$13.50

Masonry and Concrete Construction

By Ken Nolan

Craftsman Book Company,
6058 Corte Del Cedro, Box 6500,
Carlsbad, CA 92008

© **1982 Craftsman Book Company**

Library of Congress Cataloging in Publication Data

Nolan, Kenneth J., 1943-
    Masonry and concrete construction.

    Includes index.
    1. Masonry--Handbooks, manuals, etc. I. Title.
TH5311.N64   693'.1   81-22154
ISBN 0-910460-92-2   AACR2

Edited by Lois Larson

Third printing 1987

Acknowledgements

The author is indebted to the following organizations for the information and materials they provided to assist in the production of this book.

Dur-O-waL, Inc. 601 North Point Road, Baltimore, MD 21237

National Concrete Masonry Association P.O. Box 781, Herndon, VA 22070

Brick Institute of America 1750 Old Meadow Road, McLean, VA 22102

Superior Concrete Accessories, Inc. 3978 Sorrento Valley Blvd., San Diego, CA 92121

Goldblatt Tool Co. 511 Osage, Kansas City, MO 66110

Marquette Company, Gulf and Western Industries, Inc. First American Center, Nashville, TN 37238

Photo credit: Pages 67-75 *Symons Corporation* 200 E. Touhy Ave., Des Plaines, IL 60018

American Plywood Association 7011 So. 19th St./P.O. Box 11700, Tacoma, WA 98411

American Society For Testing Materials (ASTM) 1916 Race St., Philadelphia, PA 19103

Louis M. Gerson Co., Inc. Middleboro, MA 02346

Tarot Designers and Printers Inc. 4 Commercial St., Gilbertsville, NY 13776

*To my wife Barbara,
whose encouragement and assistance
helped make this book a reality.*

Contents

1 **Planning the Work**........................7
   Soil Analysis............................7
   Soil Characteristics.....................7
   Calculating Footing Sizes................8
   Elevation and Grades....................12
   Staking Out a Building..................12
   Reading a Vernier.......................14

2 **Concrete and Footings**...................15
   Components of Concrete..................15
   The Process of Hydration................16
   Proportions of Materials................17
   Placing Concrete........................18
   Steel Reinforcement.....................18
   Footings................................20

3 **Foundations**............................23
   Concrete Block Foundations..............23
   Laying a Block Wall.....................24
   Establishing the Building Lines.........24
   Laying Out the Bond.....................26
   Building the Corners....................28
   Laying the First Course.................29
   Stepped Foundations.....................35
   Pier Foundations........................35
   Grade Beams.............................36
   Slab Foundations........................37

4 **Mortar**.................................41
   Portland Cement.........................41
   ASTM Specification for Mortar...........41
   Types of Mortar and Their Uses..........41
   Properties of Mortar....................42
   Sand for Mortar.........................43
   Proportions.............................43
   Mixing Mortar...........................45
   Placement of Mortar.....................46

5 **Concrete Form Construction**.............47
   Plyform.................................47
   Jahn Forming System.....................49
   Tables for Form Design..................60
   Plate or Template Layout................61
   Other Plywood Suitable for Forming......63
   Selection of Framing for Wall Forms.....63
   Concrete Surface Characteristics........66
   Form Maintenance........................66
   Prefabricated Forms.....................67

6 **Reinforcing Masonry in Seismic Zones**...77
   Foundations.............................77
   Special Seismic Detailing...............78
   Basic Seismic Systems...................79

7 **Reinforced Masonry Construction**........81
   Basis of Design.........................81
   Reinforced Grouted Masonry..............84
   Reinforced Hollow Masonry...............90
   Reinforced Filled-Cell Masonry..........91
   Reinforced Faced Masonry................91
   Crack Control Joints....................91
   Definitions.............................93

8 **Concrete Slabs and Sidewalks**...........95
   Pouring and Tamping.....................95
   Expansion Joints........................95
   Sloping and Edging......................95
   Precast Concrete Slab Sidewalks.........98
   Concrete Slabs.........................103
   Slab-On-Ground Construction............103
   Concrete Work in Cold Weather..........107

9 **Coloring Concrete**....................109
   Iron-Oxide Pigments....................109
   Color Selection and Proportioning......110
   Mixing Colors..........................111

10 Crack Control in Masonry Walls............113
    Control Joints and Joint Reinforcement.....*113*
    Control Joint Locations.................*113*
    Control Joint Materials.................*117*
    Reinforcing Masonry: The Ivany System.....*118*

11 Brick Wall Construction..................121
    Bonds and Patterns......................*121*
    Structural Bonds.......................*122*
    Pattern Bonds.........................*127*
    Mortar Joints.........................*133*
    Mortar Joint Finishes..................*133*

12 Brick Veneer Construction................137
    New Construction......................*137*
    Existing Construction..................*141*
    Panel and Curtain Walls................*145*
    Brick-Masonry Cavity Walls.............*151*

13 Salvaged Brick..........................159
    Material Selection.....................*159*
    Physical Properties....................*159*
    Esthetics.............................*161*

14 Cleaning and Painting Brick Masonry.......163
    Cleaning New Masonry..................*163*
    Cleaning Existing Masonry..............*166*
    Removing Efflorescence and Stains........*167*
    Painting Brick Masonry.................*169*
    Surface Preparation....................*170*
    Masonry Paints.......................*171*

15 Flashing Clay Masonry....................175
    Materials............................*175*
    Placement...........................*176*

16 Chimneys and Fireplaces..................179
    Chimney Flues........................*179*
    Chimney Construction..................*181*
    Soot Pocket and Cleanout...............*182*
    Smoke Pipe Connection.................*183*
    Connection with Roof..................*183*
    Top Construction......................*186*
    Fireplace Design......................*187*
    Fireplace Construction.................*187*

17 Brick Floors and Pavements...............193
    Design Considerations..................*193*
    Material Selection.....................*194*
    Brick Paving Design Assemblies..........*197*

18 Construction Safety......................201
    Occupational Safety and Health Act........*201*
    Injury and Illness......................*201*
    Employer Responsibilities and Rights......*202*
    OSHA Standards......................*202*

Glossary......................................209

Index..217

Chapter 1
Planning the Work

Almost all residential and commercial construction must comply with local building codes. These standards or codes are regulations adopted by local governmental agencies to ensure proper construction of buildings in that particular jurisdiction.

Before starting any structure, check the code requirements of that area. The local building inspector or city engineer's office has copies of the code that is in force.

One of the first items to consider is the general definitions of the code. The code probably defines the terms listed below and illustrated in Figure 1-1.

Building Line — Line established by law, ordinance, or regulation, beyond which no part of a building, other than parts expressly permitted, shall exist.

Distance Separation — An open space between a building and an interior lot line, provided to prevent the spread of fire.

Lot Line — Line dividing one premises from another, or from a street or other public place.

Premises — A lot, plot, or parcel of land including the buildings or structures thereon.

Property Line — Line establishing the boundaries of premises.

Street Line — Line dividing a lot, plot, or parcel from a street. The street line is the line which divides the premises from the street regardless of any required setback. When a building is built to street line, the building line and the street line may coincide.

Basement — The space of a building that is partly below grade which has more than half of its height, measured from floor to ceiling, above the average established curb level or finished grade of the ground adjoining the building. Habitable space may be located in a basement provided the floor level of the basement is not more than 4 feet below the average adjoining finished grade. A basement shall be deemed a story when the floor immediately above is 7 feet or more above the finished grade. (See Figure 1-2.)

Cellars — Habitable space shall not be located in any cellar, though a recreation room is permitted. A cellar shall not be deemed a story. (See Figure 1-3.)

After the general provisions and definitions of the code you will find space requirements, structural requirements, fire-safety requirements, equipment requirements, and other provisions. Codes usually spell out each of the requirements in great detail so following them is relatively easy if you can find the provisions that apply. Codes usually show correct details for foundations in sloping ground, for example. (See Figure 1-4.)

SOIL ANALYSIS

Before building a foundation for a home or other structure, make sure you know the load capacity of the soil.

One way to avoid later problems is to get a copy of a soil survey from the U.S. Soil Conservation Service. This agency collects worldwide climatic, geographic, and soil data. It also conducts soil tests to classify soils. Soil survey reports and maps are sold by the Government Printing Office, Washington, D.C. 20402. In some areas a county extension service might also be able to help with the maps.

Soil maps consist of an aerial photograph with boundaries plotted. The map has information about the soil to a depth of approximately 4 feet and usually describes both the drainage expected from the soil and the clay content.

Soil Characteristics

Ideal foundation-bed soil will support the weight of a building, will neither swell nor shrink excessively, and will not heave from frost action. But such soils are rare.

Dry, well-compacted, sandy clay soil probably comes closest to the ideal.

Clay soils become plastic when wet. Under even moderate pressure, wet clay will squeeze from beneath the foundation. But the bearing value of clay soil can be improved if it is drained and compacted with a layer of gravel. Even dry clay can swell enough to lift a building when the clay absorbs moisture.

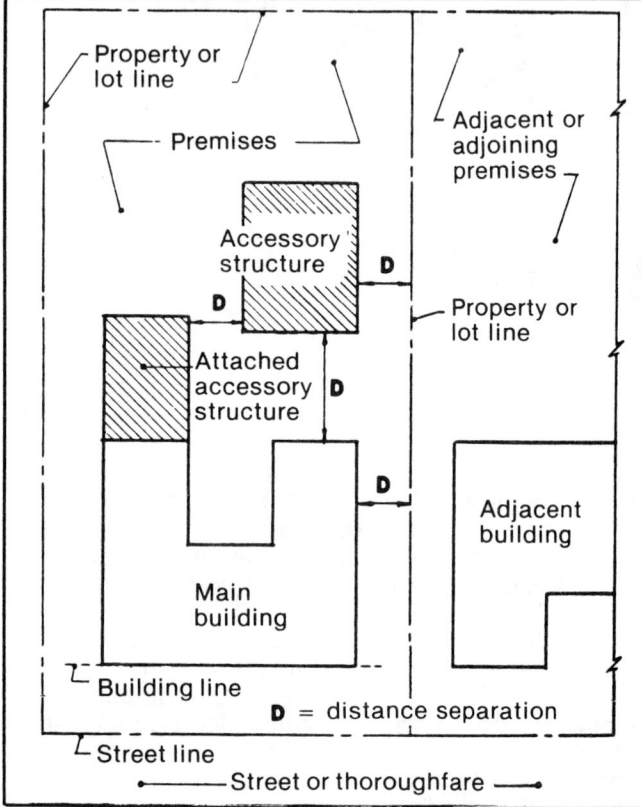

Terms used in the building code
Figure 1-1

Frost heaving can be reduced if the foundation bed is below the frostline. Sand shrinks and settles as it dries and swells or flows when wet. Either change can destroy a foundation. You can build on sand if the moisture content is constant. Otherwise, the sand under the foundation will move.

Avoid filled ground unless it is well settled. Variation in depth of fill and the different soils and waste materials used will cause the ground to settle unevenly. Different materials used for fill will have different bearing values.

Spongy or peaty soils must have specially designed foundations. A structural engineer should design both the foundation and the building.

Occasionally, a relatively thin layer of rock overlays soft clay or loose sand. This bed is unsafe for heavy buildings and for concentrated pier loads. Make sure that rock at your building site is not merely a large boulder that might loosen under the weight of the building.

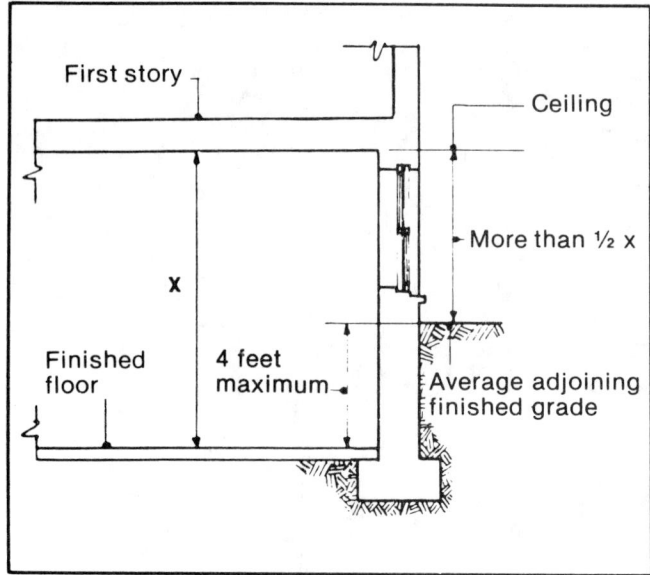

Basement location respective to grade level
Figure 1-2

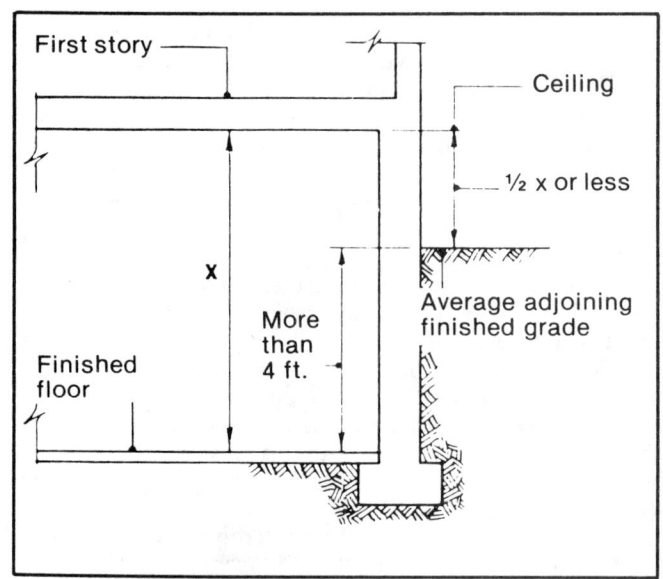

Cellar location respective to grade level
Figure 1-3

Check soil at a building site in the spring or during a wet period if possible. Problems that will complicate construction will be obvious at that time.

Table 1-1 shows accepted load-bearing capacities of soils.

Calculating Footing Sizes

After you have determined that the soil conditions will support the building, consider the size footing that is needed.

A rule of thumb is to make footings twice as wide as the finished wall and as thick as the wall is wide. That would mean an 8-inch block wall would have a footing 8 inches thick and 16 inches wide.

Before you can figure the right size of a footing for a building of a given size, you must determine the load on the footing. The structural load or building weight on a foundation includes two types of loads, *Dead* and *Live* loads. Dead load is the weight of the building itself—floors, walls, roof and foundation. The live load is the weight of whatever is in the building—people, furniture, machinery, etc. Table 1-2 shows accepted dead and live loads for estimating building weights.

Type of Soil	Bearing Capacity
Hard rock, granite, limestone	30,000
Soft rock, shale	16,000
Gravel or course sand, well consolidated	12,000
Dry, hard clay or course firm sand	8,000
Moderately dry clay or moderately dry, course sand and clay	4,000 to 6,000
Ordinary clay and sand	3,000 to 4,000
Soft clay, sandy loam or silt	1,000 to 2,000

Soil bearing capacity in pounds per square foot
Table 1-1

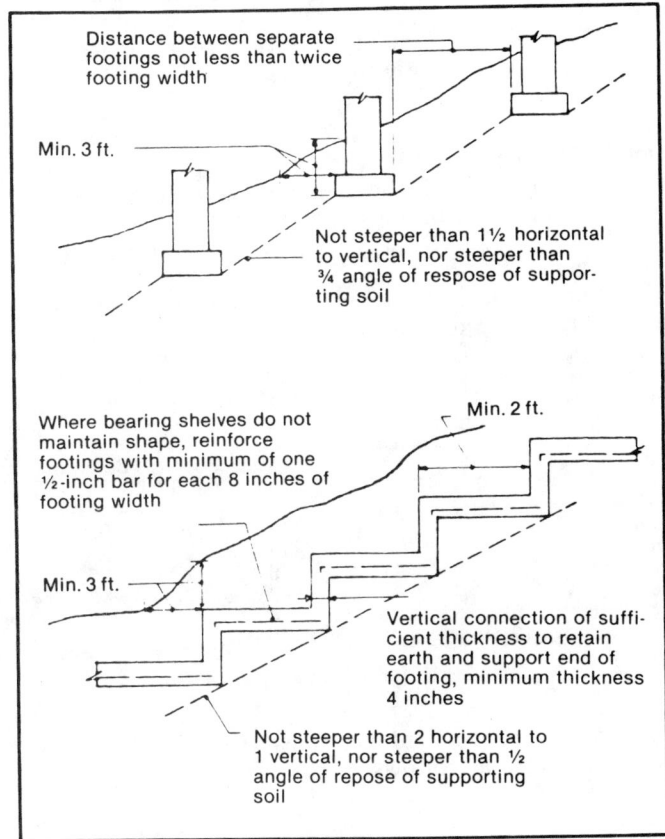

Stepping of footings
Figure 1-4

Dead Loads	Pounds Per Square Foot
Roofs: Gable, sheathed with ¾" boards, supported 2' on center, 15 pound felt, 210 pound asphalt shingles	7
Gable, Added Weight:	
Asbestos shingles	3
Built-up roof	5
Slate	7
Walls: Stud framing, plates and sills	
2 by 4s 16" on center	2
Stud wall plastered one side	10
Stud wall sheathed and sided with wood	7
Stud wall, 3/8" sheetrock both sides	6
Brick veneer	40
Concrete block 8" lightweight block	38
Concrete block 8" heavy aggregate	85
Floors: Double-on 2 by 10s 16" on center	7
Miscellaneous: Concrete or stone per 1" of thickness	12
Live Loads	
First floor in dwellings	40
Second floor in dwellings	30
Attic floor, (habitable for storage only)	20
Roofs in general	20 to 40

Dead and live loads for estimating building weights
Table 1-2

You must calculate the weight on piers and walls independently. In both cases the footing must be adequate. In general, the weight carried by a pier is the weight of the adjacent half of the beams (or girders) that are supported by the pier. This weight is represented by the shaded area of the one-story home in Figure 1-5.

Note that only one half of the total live load is used in designing the foundation of the house. It would be unusual for all of the floor area of the house to be fully loaded at the same time. (Of course, the load is used to determine the size of girders and beams because the full load might be concentrated on any one of them.) One half the total live load is also used for roofs and walls of storage buildings. The total live load *is* considered for floors in storage buildings.

A common fault is to make the footings beneath the piers too small in relation to the footings beneath the exterior walls. The following paragraphs explain how to avoid this error.

To find out how large a footing should be, you need to know two things—the bearing value of the soil (Table 1-1) and the weight the footing will have to support.

Assume that the house in Figure 1-5 has a wall load of 1,077 pounds per linear foot and a pier load of 4,664 pounds. The way to calculate these loads is shown at the end of this section.

Table 1-3 shows the width of the footing required to bear wall loads. Table 1-4 shows the size of the footing required to bear the pier loads. For example, if the soil-bearing value is 1,000 pounds per square foot, the width for the wall footing in Figure 1-5 must be 14 inches. The pier footing must be 27 inches square. If the soil-bearing value is 3,000 pounds per square foot, the wall footing

Masonry & Concrete Construction

can be less than 8 inches. But a 12-inch footing will make construction easier. Because the 12-inch footing is larger than necessary, the pier footing must also be larger than necessary to equalize the bearing capacity. Therefore, since a 12-inch wall footing will support approximately 2.8 times the wall load (2.8 x 1,077 = 3,016 pounds per linear foot), the pier footing must also be large enough to support 2.8 times the pier load (2.8 x 4,664 = 13,059 pounds per pier). The pier footing will have to be 27 inches square.

If piers (rather than a continuous wall) are used under the perimeter of the building, the combined areas of the footings beneath the perimeter piers must equal the area of the footing shown in Table 1-3. For example, the width of the wall footing on soil rated at 1,000 pounds per square foot is 14 inches. On a 48-foot long foundation, the footing would have an area of 56 square feet. If six piers were used instead of the wall, each pier would need a footing area of 9.3 square feet. The pier footings would, therefore, be about 3 feet 1 inch square.

Width of Footings (Inches)	Bearing Area S.F.	Soil-Bearing Value In Pounds					
		1,000 Lbs	2,000 Lbs	3,000 Lbs	4,000 Lbs	6,000 Lbs	8,000 Lbs
8	0.66	670	1,340	2,000	2,670	4,000	5,340
10	0.83	835	1,665	2,500	3,335	5,000	6,664
12	1.00	1,000	2,000	3,000	4,000	6,000	8,000
14	1.16	1,165	2,335	3,500	4,665	7,000	9,335
16	1.33	1,330	2,670	4,000	5,330	8,000	10,670
18	1.50	1,500	3,000	4,500	6,000	9,000	12,000
20	1.67	1,670	3,335	5,000	6,670	10,000	13,340
22	1.83	1,835	3,665	5,500	7,335	11,000	14,665
24	2.00	2,000	4,000	6,000	8,000	12,000	16,000

Safe loads per linear foot for wall footings
Table 1-3

Size of Footings Width Plus Length (In.)	Bearing Area S.F.	Soil-Bearing Value in Pounds Per S.F.					
		1,000 Lbs	2,000 Lbs	3,000 Lbs	4,000 Lbs	6,000 Lbs	8,000 Lbs
12	1.0	1,000	2,000	3,000	4,000	6,000	8,000
14	1.36	1,360	2,720	4,080	5,440	8,160	10,880
16	1.77	1,780	3,560	5,340	7,120	10,680	14,240
18	2.25	2,250	4,500	6,750	9,000	13,500	18,000
20	2.78	2,780	5,560	8,340	11,120	16,680	22,240
22	3.37	3,370	6,740	10,110	13,480	20,220	26,960
24	4.00	4,000	8,000	12,000	16,000	24,000	32,000
27	5.06	5,060	10,120	15,180	20,250	30,370	40,500
30	6.25	6,250	12,500	18,750	25,000	37,500	---
33	7.55	7,560	15,120	22,680	30,240	45,360	---
36	9.00	9,000	18,000	27,000	36,000	---	---
39	10.56	10,560	21,120	31,680	42,240	---	---
42	12.25	12,250	24,500	36,750	---	---	---

Safe loads for pier footings
Table 1-4

Unreinforced footings must be at least 6 inches thick. If the footing projects 4 inches or more beyond the wall or pier, the footing thickness must be at least 1½ times the projection. For instance, if the footing under the 8-inch foundation wall is 14 inches wide, it will project 3 inches beyond each side of the wall and must be 6 inches

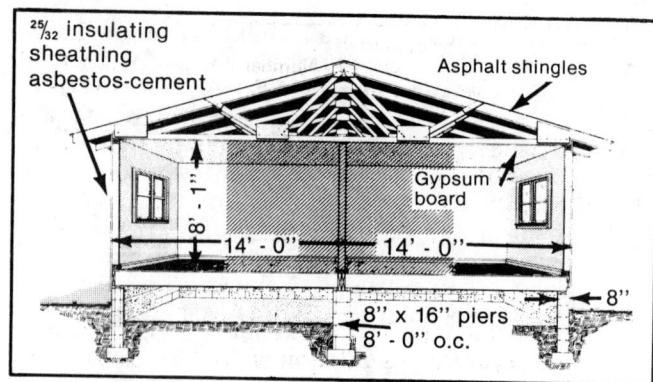

Typical one-story house with truss roof
Figure 1-5

thick. If the piers are 12 inches square, the footings will project 7½ inches and the piers will have to be at least 11¼ inches thick. Footings reinforced with steel do not have to be so thick, but a structural engineer should design them.

Estimate the dead load per linear foot of foundation wall in Figure 1-5 as follows:

	Thickness (Inches)	Pounds per Inch of Thickness	Number of S.F.	Pounds Per S.F.	Pounds of Dead Load
Roof	---	---	14 x	7 =	98
Ceiling	---	---	14 x	2 =	28
Stud wall	---	---	8 x	12 =	96
Floor	---	---	7 x	7 =	49
Foundation	8 x	12 x	4	--- =	384
Footing	12 x	12 x	½	--- =	72
Total					727

Estimate the live load per linear foot of the same foundation wall as follows:

	Number of S.F.	Pounds Per S.F.	Pounds of Live Load
Roof	14 x	30 =	420
Floor	7 x	40 =	280
Total			700

Estimate the total load per lineal foot of foundation wall as follows:

$$\text{Total dead load} + \frac{\text{Total Live Load}}{2} = \text{load per linear foot}$$
727 + 350 + 1,077 pounds per linear foot.

Each pier supports 14 feet of the width of the house in Figure 1-5 and 8 feet of the length—an area of 112 square feet.

The following shows how to estimate the number of pounds of dead load per pier:

Planning the Work

	Thickness (Inches)	Pounds Per Inch of Thickness		Number of S.F.		Pounds Per S.F.		Pounds of Dead Load
Interior wall	---	---		64	x	6	=	384
Floor	---	---		112	x	7	=	784
Pier	12	x 12	x	4		--	=	576
Footing (est.)	27	x 12	x	2.1		--	=	680
Total								2,424

The roof of the house is trussed and is supported entirely by the perimeter foundation walls. Therefore, the only live load bearing on the interior pier is the live load of the floor. Since each pier supports 112 square feet of floor, and floors have a live load of 40 pounds per square foot, the total live load on each pier is 4,480 pounds.

The estimated load on each pier is:

Total dead load + $\frac{\text{Total Live Load}}{2}$ = load per pier or 2,424 + 2,240 = 4,664 pounds per pier.

A survey is essential before you begin construction. You must build within the boundary lines established by code and on the right lot.

Even if the lot has been surveyed by a professional surveyor and even if there is a plot plan, you should verify the survey yourself. First, check the building code for the required setback from the center of the street. Then check the separation distance from adjoining buildings. Forgetting these two simple steps can result in a very expensive problem.

As a masonry contractor, most of the work you do with a level or a transit will involve relative heights: the top of the foundation wall, bearing plates, footings, etc.

For your work you will need a transit and a graduated rod or rule which is held vertical to measure heights. A common six-foot rule can be used, but it is not recommended. A leveling rod is much better because it's longer and easier to read. There are two types of rods:

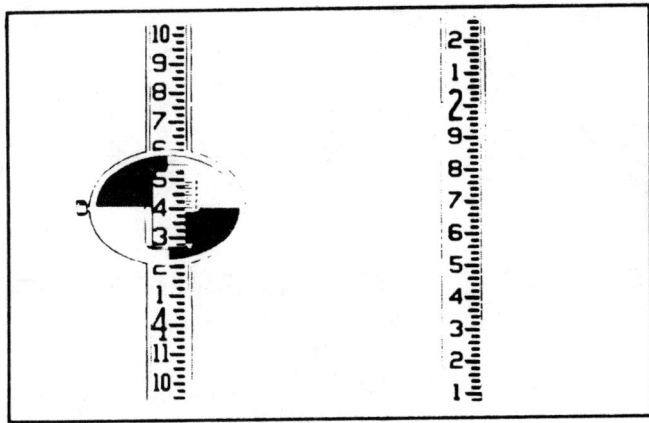

Architect's rod and engineer's rod
Figure 1-6

the architect's rod graduated in feet, inches and eighths of inches and the engineer's rod graduated in feet, tenths and hundredths of feet. (See Figure 1-6.) Both are available in 10-foot or 15-foot lengths with an oval vernier target. A snap-on target may be preferable.

The target is raised or lowered by the rod holder according to the directions of the man sighting through the transit until the horizontal line of the target is in line with the horizontal cross hair of the scope. The reading on the rod is usually taken by the rod holder. However, both types of rod may be read directly by the instrument man. These are called "self-reading rods." This method is a little faster than having the reading made by the person holding the rod.

Heights and distances on site plans are usually in decimals of a foot. Building plan dimensions are given in feet, inches and fractions of an inch. To convert dimensions, remember that 8 hundredths of a foot (0.08 foot) is about one inch. And 1/8 is about one hundredth of a foot (0.01 foot). Table 1-5 converts inches and fractions to decimals of a foot.

	Inches and Eights to Decimal of a Foot							
In.	0	1/8	1/4	3/8	1/2	5/8	3/4	7/8
0	.00	.01	.02	.03	.04	.05	.06	.07
1	.08	.09	.10	.11	.12	.14	.15	.16
2	.17	.18	.19	.20	.21	.22	.23	.24
3	.25	.26	.27	.28	.29	.30	.31	.32
4	.33	.34	.35	.36	.38	.39	.40	.41
5	.42	.43	.44	.45	.46	.47	.48	.49
6	.50	.51	.52	.53	.54	.55	.56	.57
7	.58	.59	.60	.61	.62	.64	.65	.66
8	.67	.68	.69	.70	.71	.72	.73	.74
9	.75	.76	.77	.78	.79	.80	.81	.82
10	.83	.84	.85	.86	.88	.89	.90	.91
11	.92	.93	.94	.95	.96	.97	.98	.99
Examples:	2 feet 7-1/8"		= 2.59 feet					
	8.38 feet		= 8 feet 4½ inches					

Converting inches and feet to decimal equivalents
Table 1-5

Finding Elevation Differences

To find the difference in elevation between two points which can be seen from one location, set up and level your instrument about midway between these points. Be sure that a leveling rod held on each point can be read when your telescope is level. Each point should be within 150 feet of the transit. The difference between the rod readings at the two points is the elevation difference of the two points. Figure 1-7A shows a rod reading of 69 inches at point (a) and 40 inches at point (b). Therefore, (b) is higher than (a) by 29 inches.

Suppose one of your points is below the line of sight and the other above. For example, in Figure 1-7B, the rod reading at (c) is 4 feet 6½ inches. The reading at (d), the underside of a floor beam, is 7 feet 9⅜ inches above the line of sight. This reading was made by holding the rod upside down with the foot of the rod against the beam. Point (d) is higher than (c) by 4 feet 6½ inches plus 7 feet 9⅜ inches, or a total of 12 feet 3⅞ inches.

If two points are either too far apart or at too great an elevation difference to be observed from one setup, follow the procedure shown in Figure 1-7C. Assume

Masonry & Concrete Construction

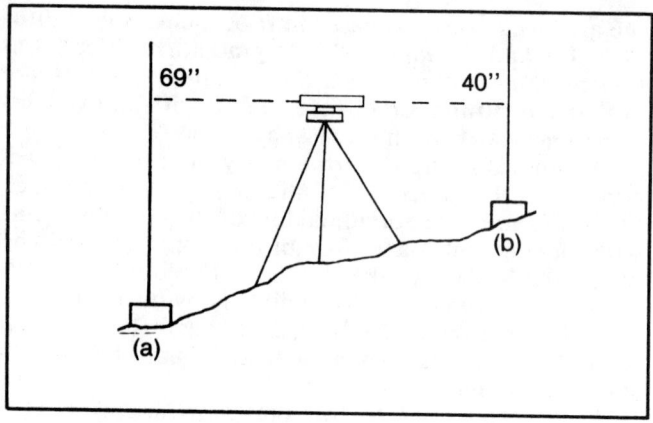

Rod reading where (b) is 29" higher than (a)
Figure 1-7A

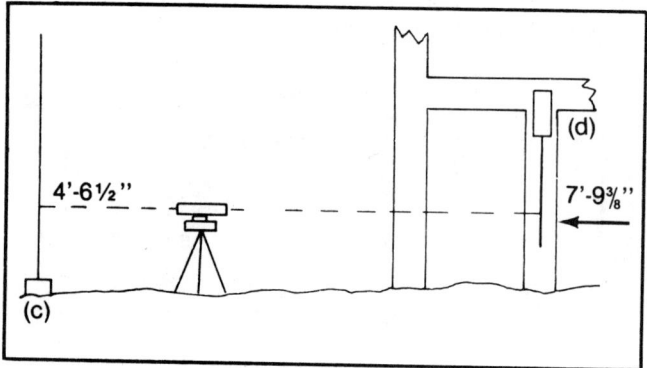

Rod reading where (d) is 12'3⅞" higher than (c)
Figure 1-7B

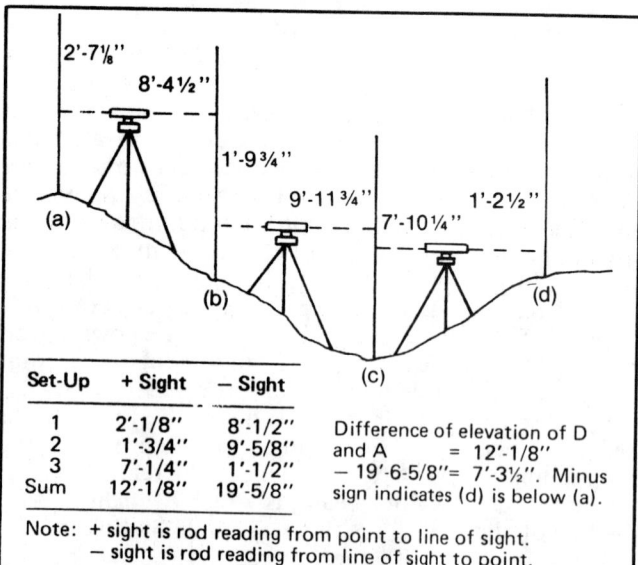

Set-Up	+ Sight	− Sight
1	2'-1/8"	8'-1/2"
2	1'-3/4"	9'-5/8"
3	7'-1/4"	1'-1/2"
Sum	12'-1/8"	19'-5/8"

Difference of elevation of D and A = 12'-1/8" − 19'-6-5/8" = 7'-3½". Minus sign indicates (d) is below (a).

Note: + sight is rod reading from point to line of sight.
− sight is rod reading from line of sight to point.

Procedure when two points cannot be observed from one setup
Figure 1-7C

that you want to find the difference in elevation between points (a) and (d). Use the terms *plus sight* and *minus sight* and make the readings at each setup as shown.

The difference of elevation between (d) and (a) is the difference between the sum of the plus sights and the sum of the minus sights. If the sum of the plus sights is larger, the final point is *higher* than the starting point. If the sum of the minus sights is larger, the final point is *lower* than the starting point.

Elevation and Grades

Most buildings you work on will have to be constructed at specific elevations. These grades are calculated from an established point of known elevation called a benchmark. Your benchmark should be a firm and definite point on the ground such as a bolt on a water hydrant, a spike in the root of a tree, a corner of a stone monument, or a chisel square on a ledge, and should be located *outside* the construction area. For a large job, several benchmarks in convenient locations are helpful. Transfer the required grades to point of construction using the "difference in elevation" method described previously. Keep a record of your survey work on each job so your values can be checked or used in later work.

STAKING OUT A BUILDING

The outline of a simple building is shown in Figure 1-8. Also shown are batter boards which help define the building lines. Each letter "N" indicates a nail driven into the top of a batter board. The various lines connecting opposite "N"s are string lines. These lines designate the outside of the foundation at a convenient height above the ground.

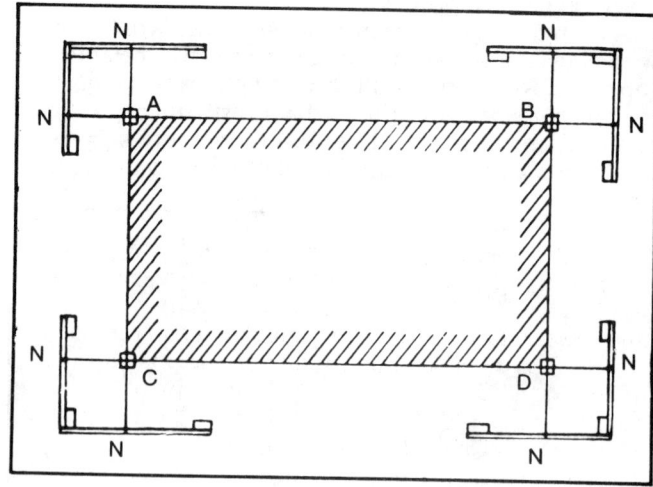

Stringline layout
Figure 1-8

Assume that the front of the building is the line between A and B, that the location of A is known, that the direction of B is known, and that all the angles are 90 degrees. Center and level your instrument over A and sight on B. Using a tape, measure the frontage of the building from A and set up point B. With the instrument still at A, turn the scope 90 degrees and set point C at a proper distance from point A. Then set the instrument over point B and take a sight on A 90 degrees to the left to find the line to D, and set D at its required distance from B. Check length CD with a string or tape to be sure that it is the same distance as AB. An excellent practice is to measure the diagonals AD and BC. If you have laid out the building correctly, the diagonal measurements will be equal.

When excavation for the building starts, points A, B, C and D will be lost. Batter boards preserve their location and establish the elevation of the building foundation. (See Figure 1-9.)

Set your batter boards about four feet outside the building lines. The exact distance depends on side conditions.

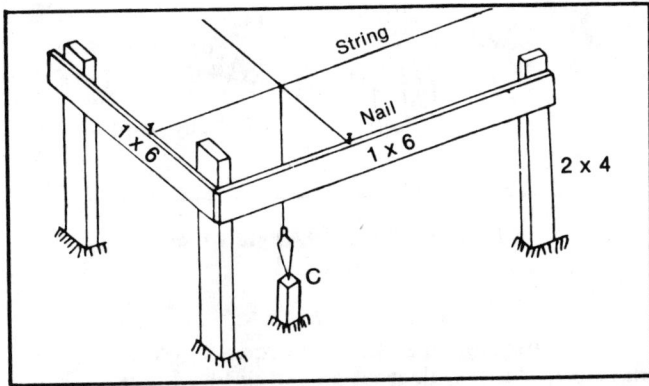

Batterboard setup at a corner
Figure 1-9

First, set posts, as shown in Figure 1-9, at all four corners. Set up your transit at a convenient location, preferably near the center of the building. The top edge of all batter boards must be at the same elevation, preferably at an even number of feet above the top of the foundation. Here's how to do it. By the leveling process previously discussed, figure the rod reading for the top of the foundation at each corner. Decrease this rod reading by the amount that the strings are to be above the foundation. Hold the rod at each batter board post and raise or lower it until the calculated rod reading is at the cross hair. Make a pencil line on the post at the foot of the rod. The pencil lines mark the tops of the batter board horizontals. The top edges of all the batter board horizontals will be at the same level.

Next, set string nails in the tops of the batter boards. Set your instrument exactly at each of the four corners and sight on the two adjacent corners, mark each board and set nails on the lines marked.

As a final step, check your work by stretching strings from nail to nail. The strings should be level and cross exactly over the corners A, B, C and D. (See Figure 1-8.)

Measuring and Laying Out Horizontal Angles

To measure a horizontal angle, set up your transit over the pivot point. Here's how that is done. Attach a plumb bob on 6 feet of string to the hook under the instrument. A large loop fastened by a slipknot works best. Adjust the plumb bob height until it is just clear of the ground. Move the entire instrument so that the bob is exactly over the pivot point on the ground. Press the legs of the tripod into the ground and lower the bob until its point is about 1/4 inch above the ground. Make final centering of the instrument over the point by loosening any two adjacent leveling screws about 1/2 turn and slowly shifting the instrument. The bob is still directly over the point on the ground. Then retighten the two screws and relevel the instrument.

To measure horizontal angles such as EFG in Figure 1-10, set up your transit over point F. Loosen the horizontal clamp screw (attached to the circle plate) and rotate the instrument until point E is in line with the vertical cross hair. Tighten the clamp screw. Turn the tangent screw until the vertical cross hair is on point E. Now by hand set the horizontal circle to read zero.

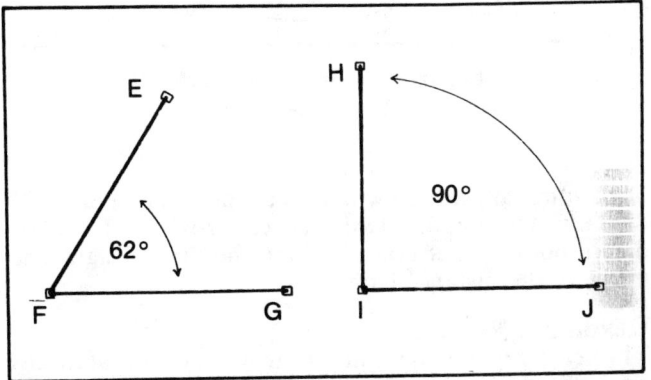

Setting horizontal angles
Figure 1-10

On some instruments, the circle is rotated. On others you set the circle to zero with a movable index. The next step is to loosen the horizontal clamp (do not touch the circle or the index) and swing the telescope until the vertical cross hair is on point G. Then tighten the clamp screw and turn the tangent screw until the vertical cross hair is exactly on point G. The horizontal index pointer will have rotated about the horizontal circle by an amount equal to the angle EFG.

Figure 1-11 shows the horizontal index and horizontal circle after measuring the clockwise horizontal angle of 62 degrees. If your instrument is furnished with a vernier instead of an index pointer, you will be able to read the angle to accuracy greater than one. The use of a vernier is explained later.

Angles of 90 degrees are most common in layout work. Assume that the 90 degree angle HIJ in Figure 1-10 is to be laid out and points H and I are known. Therefore, J is the point you want to set.

Center and level your instrument over point I. Sight the telescope on point H and set the horizontal circle to read zero degrees. Loosen the horizontal clamp and rotate the telescope until the index pointer is very close to 90 degrees. Tighten the horizontal clamp and turn the

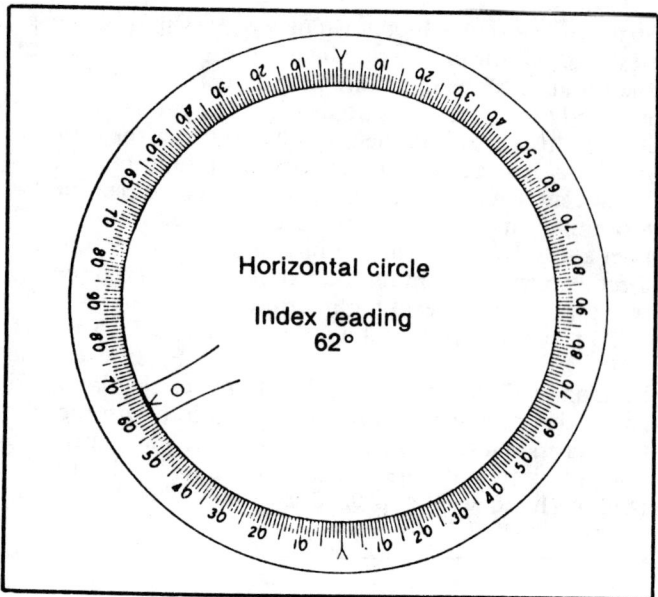

Horizontal index and circle
Figure 1-11

horizontal tangent screw until the index reads exactly 90 degrees. The line of sight (vertical cross hair) will indicate point J. Set point J along the line of sight and measure the distance from I.

Reading a Vernier

The vernier makes it possible to read very accurately any angle turned by the scope, whether to the right or to the left. For example, assume you have to turn an angle to the left (counterclockwise). First, set the circle to read zero degrees. (See Figure 1-12.)

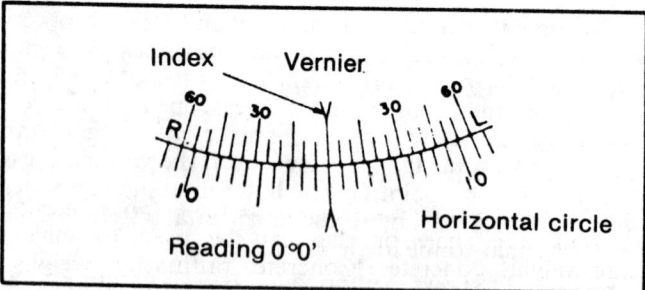

Initial setting of horizontal circle and vernier
Figure 1-12

Figure 1-13 shows what your vernier looks like after having turned the angle. Now read the angle. Remember, you have turned to the left, so you use the side of the vernier between the vernier index "V" and the letter "L." Do not be confused that the positions of "L" and "R" are reversed. Observe in this case that the vernier index has passed the 44 degree line on the circle but has not gone as far as the 45 degree line. Right away you know that the angle you have turned is greater than 44 degrees but less than 45 degrees. Now you need to know how much greater it is than 44 degrees. The vernier scale answers that question.

In the example, you add to 44 degrees the reading obtained from one of the vernier lines. Which vernier line? Look again at Figure 1-13. The fourth line from the vernier index on the top scale is lined up with one of the lines of the circle. This is the secret of the vernier. Only one line at a time can be lined up. In this case it is the fourth line from the index. We add 20 minutes to the 44 degree reading (4 times 5 minutes equals 20 minutes) because this is a 5 minute (1/2 degree) vernier. Therefore, your exact reading is 44 degrees 20 minutes.

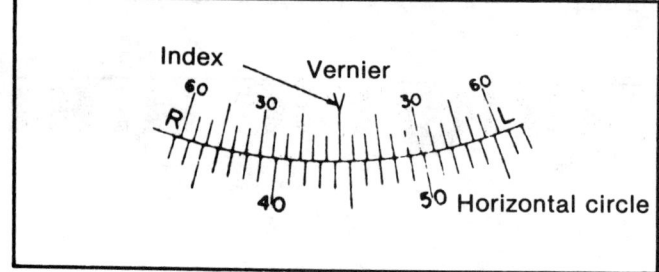

Reading 44°20' (Angle to left)
Figure 1-13

Other instruments have a vernier reading to 15 minutes (1/4 of a degree) and look like Figure 1-14.

Notice in Figure 1-14 that the index has passed the 44 degree line but has not gone as far as the 45 degree line. In this case the third vernier line from the index lines up with one of the lines on the circle. Since each vernier line represents 15 minutes, add 45 minutes to the 44 degree reading (3 times 15 minutes equals 45 minutes). The exact reading is 44 degrees 45 minutes.

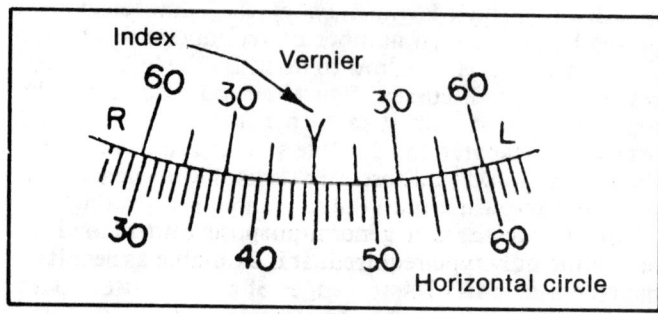

44°45' (Angle to left)
Figure 1-14

The Vertical Vernier

The vertical vernier uses exactly the same principle. The vertical vernier is below or outside the circle position rather than inside and reads angles up and down rather than left and right.

Again you have two vernier scales. The right-hand side reads angles of elevation (UP) and the left-hand side reads angles of declination (DOWN).

Chapter 2
Concrete and Footings

Concrete is a strong, economical building material that can be cast into practically any shape. Well-built concrete structures last indefinitely and require a minimum of maintenance.

Concrete has high compressive strength, and is weather and fire resistant. Its chief disadvantages are low tensile strength, high heat transmission and water vapor permeability. These may be offset, respectively, by proper use of steel reinforcement, insulation, and vapor barrier material.

COMPONENTS OF CONCRETE

Concrete is made by mixing portland cement, fine aggregate (sand), coarse aggregate (gravel, crushed stone, or other material), water and, when necessary, additives.

Portland Cement

Five types of portland cement are in general use in residential construction.

Type I Normal is a general-purpose cement and is usually the only type required. It is available as regular, air-entrained, and white.

Type II Modified generates less heat during curing than Type I and is more resistant to sulfate attack. It is used where ground water contains higher-than-normal sulfate concentrations.

Type III High Early Strength is used when strength shortly after casting is necessary. Adequate strength is obtained in one to three days.

Type IV Low Heat is used only in large masses of concrete, such as dams, where, if ordinary cement were used, so much heat would be generated that it would crack or explode. It is not used in residential work.

Type V Sulfate-Resistant should be used when concrete will be exposed to high alkali soil or water.

Aggregate

Fine aggregate (sand) consists of all grains, small pebbles, or particles of crushed stone that will pass through a ¼ inch wire screen. The sand must be clean, hard and well-graded. "Well-graded" means ranging in size from coarse to fine, excluding dust. A sample of well-graded sand is shown in Figure 4-3 in chapter 4.

Coarse aggregate, usually gravel or crushed stone or slag, should be sound, hard and clean.

The particles should range in size from 1/4 inch up to 2 inches as shown in Figure 2-1. The nature of the work determines the best size to use. No particle should be larger than one fourth the thickness of the wall or slab.

Bank or creek gravel can sometimes be used just as it comes from the pit. Usually, however, it contains too much sand and would require too much cement paste. Some supply yards sell mixed aggregate with sand and gravel combined in the correct proportion for concrete.

Lightweight aggregates such as cinders or expanded materials like shale or slag can be used to make lighterweight concrete. Concrete ordinarily weighs about 150 pounds per cubic foot. Lightweight aggregates produce concrete weighing 100 to 150 pounds per cubic foot. Concrete weighing as little as 50 pounds per cubic foot can be made with very lightweight aggregates such as pumice or expanded mica.

When using cinders as aggregate, make sure they are hard, vitreous clinkers free of soot, sulfides and unburned coal or ash. Clean them first by soaking in water for 24 hours. They should not discolor your hands when rubbed briskly. Ashes should not be used because they often contain alkali which disintegrates concrete.

Avoid using lava rock as an aggregate. It is usually too light and frothy or contains too many impurities to produce a satisfactory concrete. Lava rock from Oregon or Washington is usually satisfactory, as are rhyolite, (light-colored volcanic rock) and many of the darker basaltic lavas.

Concrete made with lightweight aggregate is excellent for fill between floor sleepers, for making precast blocks and roof slabs, and for fireproofing. It should not be used where waterproof, abrasion-resistant or high-strength qualities are needed.

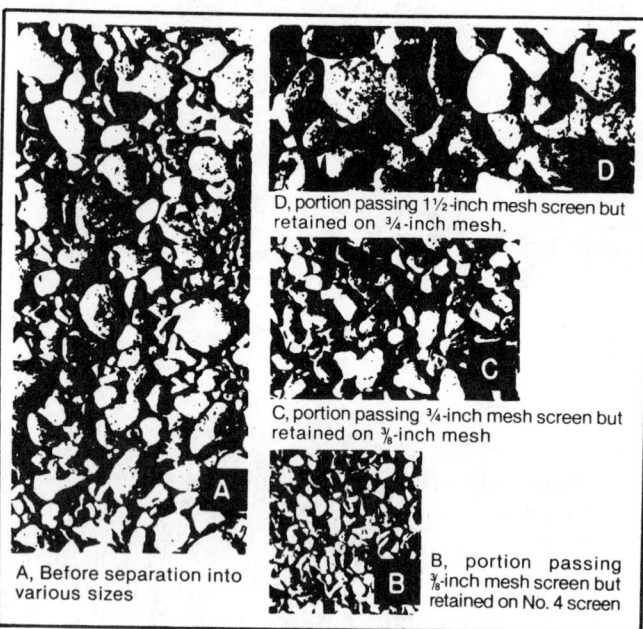

Sample of well-graded gravel
Figure 2-1

Water
Water for mixing concrete should be clean and free of strong acid, oil, alkali and organic matter.

Do not use sea or brackish water as it will weaken the concrete. Alkali salts are destructive if in excess of 0.5 percent.

In general, water that is fit to drink is suitable for concrete.

Additives
Admixtures may be put into concrete to improve workability, reduce segregation, accelerate setting or hardening, or entrain air. However, misuse of such admixtures will produce weak concrete.

Workability can be increased by adding fine materials such as powdered pumice, fly ash, and hydrated lime, but these impair the strength of the concrete. Better ways to improve workability are to add more cement to the mix, vary the mix proportions or the aggregate graduation, or to place the cement more slowly.

One to two pounds of calcium chloride per sack of cement may be added to accelerate setting and hardening, but never add more. It can be added as solution in the mixing water or as crystals in the aggregate, and should be mixed with the cement after the materials are put in the mixer.

Air
Air-entrained concrete is more durable and weather resistant than regular concrete. It should be used for all concrete work that will be subject to freezing and thawing. Five to seven percent by volume of entrained air is recommended.

Air-entrained, ready-mixed concrete is available. If you do your own mixing, use air-entrained portland cement or add an air-entraining admixture.

THE PROCESS OF HYDRATION
The concrete mix consists of cement, aggregate, and water. But it is the cement and water, reacting in a process called hydration, that form a paste or gel which binds the aggregate to the concrete. Chemicals in concrete protect the paste so it can complete its reaction.

Nevertheless, the aggregates are extremely important to the quality of the concrete's finish. Aggregates are inert materials and help extend the costly cement over as many cubic feet as possible.

The paste that results from hydration must coat each particle of the aggregate, filling all voids in the mass. The final quality of the paste depends on two things: the ratio of cement to water and the length of time hydration continues.

Control of the hydration process is vital to the quality of the concrete. Yet hydration is difficult to manage. Temperature is one problem because paste develops slowly below 50 degrees, resulting in longer and more costly supervision of the curing process.

A second control problem is moisture retention. If the moisture content of the mix is lost before hydration is complete, the concrete cannot develop the design strength. Shrinking, which leads to checking, will be one of the results. Chemicals have been developed that harden or seal concrete and thus control hydration.

Curing
The curing stage begins at the moment hydration begins and continues until hydration is complete. This is one of the most critical periods in concrete construction—and frequently one of the most neglected. The chemical reaction which takes place during the first twelve hours after the pour determines the final quality of the concrete. And, while the concrete will develop most of its ultimate strength in about seven days, the hydration process must be allowed to continue if the concrete is to reach its planned 28-day strength. So, while it's important for water to be retained in the concrete over a long period of time, it's usually impractical to leave the forms in place for even those first seven days. That is why curing methods and formulations have been devised for application to freshly formed concrete surfaces.

Of the several methods for curing concrete, most have certain disadvantages. Wet burlap or sprinkling is likely to dry out between wettings; moist earth is messy and can cause staining; waterproof paper and plastic film are expensive and costly to handle. Chemical compounds, however, offer the greatest combination of advantages—low initial cost, ease of application, reliability and lower labor costs because double handling is eliminated. These compounds form a shield, or barrier,

Concrete and Footings

Application of a chemical sealant and hardener
Figure 2-2

which minimizes the loss of reactive moisture in the concrete mixture and results in a superior cure. There are three general categories, and each has its own particular disadvantage. The categories are: resin based, which can discolor concrete; sodium silicates, most of which do not meet ASTM specifications for moisture retention; and chlorinated rubbers and acrylics, which are the best choice, though those containing chlorinated rubbers will yellow and discolor floors.

Hardening
The quality of hardened concrete is determined by its compressive strength. Filling surface voids with hard material increases its hardness. This is especially important for concrete surfaces such as floors which may be exposed to heavy traffic both during and after construction. They are applied even while the curing process is continuing. Hardening chemicals provide surfaces with superior abrasion and wear resistance and can save countless man-hours during construction by minimizing maintenance, repair and replacement of faulty surfaces.

Sealing
Concrete is porous and absorbs liquids, retains dirt and other contaminants, and is susceptible to staining. If finishes are to be applied, the surface must be clean, since even well-cured concrete tends to dust. A quality sealant—one which literally separates the concrete from direct contact with its "environment"—can quickly pay for itself in cleanup and maintenance savings.

Figure 2-2 shows spraying of a chemical called *Cure and Hard* sold by Symons Corporation. It is a concrete-curing and hardening compound which also dustproofs and creates an abrasion-resistant surface without discoloring or altering the appearance of the concrete.

PROPORTIONS OF MATERIALS
Different concrete mixes are used for different kinds of work. The mixes are designated by a three-part number. For example, 1:2:3 indicates the proportion of cement, sand, and gravel (or other coarse aggregate) used. A 1:2:3 mix indicates one part cement, two parts sand and three parts gravel.

Table 2-1 lists concrete mixes for various kinds of work and indicates the quantity of water required per bag of cement. The cement-sand-gravel proportion may need to be varied slightly to obtain a workable mix, but the water-cement ratio should never be changed. The water-cement ratio determines the quality of the cement paste, which in turn determines the strength, durability and watertightness of the concrete.

Kind of Work	Proportions Cement Sacks	Sand C.F.	Gravel C.F.	Water required per sack of cement when sand is — Wet Gallons	Moist	Dry
Very thin work 2 to 4 inches	1	2	2	3½	3¾	4½
Exceptionally watertight and abrasive-resistant work 4" to 8" thick	1	2	3	3¾	4½	5½
General reinforced and watertight work 8" to 12" thick	1	2½	3½	4½	5	6½
Mass concrete work of moderate strength and not watertight	1	3	5	5	6	7

**Trial concrete mixture for various kinds of work
Table 2-1**

Note in Table 2-1 that the water-cement ratio varies according to the moisture content of the sand. This may be determined by squeezing some sand in your hand. If the sand forms a firm ball, it is wet; if it crumbles after forming, it is moist; if it falls free from your hand it is dry.

Depending on the aggregate used, the quantities of materials required may vary as much as 10 percent. See Table 2-2 for recommended quantities.

Proportions of the Concrete			Quantities of Materials		
Cement	Sand	Gravel	Cement Sacks	Sand (damp) C.Y.	Gravel C.Y.
1	1.5	3	7.6	.42	0.85
1	2.0	2	8.2	.60	0.60
1	2.0	3	7.0	.52	0.78
1	2.0	4	6.0	.44	0.89
1	2.5	3.5	5.9	.55	0.77
1	2.5	4	5.6	.52	0.83
1	2.5	5	5.6	.46	0.92
1	3.0	5	4.6	.51	0.85
1	3.0	6	4.2	.47	0.94

**Approximate quantities of materials for one cubic yard of concrete
Table 2-2**

Concrete should be thoroughly mixed. This increases the strength of the water-cement paste and improves the workability of the concrete. For machine-mixing, mix for 5 minutes with all materials in the mixer.

Place about 10 percent of the mixing water in the drum before adding the dry materials. Then add water uniformly with the rest of the materials.

Small concrete jobs can be done with a concrete mixer. The mixer in Figure 2-3 has a capacity of six cubic feet. It can be pulled behind a truck or car to the job, and may be rented from one of the many rental yards. A mixer of this type costs about $1,500. Although the mixer may be expensive, it can pay for itself if you do a lot of small masonry jobs. Most ready-mix companies have a minimum charge for a delivery. So even if you only need a yard of concrete, you will still have to pay for 2 or 3.

PLACING CONCRETE
Place the concrete within 20 minutes after the mixing is completed. In warm weather, initial set will occur in about 20 minutes. If concrete is disturbed after initial set, it loses strength. Never rewet and remix concrete that has set before it can be placed in form. Discard it.

Place concrete as close as possible to its final position in the forms. Honeycombing and segregation may occur if it is pushed or has flowed too far from where it was first placed. Spade or vibrate the concrete as it enters the forms. A concrete vibrator (Figure 2-4) has a flexible shaft and can effectively move the concrete in the forms to fill all the voids, thus decreasing the amount of honeycombing along the edges. Vibrators are usually available at rental stores that deal in building equipment.

Do not overwork concrete in the forms. The finer material, including the cement paste, tends to work to the top, resulting in a mix that has the larger aggregates on the bottom.

STEEL REINFORCEMENT
Steel reinforcement increases the tensile strength of the concrete, which is its ability to withstand pulling and bending forces and the effects of temperature and moisture changes. The steel is placed in the forms and the concrete is cast around it. Figure 2-5 shows reinforcing rods suspended in forms.

Tying reinforcing rods can be a time-consuming job. Using the tool shown in Figure 2-6 and wire ties of the proper length will save time. (See Figure 2-7.)

The reinforcing rods in Figure 2-5 are suspended by tie wire so that the rods will be in the center of the concrete after it is poured. Tables 2-3 and 2-4 give sizes for wire ties and steel reinforcing bars.

Rod Sizes	1/4"	3/8"	1/2"	5/8"	3/4"	7/8"	1"	1-1/8"	1¼"
1/4"	3½	4	4½	5	5½	6½	7	7	7½
3/8"	4	4½	5	5	5½	6½	7	7	7½
1/2"	4½	5	5	5½	6	6½	7½	7½	8
5/8"	5	5	5½	6	6½	7	8	8	8½
3/4"	5½	5½	6	6½	6½	7½	8	8½	8½
7/8"	6½	6½	6½	7	7½	7½	8½	9	9½
1"	7	7	7½	8	8	8½	9	9½	10
1-1/8"	7	7	7½	8	8½	9	9½	10	10½
1-1/4"	7½	7½	8	8½	8½	9½	10	10½	10½

**Length of tie wires for various bar sizes
Table 2-3**

Concrete and Footings

**Small concrete jobs can be done with a concrete mixer.
Figure 2-3**

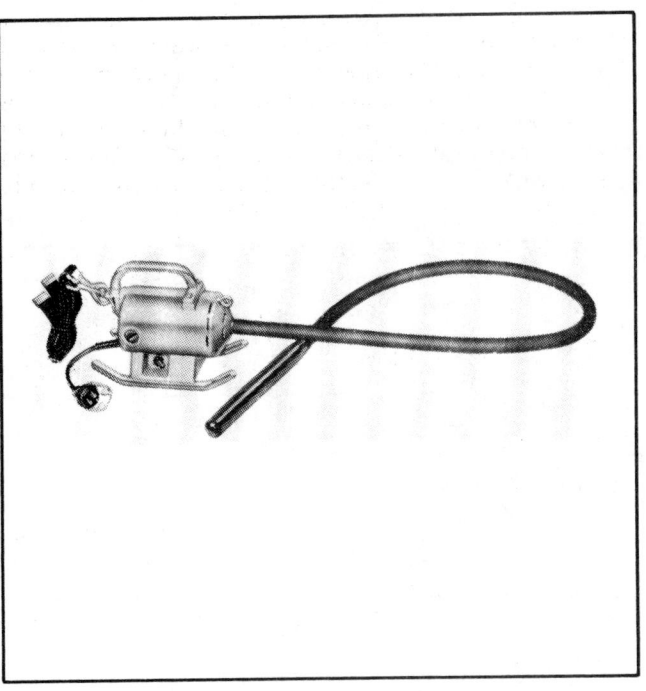

**Electric concrete vibrator
Figure 2-4**

**Steel reinforcing rods suspended in concrete forms
Figure 2-5**

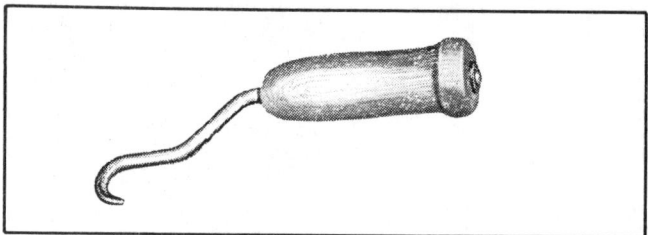

Hand tying tool
Figure 2-6

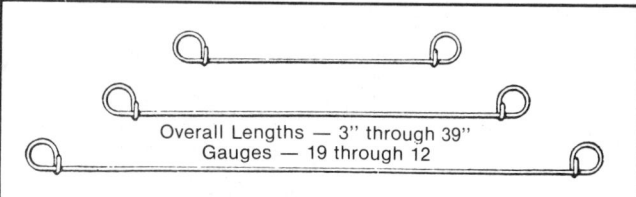

Overall Lengths — 3" through 39"
Gauges — 19 through 12

Wire ties
Figure 2-7

To make a reinforced footing at a corner, cross the reinforcing rods. (See Figure 2-8.) This produces a better bond. As shown in the Figure 2-8, the rods are suspended from the spreaders on top of the forms, and at the same height all along the footing. A short piece of rod is wired sideways to keep the rods separated. Note the board nailed to the spreaders to form a keyway in the top of the concrete footing.

Figure 2-9 shows the footing after the forms were stripped. The poured concrete wall sits centered on the keyway. This keeps the wall from slipping off the footing.

Before pouring the footing, be sure to oil the forms. Otherwise, you may have difficulty removing them when the concrete is cured. Leaving the forms on for a few days also helps the curing. A good practice after the footing is poured is to cover it with paper. This slows the curing and increases hardness. The tar paper can be used again to cover the foundation walls after a day's work.

Bar Number	Diameter Inches	Area Sq. Inches	Approximate Weight of 100 Ft Pounds
2	1/4	0.05	17
3	3/8	0.11	38
4	1/2	0.20	67
5	5/8	0.31	104
6	3/4	0.44	150
7	7/8	0.60	204
8	1	0.79	267

Size area and weight of steel reinforcing bars
Table 2-4

Correct placement of reinforcing rods at footing corner
Figure 2-8

Concrete and Footings

**Footing after forms were stripped
Figure 2-9**

**Cover footing with tar paper after it is poured to slow
curing and increase hardness.
Figure 2-10**

Chapter 3
Foundations

Each foundation for a building must be designed for the weight it will support. The well-designed and well-built foundation will have to resist the following forces:

1. The weight of the building and the supporting, or bearing, value of the soil.
2. Soil movement caused by a change in moisture content and by heaving during freezing and thawing.
3. The uplift and overturn forces of wind on the building.
4. Pressures of the outside soil on the basement walls.

Failure to resist any one of these forces will cause the foundation to crack, settle, or shift.

When you select the site for a building, do it carefully. If possible, pick a spot on a slight elevation with good drainage and a firm subsoil.

Some of the different types of foundations are:
- Continuous wall
- Basement wall
- Step foundation
- Pier foundation
- Slab foundation
- Grade beams

Foundations are usually built of either concrete or unit masonry (concrete block, brick, or stone).

In general, foundations for residential homes are concrete or block. Building stone foundations is a difficult and slow job, and locating the stone can often be a problem. They also take up twice as much space as brick or block.

CONCRETE BLOCK FOUNDATIONS

Concrete block is the most widely used material for foundations. It has many advantages including strength, good insulating qualities, and low cost.

Concrete blocks are made in lightweight and heavyweight units and come in three classes:

Solid Load-Bearing
A solid block is one with a core area of less than 25 percent. They are used in walls that have to carry a heavy load.

Hollow Load-Bearing
A hollow block is one in which the cores comprise more than 25 percent (usually about 40 percent) of the cross-sectional area.

Hollow Nonload-Bearing
Hollow nonload-bearing blocks are used mainly for partitions, and are the most commonly used of the three. They are used so frequently because they have high compressive strength, yet are lighter than solid blocks.

The American Society for Testing and Materials has classified concrete blocks into two grades:

Grade N
Grade N blocks are made for use in areas where freezing and thawing are factors for exterior walls above and below grade. They must be able to withstand 800 pounds of pressure per square inch.

Grade S
Grade S blocks are made for use above grade on interior walls or in protected exterior walls. The blocks must be able to support 600 pounds per square inch.

The ASTM makes further recommendations as to the use of these blocks in certain areas of the country:

Type I
Type I blocks have a low moisture content and are used in very dry areas of the country.

Type II
Type II blocks have no specific moisture content. They are made and used in most areas.

Masonry & Concrete Construction

Whenever possible, have the delivery truck unload the
block inside the foundation footing.
Figure 3-1

LAYING A BLOCK WALL

Careful planning helps eliminate problems and errors in building good block walls and foundations. Study the blueprints or plans carefully, and make a mental picture of what the completed job will look like. Masonry can be costly to remove and replace if you make a mistake.

Preparing the Blocks for Placement

Once the footing is in place and the concrete has had a chance to harden sufficiently, have the blocks delivered and prepare to lay the walls.

If possible, have the truck delivering the blocks drive up close to the foundation; then unload the blocks inside the foundation footing. (See Figure 3-1.) This is a money and labor saving operation.

If you aren't going to lay the blocks right away, cover them with the tar paper that you used to cover the footing. (See Figure 3-2.) This won't take long, and it can save you a lot of time and trouble later if it rains. Wet blocks are difficult to work with. They are slippery and can cut when wet. Blocks are heavy, but wet blocks are heavier still. Also, when saturated, they tend to sink in the mortar; if this happens you might have to lay some twice. Wet blocks smear the mortar and the smears show unless the walls are to be painted.

Establishing the Building Lines

In most foundations, batter boards are used to establish the building lines. To construct a straight foundation wall, stringlines must be set up along these lines. Tie a plumb bob to the intersecting points of the building lines at each corner to establish the exact corner points.

By keeping the bob about 1/4 inch from the footing, you will be able to establish the exact position of a corner. To make it easier to locate later, mark first with a lumber crayon and then make a finer mark with a pencil. Now drive in a long masonry nail at the exact corner point so that a line can be strung later. Notice that in Figure 3-3 the first block that is to be laid on the corner has the lower corner cut away so the nail won't be obstructed.

Once the corners have been marked and the masonry nails driven in, check to see if the points are accurate. This is done by measuring the diagonals as shown in Figure 3-4.

If the corners are set accurately, the diagonals will be equal and the foundation should be square. If the diagonals are different, check the measurements from corner to corner. If they are correct, then the foundation is slightly out of square. Take a large builder's square and lay it on the footing at one corner. Stretch the line to the next corner and lay the square on the line about 1/16 inch away. Stretch the next line (the one that

Foundations

**Wet blocks make the mason's job more difficult. Cover the blocks with tar paper to protect them from rain.
Figure 3-2**

**Cut away the lower corner of the first corner block so the marker nail won't be disturbed.
Figure 3-3**

To make sure the foundation is square, measure the diagonals. If they are equal, it's square.
Figure 3-4

would make a right angle) and see if it is within 1/16 inch of the other side of the square. It would only have to be off slightly to affect the diagonal measurements. Now, adjust the nails and recheck the line and the diagonal measurements.

Using a Transit. With a transit, you will be able to establish the corners of the wall. Figures 3-5 and 3-6 show the level and the transit level. The level is used primarily to lay out horizontal angles and to check differences in elevation. The transit level performs three functions that the dumpy, or plain level cannot. It lets you plumb a wall, set points in line at different elevations, and measure vertical angles.

Set the transit over a corner. To get the exact point of the corner, attach a plumb bob to the center of the bottom of the transit. By moving the legs of the transit you can set the scope almost exactly over the nail. If further adjustment is needed, the plumb bob can be placed exactly over the corner by moving the base of the scope.

Point the scope of the transit toward the far corner of the footing. With a steel tape, measure the exact distance of the wall. Mark the footing at the point of the corner where the scope meets the tape. This point marks the outside of the wall. One wall line has now been laid out. Turn the transit on the table 90 degrees and aim it at a point on the next wall line. Stretch the steel tape and mark the point where the cross hairs of the scope meet the tape. You now have two wall lines and a 90 degree corner.

When the other two sides of the foundation have been measured, you will have the four wall lines. To make sure the lines are square, check the diagonal measurements. They should be equal. Checking measurements twice doesn't take very long, and it can save your pocketbook as well as your reputation.

Laying Out the Bond

Before laying out the bond, make sure that dirt, dust, and other debris have been swept off the footing. This will make the marks clearer and easier to see. After cleaning the footing, snap a chalk line from corner to corner. Use blue chalk rather than yellow as it shows up better on light concrete.

To start laying out the bond, place a full block at any corner. Starting at this block, lay the bond out with a tape or a mason's rule. Some steel tapes come with bond marks. When using 16-inch blocks remember that for a short wall you won't have multiples of 16 inches because

Foundations

Level
Figure 3-5

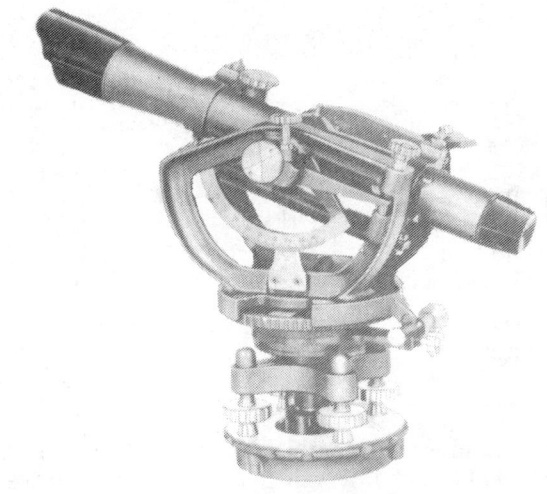

Transit level
Figure 3-6

the last block won't have an end joint. For example, three blocks lay in 47⅝ inches—not 48 inches.

In a long wall, the joint is compensated for. Figure 3-7 shows the mason marking the footing every 16 inches on his mason's rule. The modular rule has red marks for every 16 inches.

If a building is designed to be modular, all the walls should work out in full bond. By starting with a full block at one corner, you should arrive at the following corner with either a full block or a half block.

If there is a problem and the wall doesn't work out in full blocks, try to close the joints slightly to make up the difference. By closing the joints of 16 blocks just 1/8 of an inch, you can make up 2 inches. The same applies if you have to open up the joints a small amount; however, the joints should never be more than 1/2 inch wide.

If there is a problem with the size of the joints, then lay out the blocks dry along the footing to the opposite

Mark the footing every 16 inches using a mason's rule to prepare for laying the bond.
Figure 3-7

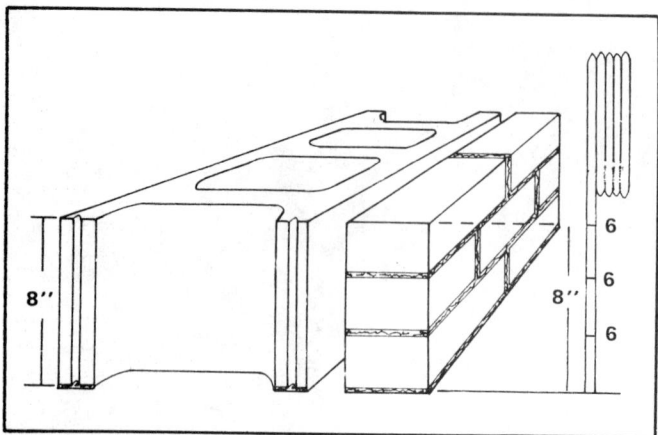

Standard modular block and brick. On the right is a rule marked at 6-inch intervals to facilitate brick placement.
Figure 3-8

corners to see how they would actually look laid in the wall.

Just as it is a bad practice to open up joints over 1/2 inch, it is also a bad practice to lay a "plug" in a wall. A plug is a piece of block smaller than a half. One way to avoid using a plug is to make two or three 12-inch pieces and lay them one after the other in each course. Keep them at the same spots in the wall for a good appearance and for strength. Blockwork should never be "jack-over-jack," meaning joints are directly over each other. This may cause the wall to crack.

If the bond works out to an 8-inch piece, reverse the corner block at the next corner. If there has been a mistake, the corner should be torn down and the blocks replaced properly—even if the wall has been built scaffold-high.

If the wall is an inch or so short of working out in bond, cut off a piece of a block. Most stretcher blocks have ends extended on each side, sometimes called "ears." By chiseling the ears off a block, you will save about one inch in the length of the wall. But this is time-consuming, and can often be avoided by careful planning.

Modular Construction. When you look over the plans, see if the dimensions are divisible by 4. If so, the building is laid out in modular dimensions. Modular construction is based on the 4-inch module, and is accepted by the American Standard Association. The A.S.A. determined that if all building materials were made to fit the 4-inch modular grid system, the construction industry would benefit from the savings in waste and labor. Consequently, almost all masonry materials fit the modular grid system, except non-modular brick.

The biggest advantage in modular construction is that few, if any, small pieces are involved, making it a fast and inexpensive way to build.

Figure 3-8 shows how brick and block fit the modular system.

Building the Corners

When the building lines have been established and the bond has been laid out, you are ready to construct the corners.

Set up the transit level or dumpy level in the center of the foundation, as shown in Figure 3-9. Have a helper

Set up the level in the middle of the foundation or at a point where you can see all the corners.
Figure 3-9

Hold the rod with the fingertips so the man with the scope can see the markings.
Figure 3-10

stand with the mason's rule or a rod at each corner. The rod should be held with the fingertips, as shown in Figure 3-10. This allows the man on the scope to get an accurate, unobstructed fix on the target. Have the rule or rod held on the center of each corner, and read all corners before starting the blockwork. Record the readings and determine which corner is high and which is low. Then adjust the first course to make all the corners the same height.

If you have to set up the scope some distance away or if the building is rather long, hand signals may be necessary. Figure 3-11 shows some of the most commonly used hand signals. The rod and level team should become familiar with these signals to facilitate leveling and to prevent errors in communication.

After you have set the first corner blocks at the same elevation, start building the corners. Follow the chalk lines laid down previously and lay three or four blocks each way. To make sure you are proceeding in a straight line, stretch a stringline to the next corner and lay the blocks along this line. This procedure is called "ranging the wall," and if it isn't done, you might get a "wow" or a bow in the wall.

Figure 3-12 shows the beginnings of a foundation. Notice the built-up corners and the conveniently placed material. Most blocks are wider at the top than at the bottom so they can hold more mortar and be handled more easily. Instruct the laborers to place the blocks with the wide side up.

Laying the First Course

When laying the first course of the foundation, use a full bed of mortar. (See Figure 3-13.) This will help prevent water leaks and maintain a good bond with the footing. But never put down a mortar bed thicker than 1 inch. A mortar bed that is too thick won't hold the weight of a block very well and takes a long time to stiffen. If there is a low spot in the footing, use pieces of block embedded in mortar to get the right elevation. Similarly, if you encounter a high spot in the footing, cut the block to make the first course level. This will prevent lost time and other problems that may develop, such as a hump in the wall.

Corners are usually built seven courses high, commonly called "scaffold-high." This is because a scaffold is usually required after seven courses have been laid.

Masonry & Concrete Construction

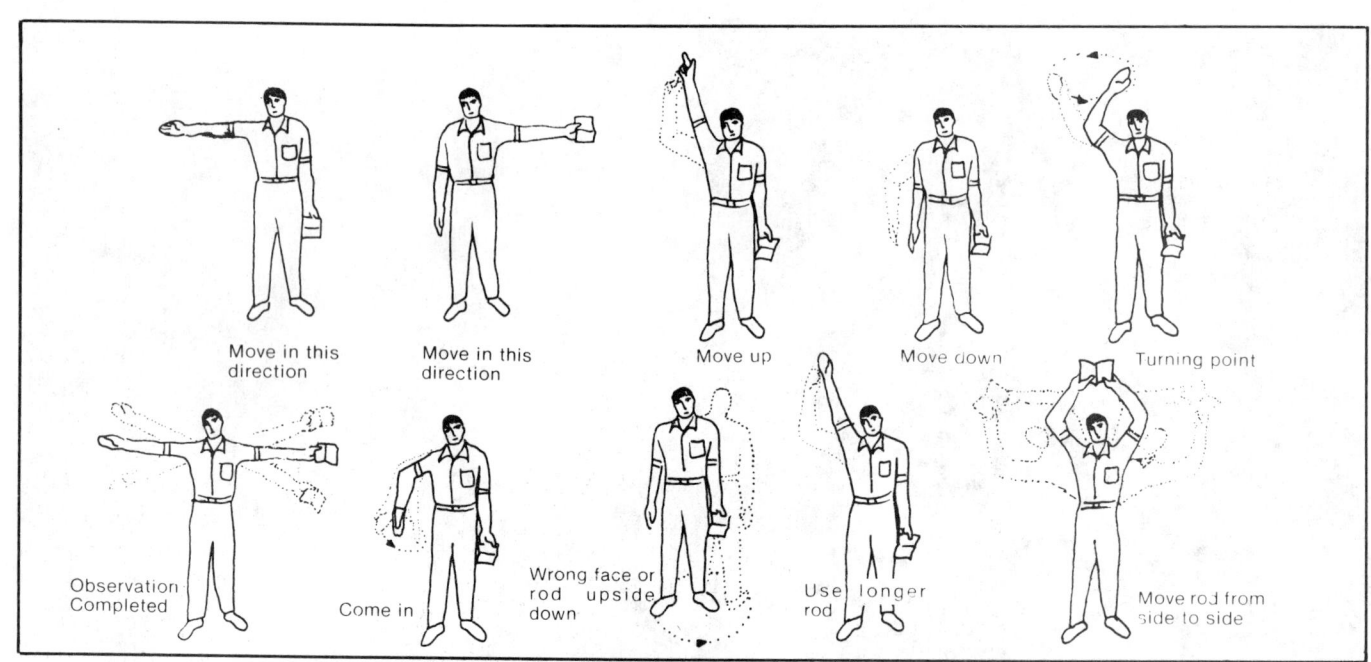

Commonly used hand signals
Figure 3-11

The beginnings of a foundation. Notice the built-up corners and conveniently placed blocks.
Figure 3-12

A good bed of mortar for the first course of block. Notice how the footing has been cleaned off prior to laying the first course.
Figure 3-13

In some situations, you can use a foot plank to lay an eighth or extra course. (See Figure 3-14.)

Using a Story Pole. Before building the corners, consider using a story pole. The story pole is used by some masons to keep their rule clean and to help keep the elevations accurate. (A mason's rule bends easily when opened all the way.)

Make a story pole out of a good 1 x 4 or 1 x 3 board about 8 or 10 feet long. Mark one end of the pole with a small square and cut it off. From this point, mark the pole every 8 inches. The 8 inches is the height that blocks are usually laid, including the mortar bed. Figure 3-15 shows a story pole.

Sometimes the pole won't be long enough to have full markings all the way up, and the marks could be read incorrectly if the pole were held upside down. Therefore, it is a good practice to place an arrow beneath each mark indicating which way is up.

One other problem you may encounter is that the point where the pole is held at each corner isn't the same for all the corners. Assuming each corner block has been laid at the same elevation, you could avoid this problem by placing a piece of hacksaw blade in the joint (on the top of the first block) between the first and second courses. Set the pole on the hacksaw blade to check heights. Remember to cut out the piece of blade before leaving the wall as completed.

Keep the story pole in a dry place when not in use to prevent warping and possible breakage. On a large job you might want to have more than one story pole. If so, make several at the same time to help ensure consistency and prevent error.

If you take care to get the first course level all the way around the footing, you will save time later. Stretch a good mason's line (nylon) between corners, and pull the line tight. (Nylon will stretch and can take a strong pull.) If the wall is long, the line might sag. If it does, sight the line at eye level or measure its height in the middle of the wall.

Using a Trig. You may want to use a trig to correct a sagging line. A trig is a paper-thin metal clip about 4 inches long with a slot to hold the line.

Figure 3-16 shows a metal trig on a block laid in the center of the wall and held in place by a stone. The trig

Masonry & Concrete Construction

The foot plank permits laying of an extra course.
Figure 3-14

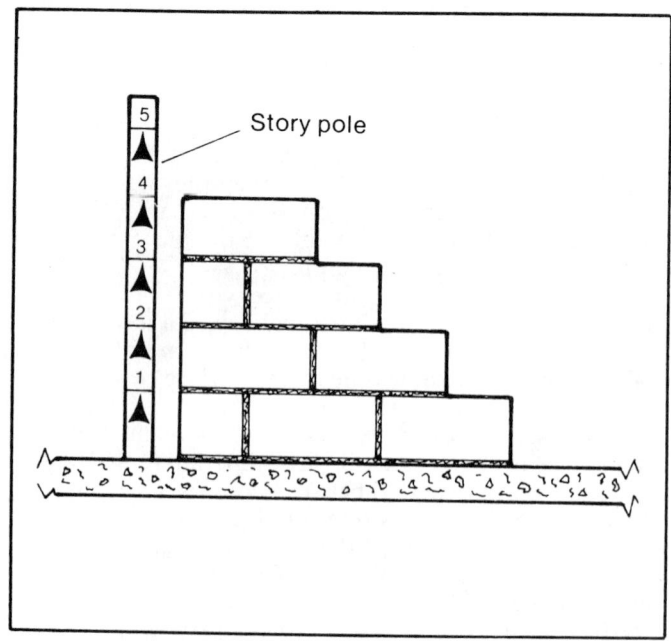

Number the story pole to show the courses of block. Arrows under the numbers show which way is up.
Figure 3-15

A metal trig keeps the line from sagging or swaying.
Figure 3-16

The closure block, with end joints filled in solid
Figure 3-17

is laid on a block to keep the line from sagging or swaying. If the metal trig isn't available, you can improvise by using a matchbook cover or a piece of paper. The trig block should be a standard 8-inch block laid level and plumb. Stand at one corner and sight the line across the trig to the opposite corner. This will determine if there is a hump in the wall.

With a builder's level you can set the trig block at the same elevation as the corner blocks. Then all you will need to do is measure the height of each succeeding trig block at 8 inches.

The Closure Block. A problem will arise when you are ready to lay the last block, commonly called the closure block, in the wall. The closure block can't be pushed against its adjacent blocks, so the end joints can't be compressed. To get a satisfactory bond, you will need to fill the ends of the block with mortar. (See Figure 3-17.) Some State inspectors require this, and it is a good habit to develop.

Stocking the Block
With the first course in, plan the stocking of the block. By counting the number of blocks in the first course, you can determine the number of blocks required for the remaining courses. Have these blocks placed (stocked) near the footing of the wall.

Be careful when stocking up. Concrete block chips very easily. Filling in these chips is a costly, time-consuming job, and usually leaves an unsightly finish.

Also be careful not to smear the blocks. Even if the blocks are painted later, the smears may still show.

Before stocking the blocks, decide where the work on the wall is to be done. If the wall is to be built entirely from the inside, try to stock all the block inside the building lines. If this isn't possible, leave a V opening in one wall through which to bring your materials. This opening can be filled once the walls have been completed.

The location of the mixer can also save time and money. Placing it close to the work can save the laborers a lot of walking.

Pilasters
Foundations are often constructed with walls that are long and not tied with any partitions or columns. These walls need some type of reinforcement to keep from cracking. Wire reinforcement, which is discussed in Chapter 5, is one common type. Pilasters are another. (See Figure 3-18.) A pilaster is constructed in a wall to help withstand horizontal pressures caused by backfilling against a hill or incline.

Construct the pilaster as the wall goes up. Figures 3-18 and 3-19 show how a 10-inch block wall pilaster is constructed. In this course, the ten-inch jamb or square block is placed on the pilaster with a six-inch block behind it. The bond continues on through the pilaster.

Figure 3-19 shows the six-inch block laid on the pilaster with the ten-inch stretcher blocks laid in the normal position in the wall. The pilaster is braced against outside pressure, but is not visible from outside of the wall. Some building codes do specify, however, that pilasters project at least 4 inches from the face of the wall.

Masonry & Concrete Construction

Ten-inch jamb or square block placed on the pilaster with a six-inch block behind it. The bond continues through the pilaster.
Figure 3-18

**Six-inch block pilaster
Figure 3-19**

STEPPED FOUNDATIONS

A stepped foundation is a variation of a continuous wall foundation and is used where the ground slopes or where there is only a partial basement.

The foundation is stepped down gradually to keep the footing on solid ground and to avoid undermining the higher part of the foundation with excavations at the time of construction. (See Figure 3-20.)

Where the foundation is built on sloping rock, cut level steps. These steps prevent the foundation from slipping. Slight slopes can often be chipped or doweled. If the surface is loose or decomposed due to weathering, it must be cut away.

When building a stepped foundation always make the horizontal distance, H, over 2 feet, and the vertical step, V, less than three-fourths the horizontal distance. For average soil, a vertical step of not more than 2 feet in a horizontal distance of 4 feet is generally satisfactory.

There should be a projection on the vertical part of the step. This projection should be as wide as the footing on the horizontal part of the step and at least 6 inches thick. (See Figure 3-20.)

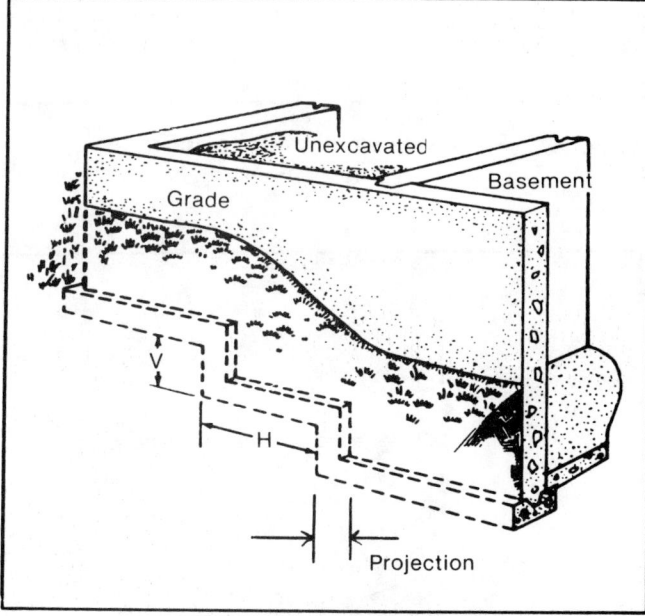

When building a foundation on sloping rock, cut steps to prevent the foundation from slipping.
Figure 3-20

PIER FOUNDATIONS

Pier foundations are more economical than continuous wall foundations. Standard piers are built of brick, concrete, rubble stone, and hollow masonry filled with concrete. More slender piers are made with reinforced concrete.

Pier spacing is largely a matter of economy and building weight. Since beam strength varies inversely with the square of the beam span, a beam spanning 6 feet is four times as resistant to bending as a beam spanning 12 feet. Closer spacing of piers is more economical than using heavier beams.

Concrete and concrete block piers 3 to 4 feet high must be reinforced at each corner with 3/8-inch vertical steel rods. The rods should be at least 2 feet long and hooked at the bottom. Figure 3-21A shows the rods buried 5 inches into the footing to bond the pier to the footing; the rest of the rod is extended into the pier.

Concrete and concrete block piers 4 to 6 feet high need additional reinforcing as shown in Figure 3-21B. Seek the advice of a structural engineer for larger piers.

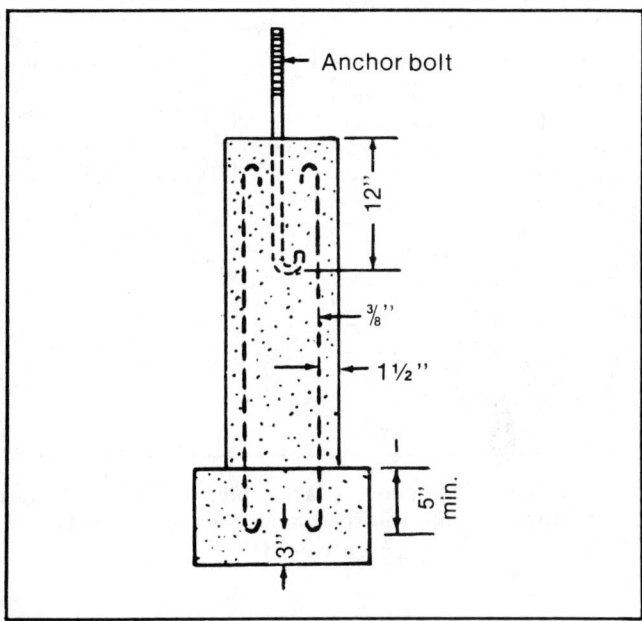

Reinforcing and anchoring short piers
Figure 3-21A

It is easier to reinforce 4- to 6-foot piers with two-piece reinforcing rods than with one-piece rods that run the full height of the pier. Set short dowels of reinforcing rod into the footing so that they protrude 18 inches. (See Figure 3-21B.) They should align with the corresponding pier corner-reinforcing rod. Wire the dowels to the pier corner rod.

Anchor the building to the piers with 5/8-inch bolts. Set the bolts with a hook bent in the lower end 12 to 18 inches into concrete piers and 3 feet into piers of unit masonry. They should extend above the pier far enough to fasten down the building sills or girders. Figure 3-21C shows how wood posts are anchored to concrete piers.

Buildings with pier foundations are more susceptible to wind uplift than buildings with other types of foundations. This is because the piers are often not heavy enough or sufficiently anchored to the ground.

Piers made of masonry may not be taller than ten times their smallest dimension. When using structural clay tile or hollow concrete masonry units for isolated piers supporting beams or girders, fill the cells and spaces with concrete or type M or S mortar whenever the unsupported height exceeds six times its smallest dimension. Loads supported by piers should be centered directly on the piers.

Masonry & Concrete Construction

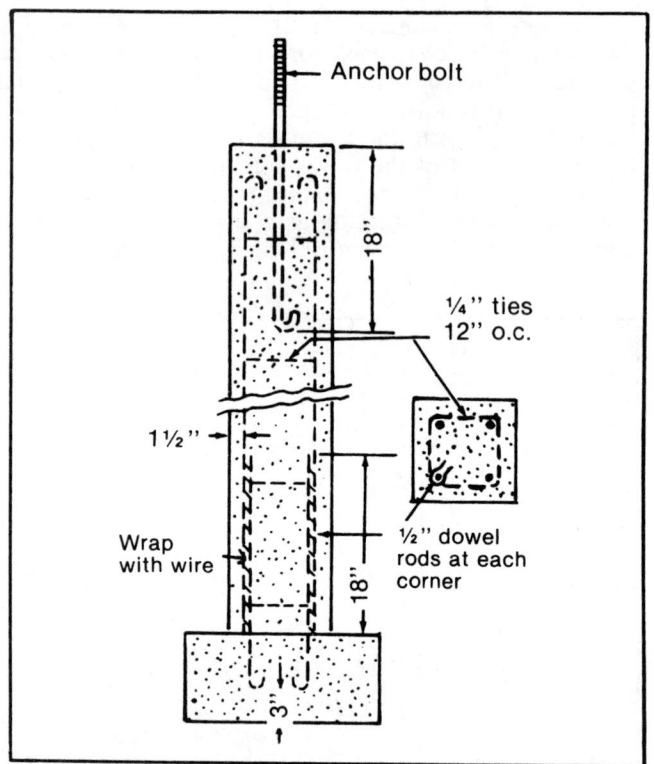

Reinforcing and anchoring larger piers. Note the dowels wrapped to the reinforcement with baling wire.
Figure 3-21B

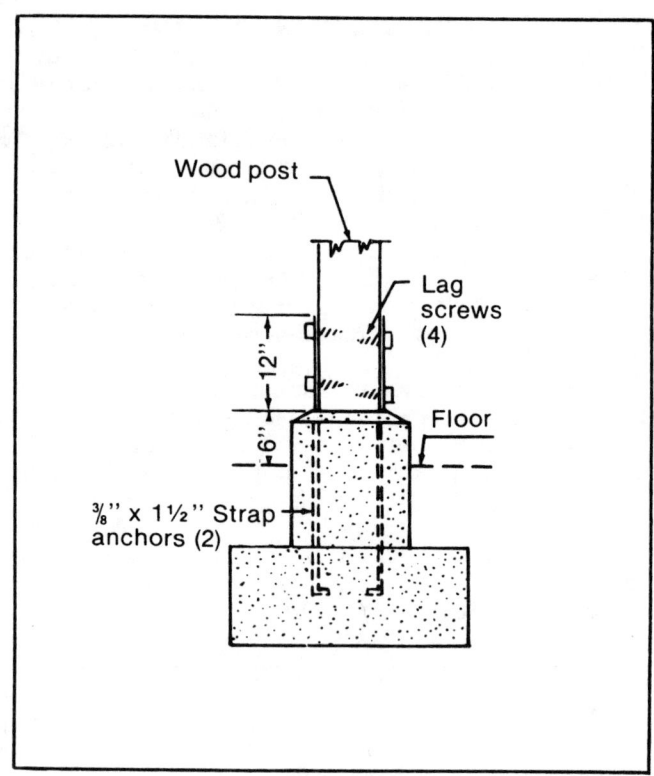

Anchoring wood posts to concrete
Figure 3-21C

GRADE BEAMS

A grade beam is a rectangular, reinforced concrete beam that serves as a continuous foundation. It does not extend into the soil more than a few inches, but is supported on reinforced concrete piers.

Where soft soil overlays a more compact soil, a good foundation can often be secured with a grade beam. Drill holes through the soft soil and fill with reinforced concrete piers. These piers support the reinforced beam on the soft soil.

A grade beam itself extends about 8 inches above grade and about the same distance below grade. It is placed over loose fill such as gravel, cinders, or similar porous material that will drain water from beneath the beam.

The concrete piers are 10 inches in diameter. Space them 8 feet on center for a one-story building. The piers are reinforced with 5/8-inch reinforcing rod that extends through the grade beam.

The cross-sectional size of the grade beam for the average one-story house should be about 8 inches by 16 inches. The size for buildings other than homes depends on the weight of the structure and the length of the beam. These sizes should be determined by an engineer. Before agreeing to build with a grade beam, check the local building codes or building inspector's office to see if it is allowed in the area. In cold climates they are not usually allowed. Figures 3-22A through D show the types of grade beam construction.

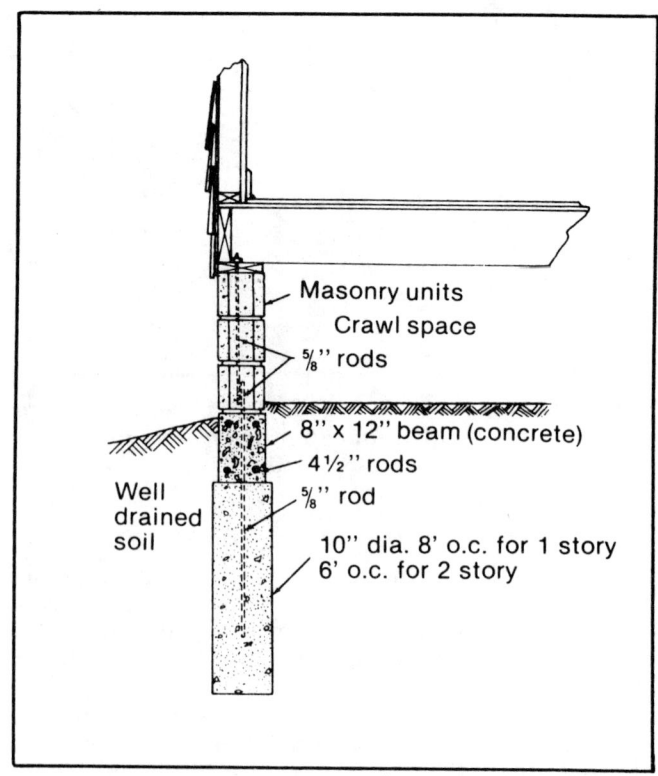

Grade beam for a frame house with crawl space
Figure 3-22A

36

Foundations

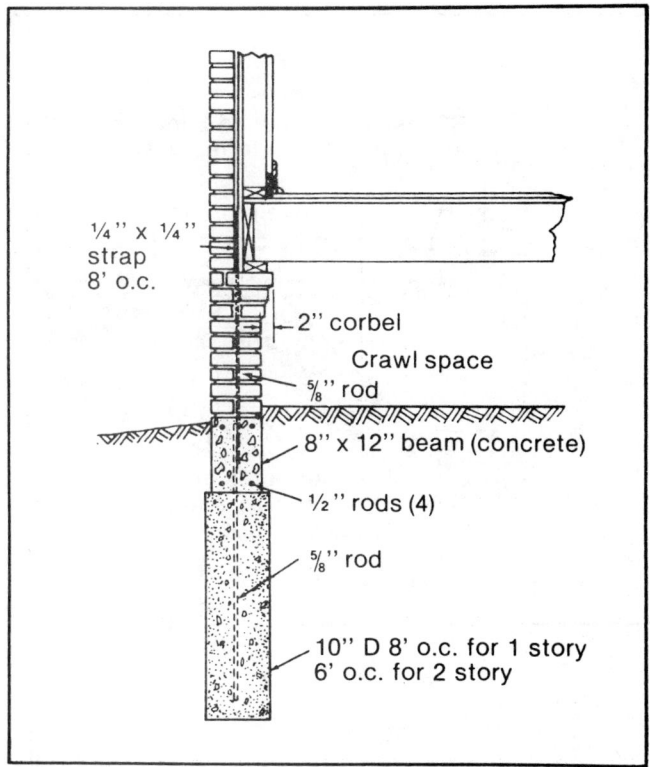

Grade beam for a brick veneer house with crawl space
Figure 3-22B

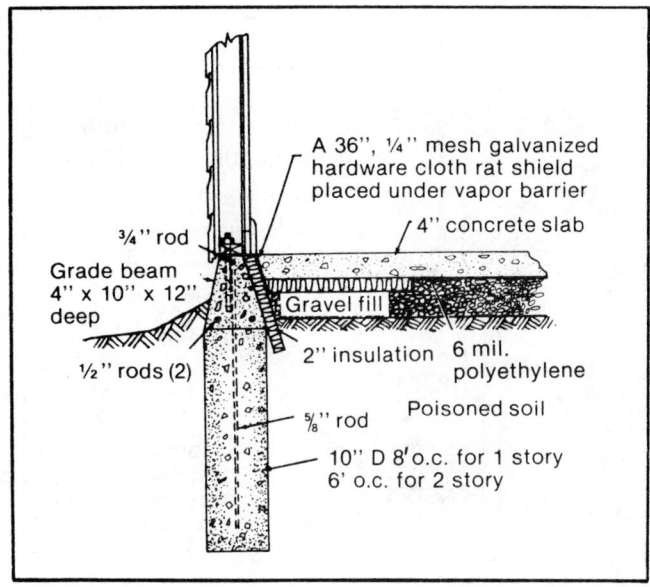

Grade beam for a frame house with slab floor.
Figure 3-22C

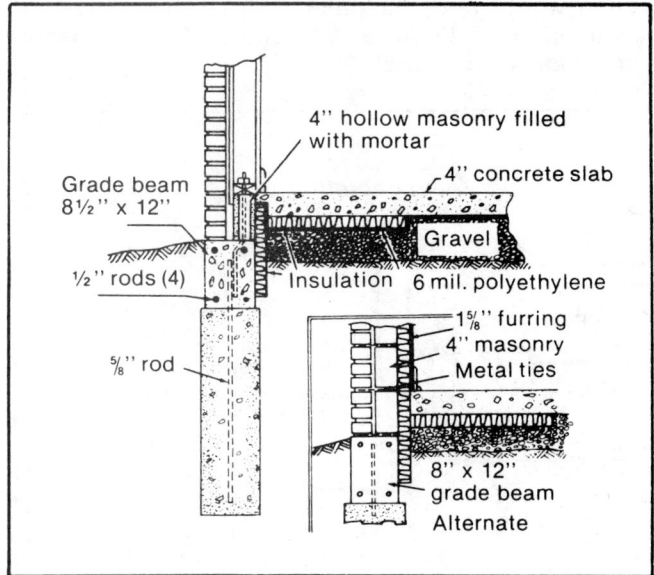

Grade beam for a brick veneer house with slab floor.
Figure 3-22D

SLAB FOUNDATIONS

Slab foundations differ from slab floors. Slab floors are cast independently of the foundation and are usually isolated from the foundation with rigid insulation.

Slab foundations are both foundation and floor and are cast as one reinforced unit. The building is anchored to the slab with 5/8-inch bolts that are set in the slab when it is cast. The weight of the slab anchors the building to the ground. The design of a slab foundation—its dimensions and reinforcing—depends on the weight of the building, the type and drainage of the soil, and the climate. The slab for even light structures should be designed by a structural engineer familiar with the local soil conditions.

Compact the soil under the slab to prevent it from settling and cracking. If fill is required it must be allowed to settle thoroughly and be as firm as undisturbed soil before the slab can be poured. Cover the soil under the slab with 4 or 5 inches of gravel to give a well-drained subgrade for the slab.

For protection against frost, the adjoining ground should slope away in all directions, and the underlying soil should be sand or gravel to reduce to a minimum the heaving action due to frost. Avoid silty sand or clay soil. For a perimeter wall foundation, the bottom of the footing should be below the frost line. If necessary, install footing drains or use some other means to reduce the height of the water table below the slab.

Concrete for use in slabs on soil should have a minimum compressive strength at 28 days of 2,000 psi and should be either controlled concrete or average concrete. If possible check the local building codes for psi requirements. Average concrete must be in the proportions of 1 part portland cement to not more than 6 parts of combined volumes of fine and coarse aggregate, and not more than 7½ gallons of water for each 94-pound bag of cement. It should have a minimum of 5 sacks of cement per cubic yard.

Concrete reinforcement should conform to ASTM Standard Specifications for Welded Steel Wire Fabric For Concrete Reinforcement. The reinforcement should consist of a minimum of 20 pounds for every 100 square feet, distributed evenly in both directions, and lapped a

Masonry & Concrete Construction

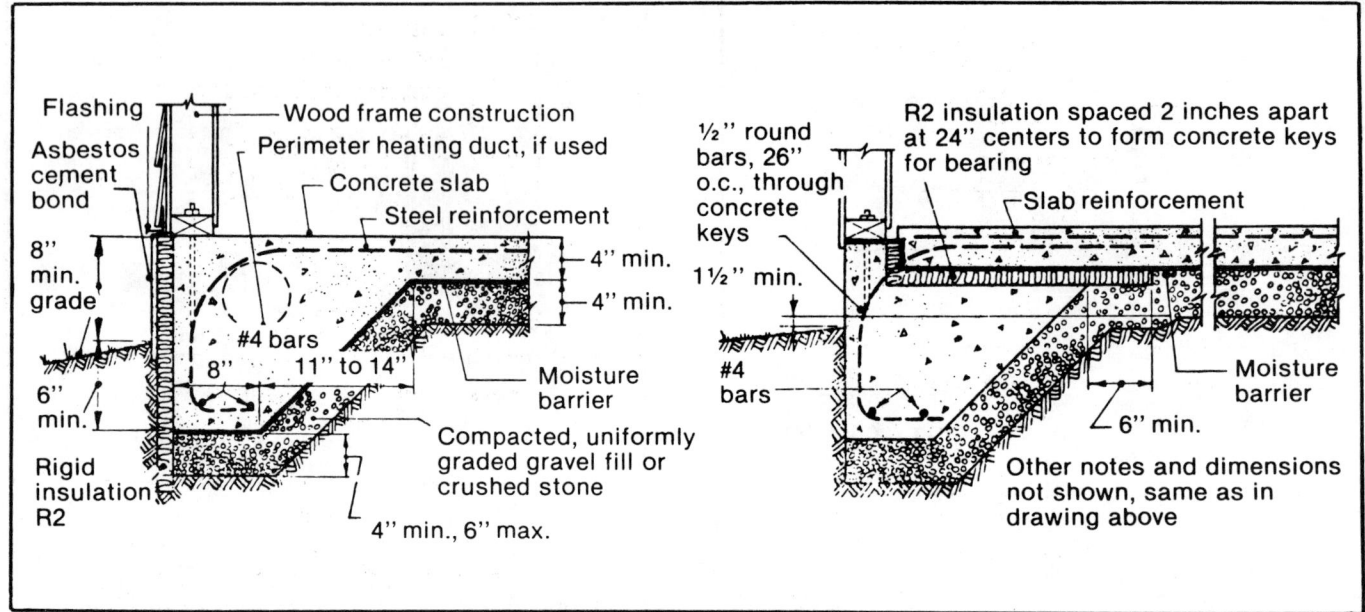

Floating slab foundations
Figure 3-23

minimum of 6 inches at all edges. When using plain or deformed bars, reinforcing should be structural grade or better. A slab should have a minimum of 30 pounds for every 100 square feet, distributed evenly.

Moisture barriers in slabs should be either membrane waterproofing or 35-pound minimum asphalt or coal tar pitch roofing felt. It should be lapped 6 inches at all edges.

In dwellings, slabs on soil below grade subject to ground-water accumulation should be waterproofed in accordance with the illustration in Chapter 5, on waterproofing masonry.

Figures 3-23 and 3-24 illustrate two types of slab-on-ground wood frame buildings with heights of 12-foot maximum from floor to eave and 20-foot maximum from floor to gable peak.

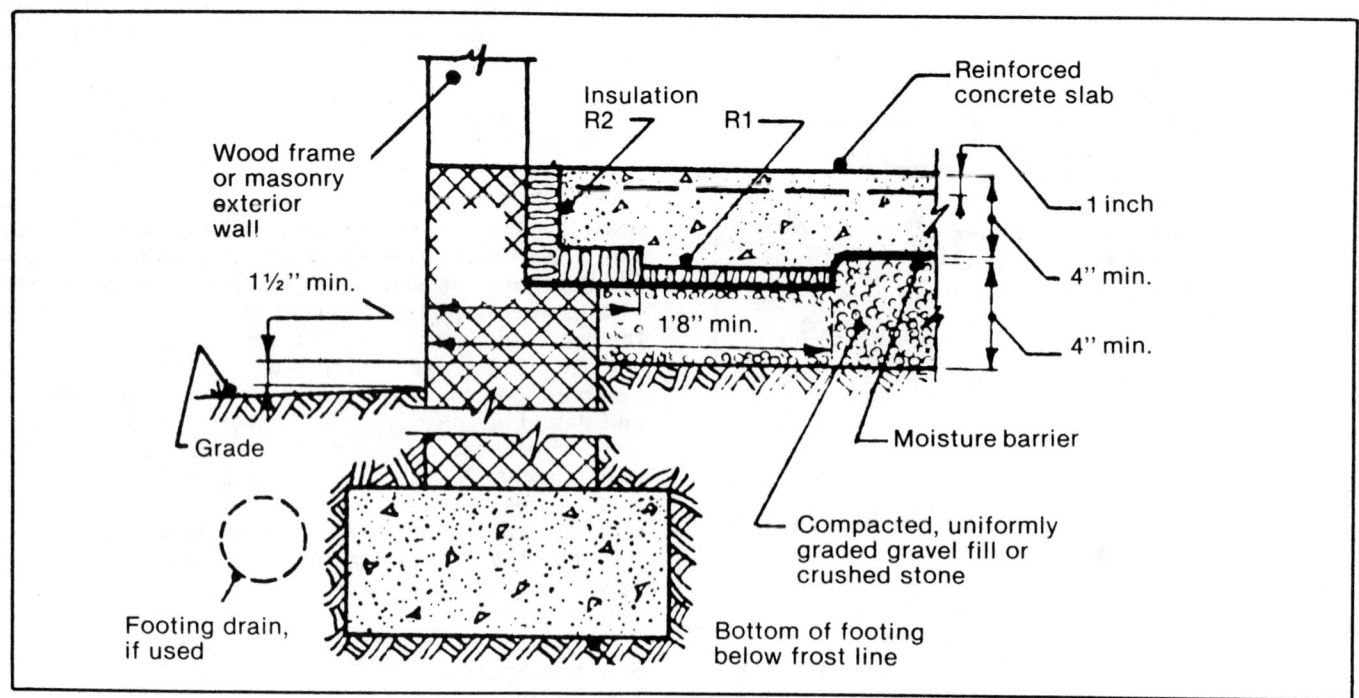

Perimeter wall foundations
Figure 3-24

Foundations

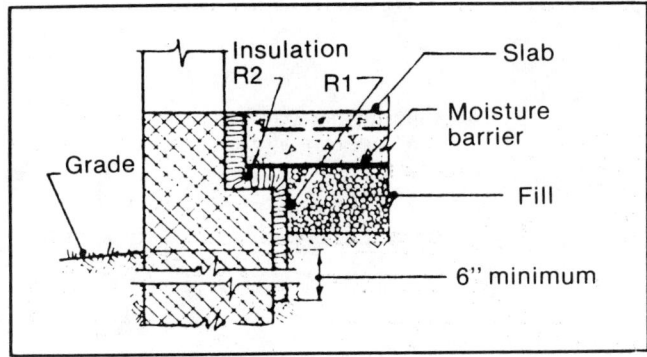

Cellular glass enclosing sealed-in gas
Figure 3-25

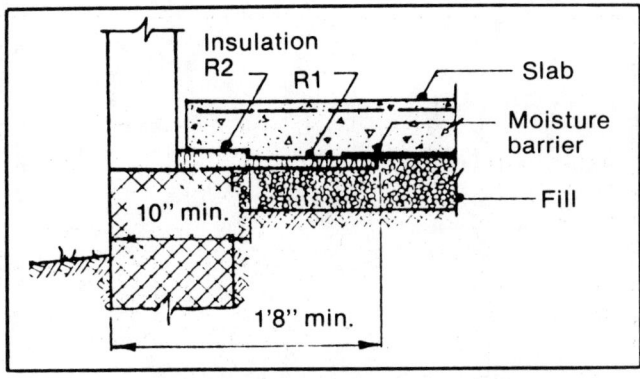

Glass fiber with a plastic binder
Figure 3-26

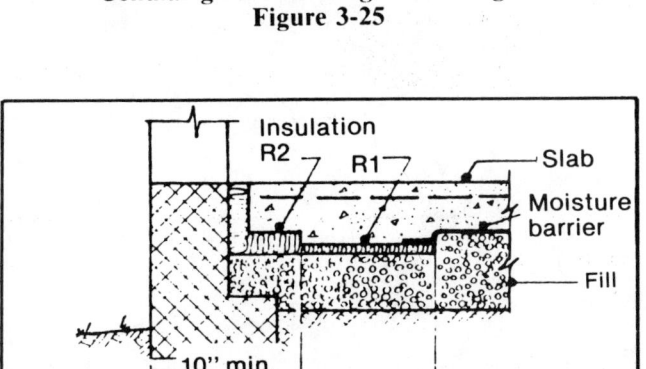

Cane or wood fiberboards
Figure 3-27

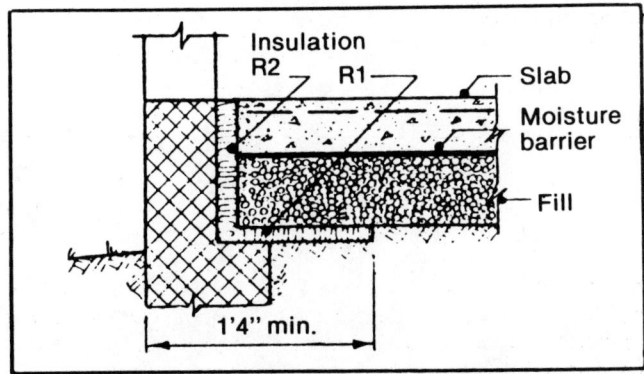

Hard cellular rubber enclosing sealed-in gas
Figure 3-28

Slab-On-Ground Construction: Perimeter Wall Foundation

Along with the explanation of proper slab-on-ground construction, Figures 3-25 through 3-28 show the proper method of insulation. In dwellings, insulation should be the rigid type. The following materials and minimum thicknesses are recommended:

Cellular glass enclosing sealed-in gas, 2 inches thick for R1 and R2. Once cut to size, it is dipped in roofing pitch or asphalt. Use wire ties to attach in a vertical position. (See Figure 3-25.)

Glass fibers with a plastic binder, ¾ inch thick for R1 and R2. Once cut to size, it is dipped in a roofing pitch or asphalt. (See Figure 3-26.)

Cane or wood fiberboards, ½ inch thick for R1, $25/32$ inch thick for R2. Once cut to size, it is dipped in roofing pitch or asphalt to form a heavy coat. (See Figure 3-27.)

Hard cellular rubber enclosing sealed-in gas, ½ inch thick for R1 and R2. For attaching to masonry in a vertical position, coat with asphalt or pitch, or use cement keys or metal ties. (See Figure 3-28.)

Chapter 4
Mortar

Mortar used in masonry work has four functions:
1. To bind together the units — the blocks or bricks — in a stable structure.
2. To present an effective barrier to the infiltration, absorption, or passage of moisture into, through, or out of the wall.
3. To resist weathering.
4. To present a neat, uniform appearance.

PORTLAND CEMENT

Portland cement is the binding agent in mortar, having replaced lime mortar and natural cement. The ASTM (American Society for Testing and Materials) has made specifications for different types of portland cement. Of these types, Type I and Type III are used most often in the making of mortar. The other types of portland cements are usually reserved for concrete.

Type I portland cement is general purpose and used in most situations. Type III is a high early-strength cement and is used when fast setting is needed, as in cold weather.

When portland cement is used to make mortar, another ingredient, hydrated lime, must be added. Hydrated lime serves to keep the mixture from becoming stiff, sticky and hard to use by holding water in the mix.

Most mortar used today for laying brick or block is made of masonry cement. Masonry cement is factory made and composed of portland cement, lime, and some admixtures. This premixing at the factory results in a high quality and a good appearance because the materials are mixed before they are put into bags. Another advantage is that when masonry cement is used, there are fewer materials on the job and fewer chances of error in proportioning the materials.

ASTM SPECIFICATION FOR MORTAR

Many buildings that are built to an architect's specifications must be made of mortar that conforms to ASTM specification C270, "Mortar for Unit Masonry."

In 1954 the ASTM committee on Mortars for Unit Masonry changed classifications of mortar types from A-1, A-2, B, C, and D to M, S, N, O and K. The main reason for the change was that in the United States, the term A-1 has become synonymous with the "best" or "top quality," and some committee members believed that this designation for high early-strength mortar was misleading. The new classification was apparently developed by dropping the alternate letters A, O, W and R from the words "mason work."

Recommended Applications of Types of Mortar:

Type M A high-strength gray masonry cement for load-bearing walls subject to extremely heavy loads, violent winds, seismic loads, and severe frost action. For use in high-rise structures, free-standing structures, isolated piers and other load-bearing structures.

Type S A medium-strength gray masonry cement for load-bearing walls at or below grade, nonload-bearing exterior walls above grade, retaining walls, foundation walls, foundations, sewers, brick pavements, walks, patios, manholes, stucco, plaster, and all other uses subject to heavy structural loads.

Type N A waterproofed gray masonry cement for use in mortar for laying brick, concrete block, tile, stone and glass block. Used in nonload-bearing interior and above grade exterior walls.

Type O A masonry cement that is low in strength, suitable for interior use in nonload-bearing partition walls. The walls should not be subject to freezing temperatures.

Type K This type of mortar is very low in bond strength and in compressive strength. It attains only 75 psi after 28 days. Used in decorative work that isn't subjected to any load other than its own. It isn't used very often.

PROPERTIES OF MORTAR

Mortar must develop certain properties in order to produce good functional masonry work. The properties and the means by which they are developed are as follows:

Plasticity
The unhardened mortar must be very plastic, smooth and workable so that it may be spread with a minimum of effort to cover the surfaces to be coated and bonded.

Body
The mortar must possess sufficient body and cohesiveness, and must be able to adhere to itself and to all masonry surfaces with which it comes in contact. It must resist being squeezed out of the joints and prevent uneven settlement as additional courses are added to the new wall or column being built.

Water Retention
The mortar must possess high water retention: the ability to resist loss of water to highly absorbent masonry units being bonded, and to resist "bleeding" of water when the mortar is in contact with nonabsorbent units, such as glass blocks. Loss of water due to poor retention results in rapid loss of plasticity in the mortar. This may seriously reduce the completeness and effectiveness of the bond and strength development.

Bond
The essence of masonry construction is that the hardened mortar binds the units together and does not just hold them apart. To make a thorough bond, coat all contact surfaces of the unit fully and evenly with mortar so that there is an intimate and complete contact. Mortar must be highly plastic and must retain that plasticity through water retention and proper chemical set retardation until the block above is pressed down on it. It must, however, have adequate "body" to resist being squeezed from the joints under the load of later courses.

Strength
An important strength in masonry work is resistance to tension and buckling. Distress due to crushing from compressive loads is virtually unheard of. In most cases in which ruptures of any sort occur, the unit and the mortar remain essentially undamaged. The failure occurs in the bond between the unit and the mortar.

Water Repellency
As important as binding qualities is the effectiveness of the bond between the unit and the mortar in resisting moisture. Leakage, dampness, or other manifestations of moisture in masonry work that are not due to faulty flashing or poor design can always be traced to poor bonding. The qualities which promote good structural bond also contribute to watertightness and water repellency.

Yield
The cost of masonry construction is largely dependent on the mortar's yield, or extendibility — the number of masonry units which can be laid with a given quantity of mortar while retaining the necessary properties given above.

Color
Uniform, harmonious color is most desirable in exposed surfaces of masonry work. First of all, the mortar itself must be uniform in color. Using cement from one bag ensures uniformity of color.

Cement is available in various colors. A quick check with a local material supplier will help. The more common colors like gray come in bags. Marquette Cement Company puts out a product called *Dark* which is sold in bags. It is a waterproofed black masonry cement for use in mortar for laying brick tile or stone where a certain aesthetic effect is desired. Due to the minute particle size of the coloring material and a unique dispersing system developed by Marquette, this cement is one of the darkest-colored masonry cements on the market. Marquette also offers a Buff and Dark Brown that, like the Dark, are 750 psi (Type N). To locate a dealer near you, write Marquette at First American Center, Nashville, Tennessee 37238.

Buying pre-colored mortar instead of using coloring agents has many advantages. Pre-colored masonry cement will remain uniform in your batches whether you mix a partial or a whole bag with sand, whereas coloring agents may produce variations in color and make a good job look bad. Also, adding too much coloring agent will affect the strength of the mortar.

Admixtures
Use of the modern portland-cement-based masonry cement eliminates the need for special combinations of material. In fact, these masonry cements develop the best qualities when no admixture, waterproofer, plasticizer, or other material is added except good, well-graded mason sand and clean water. Also, the use of portland-cement-based masonry cement is comparable in cost to using mortars containing such substitutes as cement lime and cement lime plus admixture.

Efflorescence
The biggest mortar problem in masonry construction is efflorescence. Efflorescence is deposits of salt on the surface of masonry walls. It is usually white, chalky and when dry, can be brushed off. Efflorescence usually occurs under two conditions: when the wall contains some soluble salts, and when there is moisture in the wall to make the salts "bleed." If you eliminate these two problems, you eliminate efflorescence.

It is believed that as many as ten different chemical (salt) combinations cause efflorescence. One chemical that is used in the construction of walls is calcium chloride. It is used during cold weather to accelerate the setting up of the mortar to decrease the likelihood of the mortar freezing. The proper proportion is 2 percent of the weight of the mix. Adding too much of the chemical can create problems later and may lead to efflorescence. Mortar doesn't start to freeze until around 28 degrees,

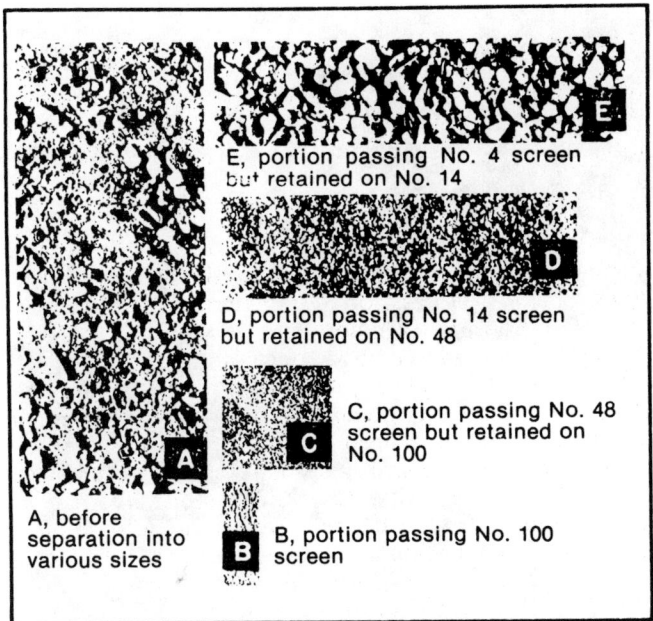

Sample of well-graded sand
Figure 4-1

because the chemical reaction of the mortar setting up generates a slight amount of heat.

The following are some precautions you can take to prevent efflorescence:

1. Use clean water, never salt water.
2. Try to buy washed sand.
3. Keep materials covered and dry.
4. Cover the walls at the end of each workday or if it begins to rain.
5. Use good mortar.

SAND FOR MORTAR

Sand is the cheapest and most widely used material in masonry construction, yet its importance is often overlooked. To make strong mortar, the sand must be clean, free of impurities, and well graded. (See Figure 4-1 and Table 4-1.)

Most masonry supply dealers have washed sand for masonry work. It's always a good idea to check the quality of the sand, especially if you are buying from an unfamiliar source. This can be done with the silt-and-clay content test and the colorimetric test. These tests will determine if there is any organic matter in the sand.

The test for silt or clay content is as follows:

1. Put 2 inches of the sand in a quart fruit jar. Add water until the jar is 3/4 full. Screw on the cover, and shake the jar vigorously until the sand is thoroughly washed. Let the contents settle for about 12 hours. The silt will be deposited in a layer above the sand.
2. Measure the layer of silt. If it is more than 1/8 inch thick, the sand is not clean enough for mortar.

The test for organic matter is done as follows:

1. Dissolve a heaping teaspoonful of lye in one-half pint of clear water. Any household lye that contains at least 94 percent sodium hydroxide is suitable. Or you can use just sodium hydroxide which can be purchased at any drugstore.
2. Pour the solution into a glass jar containing one-half pint of the sand. Cover the jar and shake it vigorously for one or two minutes.
3. Let the contents stand for several hours. The color of the liquid will indicate whether the sand contains harmful amounts of organic matter. A clear color indicates clean sand. A straw color indicates some organic matter, but not an objectionable amount. Darker colors indicate an excessive amount of organic matter, in which case the sand must be washed and retested before use.

Caution: Measure the materials carefully in making the test because variations in the concentration of the solution can alter the color. Avoid spilling the solution as lye and sodium hydroxide are highly caustic.

Even though sand is the cheapest building material, don't waste it. One way of saving sand is to place tarps on the ground and have the sand dumped onto the tarps. (See Figure 4-2.) Using tarps also keeps stones from inadvertently being picked up with the sand and mixed in the mortar. The tarps are part of inventory and can be used repeatedly.

After the foundation is built, draw the tarps toward the center and save the leftover sand for another job. (See Figure 4-3.) This also makes cleanup easier.

	Percentage Passing	
Sieve Size	Average	ASTM Limits
No. 4	100	100
No. 8	100	95-100
No. 16	80	60-100
No. 30	50	35-770
No. 50	25	15- 35
No. 100	7	0- 15

Grading of mason's sand
Table 4-1

PROPORTIONS

Mix mortars according to ASTM specifications for Unit Masonry (ASTM C 270-80a). (See Table 4-2.)

In most cases, mortar (Type N) consisting of one part masonry cement (a 70-pound bag is equivalent to one cubic foot) mixed with three parts (cubic feet) of mason's sand will produce desired strengths (minimum of 750 psi at 28 days) and plastic-placing properties with a minimum of shrinkage and waste. Richer mixes (Types S and M — Table 4-3) containing less sand are desirable only when very high strengths (1800 psi and 2500 psi minimum strengths) are required for special structural purposes. However, using poorly graded sand sometimes necessitates the use of rich mixes to attain the desired plasticity. Make every reasonable effort to obtain mason's sand which meets ASTM Specification C144, "Aggregates for Masonry Mortar."

Masonry & Concrete Construction

**Place tarps on the ground before dumping sand.
Figure 4-2**

**When the foundation is finished, collect leftover sand
for future use.
Figure 4-3**

	Parts By Volume Of -			
Mortar Type	Portland Cement	Masonry Cement	Hydrated Lime	Aggregates measured in a damp loose condition
M	1	1	--	Not less than 2¼ and not more than 3 times the sum of the volumes of the cements and lime
	1	--	¼	
S	½	1	--	
	1	--	Over ¼ to ½	
N	--	1	--	
	1	--	Over ½ to 1½	
O	--	1	--	
	1	--	Over 1¼ to 2½	
K	1	--	Over 2½ to 4	

ASTM specifications for unit masonry
Table 4-2

Mortar Type	Average Compressive Strength at 28 Days
M	2500 p.s.i.
S	1800 p.s.i.
N	750 p.s.i.
O	350 p.s.i.
K	75 p.s.i.

Compressive strengths for mortar type
Table 4-3

Tables 4-4 and 4-5 show amounts of materials necessary to make mortar for 100 units and for 100 square feet of wall. (Note: There is no allowance for waste in Table 4-5.)

Quantities for a 1:2½ mix can be calculated from a 1:3 mix for corresponding conditions of joint and wall thickness.

1. Multiply the cement content by the weighted factor of 1.1553.
2. Multiply the new cement content by 2½ for the sand volume.
3. Be careful when using damp sand. Adding just 5 to 10% more moisture can cause the sand to bulk, leaving a smaller amount of sand per cubic foot.

Concrete or Clay Unit Size	Wall Thickness	No. of Units	Mortar C.F.	Masonry Cement Bags	Sand C.F.
5″ x 8″ x 12″	8″	220	3.7	1.2	3.7
8″ x 4″ x 16″	4″	110	2.6	0.9	2.6
8″ x 8″ x 16″	8″	110	2.6	0.9	2.6
8″ x 10″ x 16″	10″	110	2.6	0.9	2.6
8″ x 12″ x 16″	12″	110	2.6	0.9	2.6

Material required for 100 square feet of wall using a 1:3 mix, ⅜-inch thick joint
Table 4-4

Concrete or Clay Units	Wall Thickness	Mortar Required C.F.	Masonry Cement Bags	Sand C.F.
5″ x 8″ x 12″	8″	1.8	0.6	1.8
8″ x 4″ x 16″	4″	2.4	0.8	2.3
8″ x 8″ x 16″	8″	2.4	0.8	2.3
8″ x 10″ x 16″	10″	2.4	0.8	2.3
8″ x 12″ x 16″	12″	2.4	0.8	2.3

Material required for 100 units 1:3 mix dry material, ⅜-inch thick joint
Table 4-5

MIXING MORTAR

Very little has been written about the mixing of mortars, yet it is an important link in the preparation of good mortar. Mason's sand and the extremely fine cement must be mixed thoroughly to distribute the cement evenly throughout the batch. It is also important to use the right amount of water and to mix it thoroughly into the batch to achieve the desired plasticity.

Hand Mixing

The old method of mixing mortar with a mason's hoe in a mortar box or wheelbarrow is still used for small jobs, but it is difficult and impractical to use for large projects. When using this method, mix the sand and the cement first until a uniform gray color is obtained, then add the water a little at a time. If the mix should turn out soupy, add more sand and cement at the same ratio as in the original mix.

Machine Mixing

There are many excellent mechanical mixers available which remove the drudgery and produce uniform, well-mixed batches of mortar.

In machine mixing, place approximately half of the required water in the machine. Add about half the required sand as the mechanical mixing action continues. Then add the cement and the remainder of the sand. As the mixing continues and the mass stiffens, add the remainder of the water. The mixing action should continue at least three minutes after all the materials are in the mixer. A longer mixing time of five minutes can increase yield, workability, water retention, and board life.

One tip to increase efficiency is to place the small mixer on cement blocks to make it high enough to dump into a small mortar box. (See Figure 4-4.) Using a mortar box allows more than one batch to be mixed at a time and still allows the mixer to be cleaned before the mortar is used up. A few extra minutes spent cleaning the mixer at the end of each workday will make the mixer last much longer and will also make the mixing of the mortar easier the next day.

Retempering

By mixing in more water, retempering may be accomplished at any time within the first 2½ hours after mixing without damage to the ultimate properties of the

**By standing the mixer on cement blocks so it can dump into a mortar box, a builder can mix more than one batch at a time.
Figure 4-4**

mortar. The mortar should always contain as much water as it can carry while maintaining adequate "body" and water retention. Only wet plastic mortar can develop good bond with absorbent masonry units. Therefore, retempering should be done as frequently as needed; however, it is preferable to keep mortar batches to a size which can be used in the work in a reasonably short time.

The length of time that mortar will retain its full plasticity before retempering is needed depends on the natural setting characteristics of the cement, the temperature of the mixture, and the temperature and humidity of the surrounding air, as well as the amount of sunlight and wind to which the mortar is exposed. Mortar will retain plasticity longest on days that are cool, damp, cloudy and still. Plasticity will disappear most rapidly on hot or cold days that are dry and windy when the water used in the masonry cement is heated above 80 degrees.

Where practical, cover the mortar with wet burlap to protect it from premature loss of plasticity. Mortar left overnight should never be retempered.

Masonry cement is manufactured to develop setting properties which blend best with most masonry operations — that is, to set fast enough to permit speed in construction and slow enough to permit reasonably large batches to be mixed and used prior to stiffening. These characteristics are set for normal weather conditions. In hot, dry weather, use smaller batches and damp sand to prevent frequent retempering. In winter, keep the mortar warm to prevent unduly delayed setting. But use care! Hot mortar may flash-set any time of the year and chemical reactions are faster at high temperatures than they are at low temperatures.

PLACEMENT OF MORTAR

Workmanship is the key to successful, well-bonded, water-repellent, crack-free masonry construction. When you consider that each individual block or brick and trowelful of mortar is placed by hand, the importance of good workmanship is evident. This type of construction is one of the few that have resisted mechanization. Artisans must be top notch, yet adequate workmanship is frequently lacking due to emphasis on speed, skimping on materials, or failure to appreciate the importance of the manual operations involved. Take the time and care to do the job properly.

Chapter 5
Concrete Form Construction

Building contractors have the option of using prefabricated forms or of constructing their own on the job site. If you decide to construct your own forms you will probably use plywood for the construction material. Virtually any exterior-type APA plywood can be used for concrete formwork because all panels with that identification are manufactured with waterproof glue.

PLYFORM

Plyform is a plywood product designed specifically for concrete forming and is the recommended panel for most general forming uses. It is an exterior-type plywood composed of certain wood species and veneer grades to ensure quality and durability. All Plyform panels bear the APA trademark of the American Plywood Association.

Unless otherwise specified, Plyform panels are sanded on both sides and oiled at the mill. Face-oiling reduces moisture penetration and keeps the concrete from sticking to the forms. Unless the mill-oiling is reasonably fresh when the panels are first used, the plywood may require additional oiling. Many users apply a top-quality edge sealer before the first pour. With proper care it is not uncommon for Plyform panels to last through as many as ten or more pours.

Plyform is available in two classifications: Plyform Class I and Plyform Class II. Of the two, Class I is the stronger and stiffer panel. Either type may be ordered with High Density Overlaid surface on each side.

HDO Plyform

High Density Overlaid Plyform Class I and Class II meet the same general specifications as Plyform Class I and II. Both classes of HDO Plyform have a hard, smooth semi-opaque surface of thermosetting, resin-impregnated materials that form a durable, continuous bond with the plywood. Though the abrasion-resistant surface does not require oiling, many users wipe the panels lightly with oil or other release agents before each pour to ensure easy stripping.

To begin a plywood-formed concrete foundation, set and brace one corner first. Then place the succeeding panels and add bracing to hold them in place.

Notice in Figure 5-1 how the keyway formed in the footing is in the center of the concrete wall line. This serves to lock the wall and the footing together.

The APA Standard requires panels to be square within 1/64 inch per linear foot for panels 4 feet by 4 feet or larger. Panels must be manufactured so that a straight line drawn from one corner to an adjacent corner will fall within 1/16 inch of the panel edge. Figure 5-2 shows a plywood concrete form braced by walers and strongbacks.

The tolerances and requirements set by the American Plywood Association ensure quality and minimize the time and labor required to build forms. Good construction practices dictate an awareness of these tolerances at the jobsite. In an extreme case, two 3/4-inch sanded panels, both within manufacturing tolerances, could form a joint with a 1/32-inch variation in surface level from panel to panel. Realignment of panels or the addition of shims are quick, easy solutions.

HDO plyform is usually specified when the smoothest concrete finishes are desired because the panel has a hard, smooth surface. Both sides are moisture resistant and can be used with equal effectiveness. With reasonable care, HDO Plyform can normally be used for as many as 50 pours. Some concrete-forming specialists have used the same panels for as many as 200 pours or more with good results.

Structural I Plyform

This panel is stronger and stiffer than Plyform Class I

Masonry & Concrete Construction

The corners are the first part of a plywood-formed footing
Figure 5-1

Plywood concrete form braced by walers and strongbacks
Figure 5-2

Concrete Form Construction

Grade-Use Guide for Concrete Forms*

Use these terms when you specify plywood	Description	Typical Trademarks	Veneer Grade Faces	Veneer Grade Inner Plies
B-B PLYFORM Class I & II** APA	Specifically manufactured for concrete forms. Many reuses. Smooth, solid surfaces. Mill-oiled unless otherwise specified.	B-B PLYFORM CLASS I EXTERIOR APA PS 1-74 000	B	C
High Density Overlaid PLYFORM Class I & II** APA	Hard, semiopaque resin-fiber overlay, heat-fused to panel faces. Smooth surface resists abrasion. Up to 200 reuses. Light oiling recommended between pours.	HDO PLYFORM EXT-APA PS 174	B	C Plugged
STRUCTURAL I PLYFORM** APA	Especially designed for engineered applications. All Group 1 species. Stronger and stiffer than PLYFORM Class I and II. Recommended for high pressures where face grain is parallel to supports. Also available with High Density Overlay faces.	STRUCTURAL I B B PLYFORM CLASS I EXTERIOR APA PS 1-74 000	B	C or C Plugged

* Commonly available in 5/8" and 3/4" panel thicknesses (4' x 8' size).
** Check dealer for availability in your area.

Grade-use guide for concrete forms
Table 5-1

or II and is designed specifically for engineered applications. HDO Structural I Plyform contains all Group I wood species and is often recommended for high pressures where face grain is parallel to supports. It is also recommended where several reuses are required.

Related Grades

Additional plywood specifications designed for concrete forming include special overlay panels and proprietary panels. These panels are designed to produce a smooth uniform concrete surface and are generally mill treated with a form-release agent. Some proprietary panels are made of Group I wood species only and may have thicker face and back veneers than those normally used. This provides greater parallel-to-face grain strength and panel stiffness. Faces may be specially treated or release coated. Check with the manufacturer for design specifications and surface treatment recommendations. Table 5-1 is a grade-use guide for plywood forms.

Plywood Tolerances

Plywood is an engineered product, manufactured to exacting tolerances under U.S Product Standard PS 1. A tolerance of plus 0.0 inch and a minus 1/16 inch is allowed on the specified width and/or length. Sanded Plyform panels are manufactured with a thickness tolerance of 1/64 inch of the specified panel thickness for 3/4 inch and less, and plus or minus 3 percent of the specified thickness for panels thicker than 3/4 inch.

Overlaid Plyform panels have a plus or minus tolerance of 1/32 inch for all thicknesses through 13/16 inch. Thicker panels have a tolerance of 5 percent over or under the specified thickness.

JAHN FORMING SYSTEM

Though there are many different brands of concrete form accessories, the "Jahn" forming system is probably the most common.

5/8-inch or 3/4-inch plywood is used. Although 5/8-inch is slightly lower in cost, 3/4-inch is stronger, lasts longer, uses fewer ties and brackets and is more economical in the long run.

A 5/8-inch eccentric take-up on brackets allows use of a 4¾-inch wall tie with either 3/4-inch or 5/8-inch plywood.

The only preparation required with the Jahn forming system is gang drilling the plywood. This operation is shown in Figure 5-3. See Figure 5-11 for spacing.

Drill holes 1/8 inch larger than tie-end. This will usually be 9/16 inch or 5/8 inch. (See Figure 5-4.)

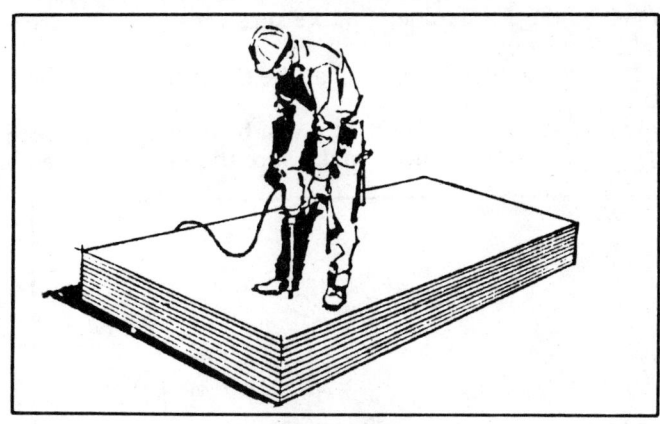

Gang drilling plywood
Figure 5-3

Masonry & Concrete Construction

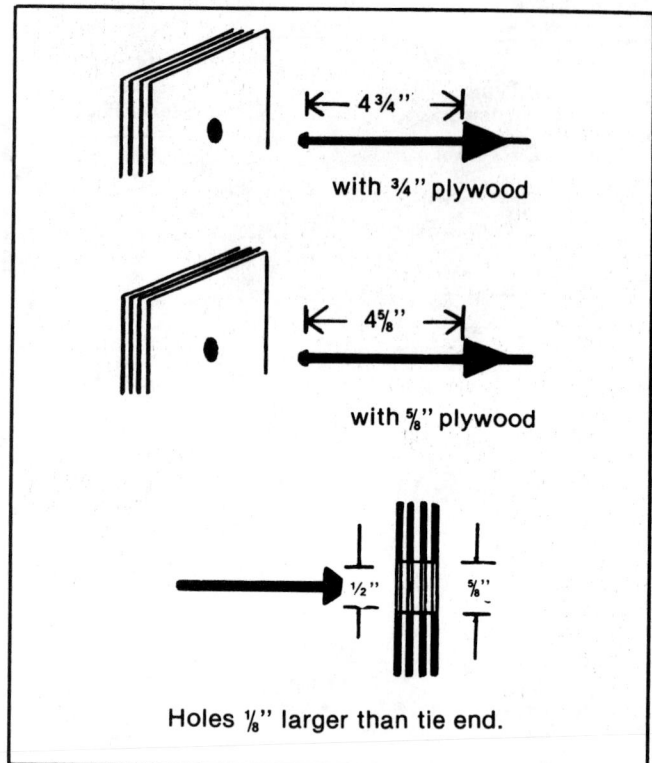

Tie and hole sizes
Figure 5-4

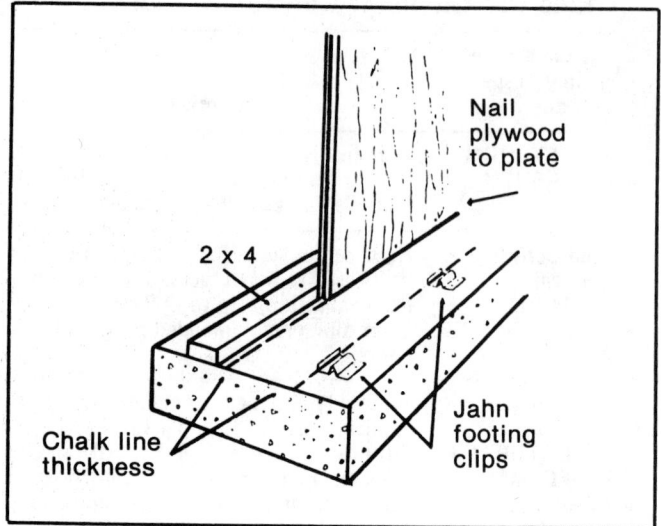

Jahn footing clips used with 2 x 4-inch plate
Figure 5-6

Procedure

Footing Plates and Clips Good forming requires good level footings or subgrade. Snap a chalk line behind the plywood at the wall line.

Nail down Jahn footing clips (Figure 5-5) 2 per plywood sheet at 2-foot maximum spacing, or nail down a 2- by 4-inch plate as shown in Figure 5-6. On heavy pours put plates on both sides.

Panel Erection Figure 5-7 shows the correct positioning of the panels. (A) First, set the plates or clips. (B) Then plumb and nail the first sheet of plywood to the plate and brace temporarily. (C) Now erect additional sheets, nailing them to the plate and holding temporarily with Jahn Ply Holders. Make sure vertical joints are tight and snug.

Be sure the panels are level at the top if they are to be staked.

Once the first side is set up it can be oiled. When this is done check the panels. Make sure the ties have been

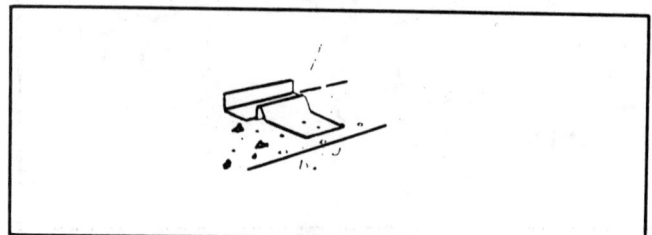

Jahn footing clip
Figure 5-5

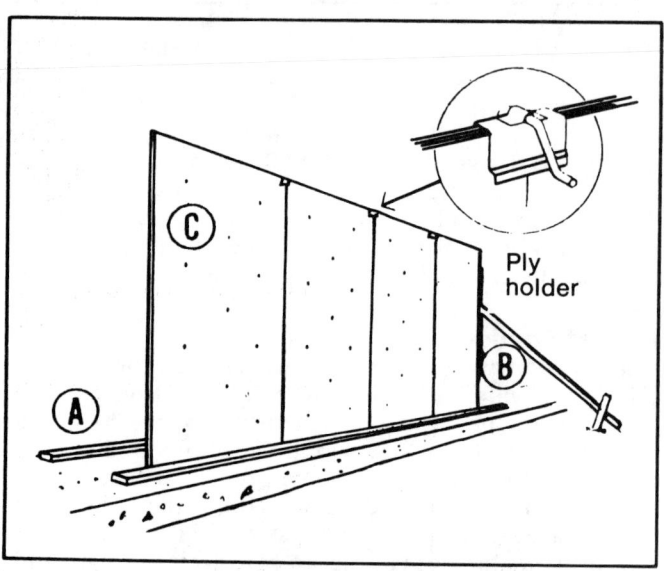

Panel erection
Figure 5-7

inserted properly and the plywood forms have been placed over the keyway in the footing. (See Figure 5-8.) Check the forms for plumb and make sure they are well braced and safe before ordering concrete.

Installing Ties and "A" Brackets

1. Snap Ties—Any standard 4¾-inch end 3000 lb. snap tie will work with the Jahn "A" Bracket and 3/4-inch plywood. For 5/8-inch plywood use 4⅝-inch end ties. Two men can do this job quickly and efficiently if one inserts the ties and the other installs the Jahn "A" Brackets. (See Figure 5-9.)

The Jahn "A" Brackets (Figure 5-10), with the eccentric take-up, are the only brackets that will not shake loose under internal vibration. Their features include the following:

Concrete Form Construction

Plywood forms being set up for a concrete foundation
Figure 5-8

Installing ties and "A" Brackets
Figure 5-9

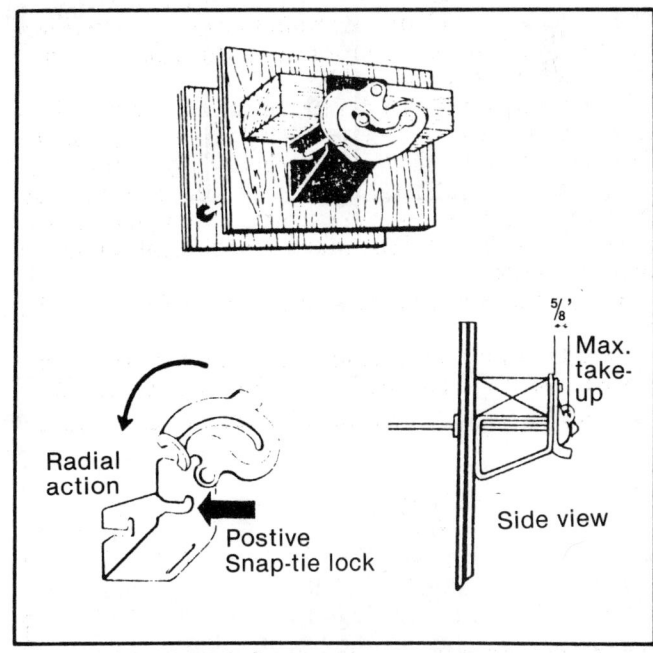

Jahn "A" Bracket with eccentric take-up
Figure 5-10

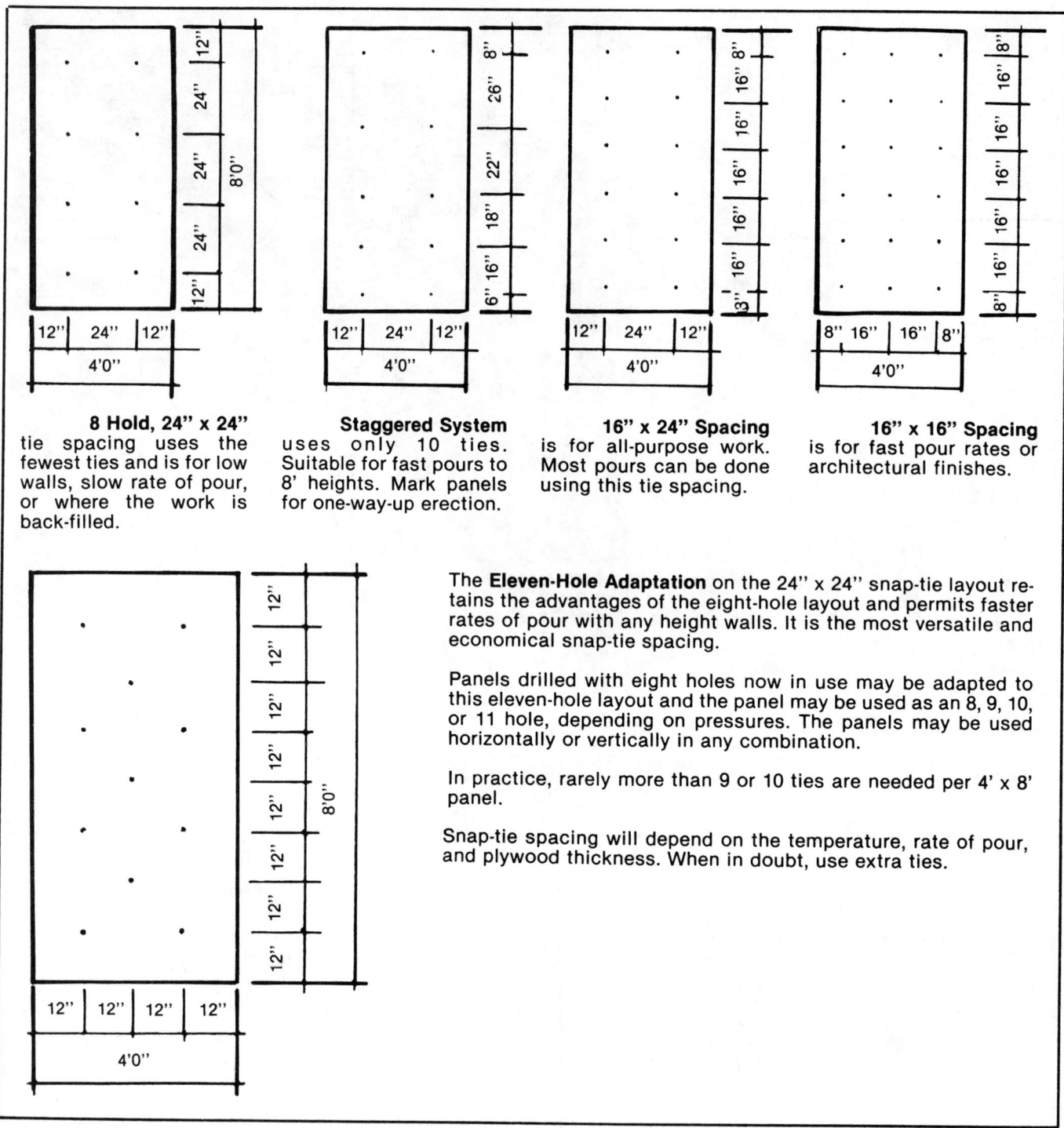

Typical snap-tie layouts
Figure 5-11

- They can be installed either before or after the walers are in place
- The bracket slots slip easily over the tie end
- The tie seats positively in the slot end
- The radial action of the eccentric has a drawing instead of a shearing action against the tie head
- The offset fastening of the eccentric on the body helps keep the eccentric from loosening under vibration
- The pressure of the bracket body is applied against the 2 by 4 instead of against the plywood

The snap tie spacing layout you use will depend on the temperature, rate of pour, plywood thickness, and nature of the job. Figure 5-11 shows typical snap tie layouts.

Concrete Form Construction

2. **Special Tie Spacing Layout**—The special tie-drilling layout shown in Figure 5-12 allows single 2 by 4 stud coverage of the vertical panel joint whether the joint is continuous or the panels are staggered. This layout may be used with straight or radius wall work. CAUTION: Be sure to start drilling the tie holes 1⅛ inches in from the right edge of the panel. This applies to both interior and exterior wall forms.

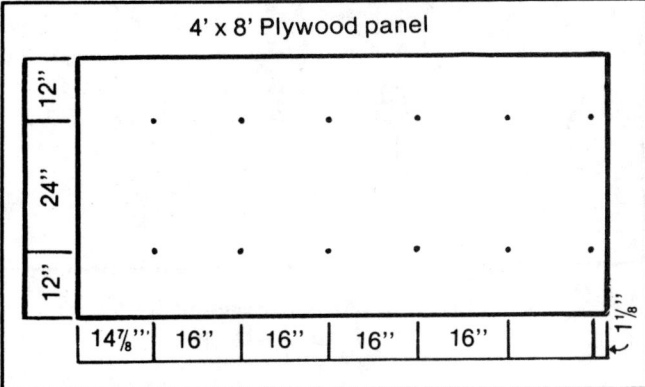

Special tie spacing layout
Figure 5-12

3. **Proper "A" Bracket Installation**—Figure 5-13 illustrates each step in "A" Bracket installation procedure with waler not already in place.
• Place an "A" Bracket on the tie end by slipping the slotted bracket body over the end of the tie (A).
• Slip the eccentric loosely over the tie end so the tie fits in the slot (B).
• Drop the 2- by 4-inch waler in place. Seat the waler by hand or with a hammer, if necessary (C).

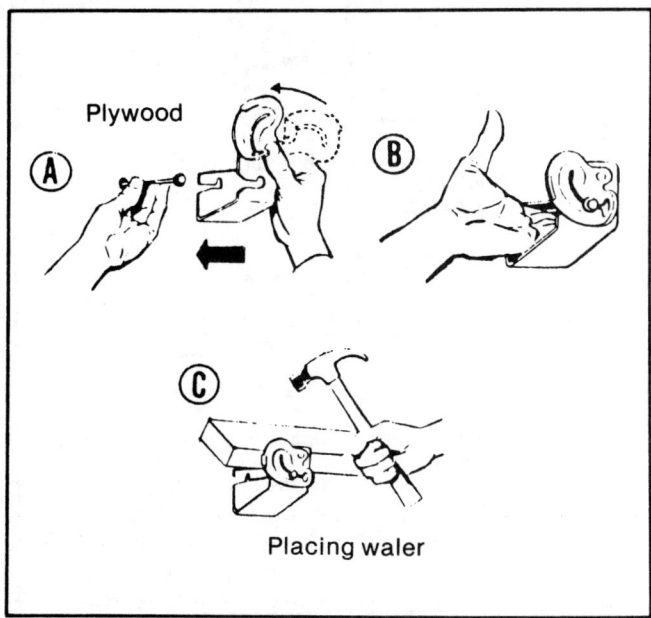

Proper "A" Bracket installation without waler in place
Figure 5-13

If the waler is already in place, install the "A" Brackets according to the procedure shown in Figure 5-14.
• Slip the back slot of the "A" Bracket over the tie directly behind the tie head (A).
• Push the bracket back toward the plywood until the tie head emerges through the hole in the front slot (B).
• Swing the eccentric over the tie end and secure it to the tie (C).

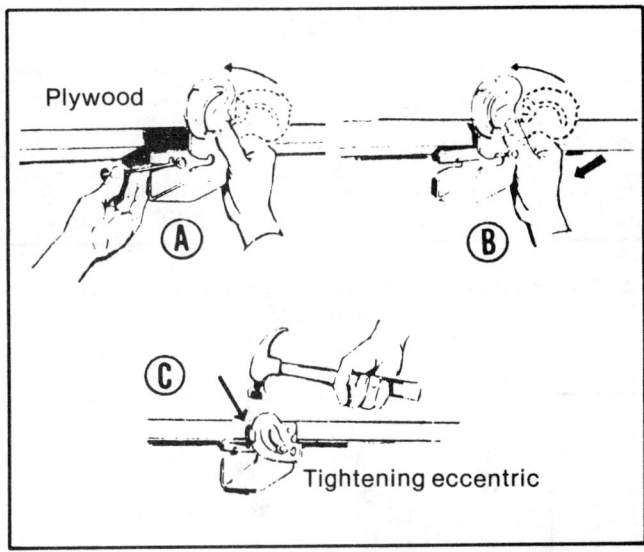

Proper "A" Bracket installation with waler in place
Figure 5-14

Forming Low Walls The method for forming low walls is illustrated in Figure 5-15.
• Install the top waler in the "A" Brackets first (A).
• Tighten the waler by striking the eccentric lip with a hammer (B).
• Make sure the waler is plumb and in proper alignment (C).
• Note that brackets may be installed either before or after the walers (D).
• Working from the top of the wall to the bottom, install the brackets and walers, tightening as you go (E).
• Break joints should occur at "A" Brackets or where there is no vertical joint in the plywood (F).
• Scabs may be installed at any waler joint using "C" Brackets (G).

Inside Walls Most of the second wall is constructed from the outside. As shown in Figure 5-16, this can best be done by two men slipping the plywood over the tie ends. Starting at the bottom move the panel from side to side or up and down to align the holes in the plywood with the tie ends.

The Jahn "C" Bracket is used in constructing vertical strongbacks where sturdy reinforcement is needed, such as in high walls or foundations. (See Figure 5-17.) The "C" Bracket is also designed for use with loose 2 by 4's and standard 4 by 8 plywood panels. A standard long-end snap tie is used with this bracket. The "C" Bracket and double waler method can be used to support a horizontal seam.

F Waler joints
A
B
C
D
E
G Scabbed joints

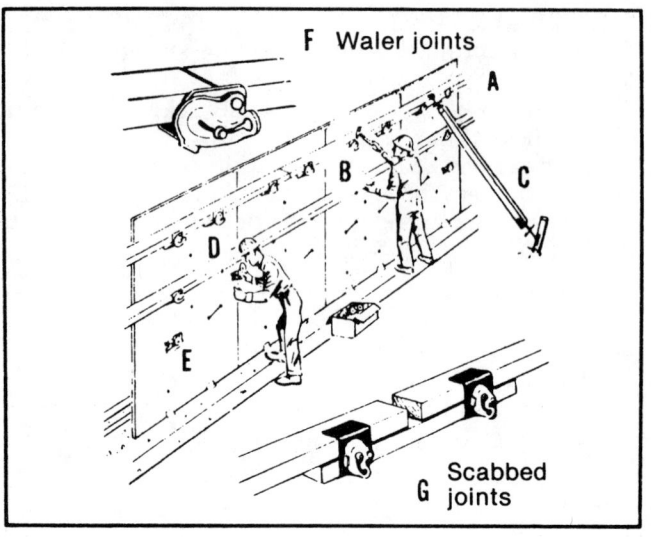

Procedure for erecting forms for low walls
Figure 5-15

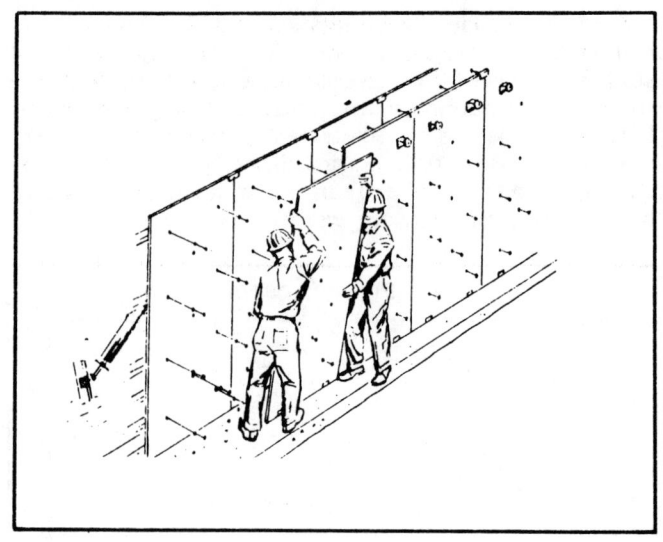

Erecting panel for second wall
Figure 5-16

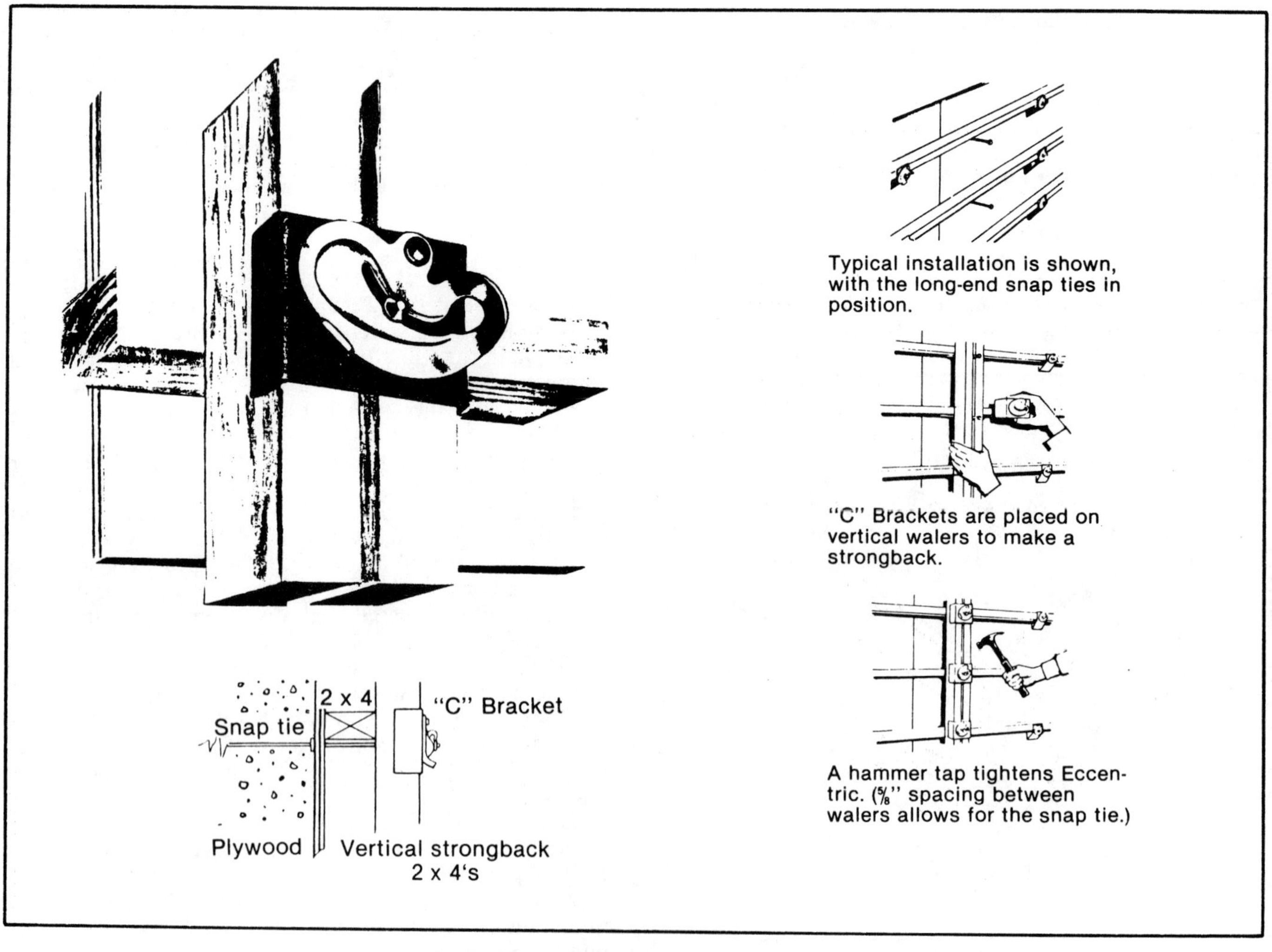

2 x 4
"C" Bracket
Snap tie
Plywood
Vertical strongback
2 x 4's
Typical installation is shown, with the long-end snap ties in position.
"C" Brackets are placed on vertical walers to make a strongback.
A hammer tap tightens Eccentric. (⅝" spacing between walers allows for the snap tie.)

Use of Jahn "C" Bracket in constructing vertical strongbacks
Figure 5-17

Outside Corners To eliminate cutting full panels, start a corner on the inside wall first, using full 4-foot plywood panels. When the outside wall is erected, install full sheets in line with the inside ones and use special filler panels to fill out the exterior corner. (See Figure 5-18.) These fillers should be the same width as the wall thickness. The Jahn Cornerlock is used to secure the walers to the outside corners.

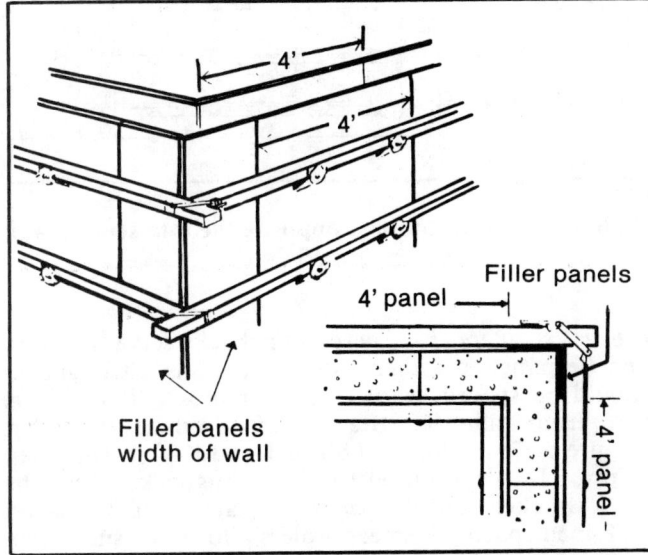

Use of filler panels to fill out exterior corner
Figure 5-18

The cam action of the Jahn Cornerlocks draws the walers together and eliminates costly overlapping, blocking and nailing. Place one waler flush at the corner and slip the Cornerlock into place with its handle perpendicular to the waler. (See Figure 5-19). Drive two nails through the holes on the clamp and pull the handle around 90 degrees. The result will be a snug, tight outside corner. (See Figure 5-20.)

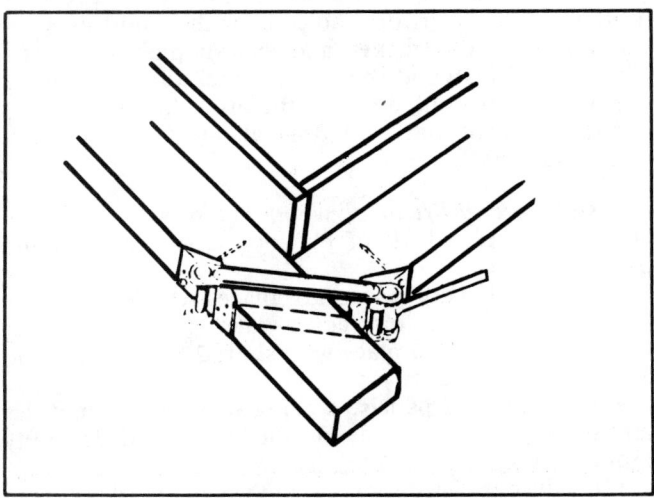

Placement of Jahn Cornerlock
Figure 5-19

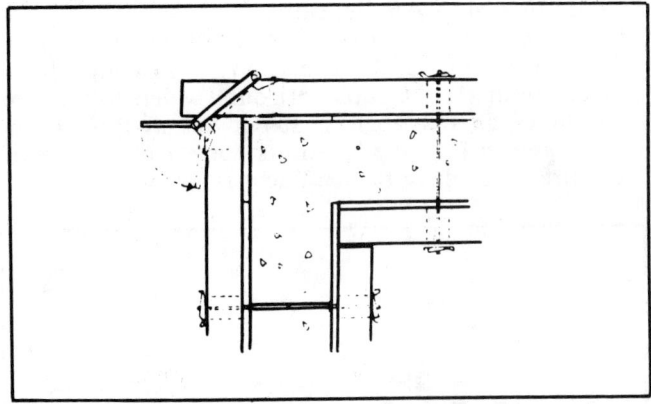

Clamping Cornerlock
Figure 5-20

Attaching Plyform to Framing For most forms, Plyform is attached to the framing with as few nails as possible. For slab forms, each panel must be at least corner nailed. Use 5d nails for 5/8-inch Plyform and 6d nails for 3/4-inch Plyform. In special cases, such as gang forms, additional nailing may be required. Do not butt panels too tightly, especially on the first pour. Figure 5-21 shows a typical form design made from plywood.

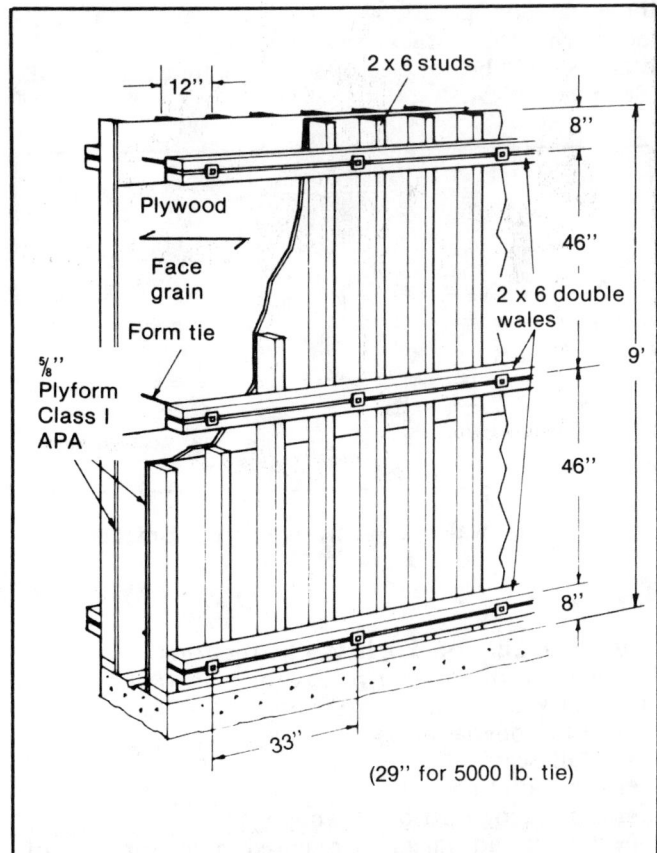

Typical plywood form design
Figure 5-21

Special Forming Applications

Step forming Place forms for steps as shown in Figure 5-22. Using Jahn "C" Brackets and Tie Extenders for a supporting stiffback allows horizontal walers to run free if tie holes do not align at stepdowns in foundation walls. Inset in Figure 5-22 shows correct placement of "C" Brackets where tie alignment is fairly close.

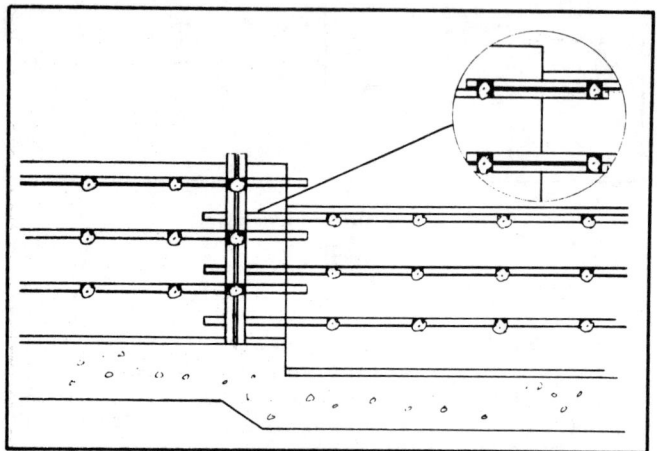

Step forming
Figure 5-22

Three-way wall forming In constructing forms for three-way configurations, horizontal walers can be attached with "C" Brackets as shown in Figure 5-23. The walers should be doubled opposite the intersecting wall. (See Figure 5-24.)

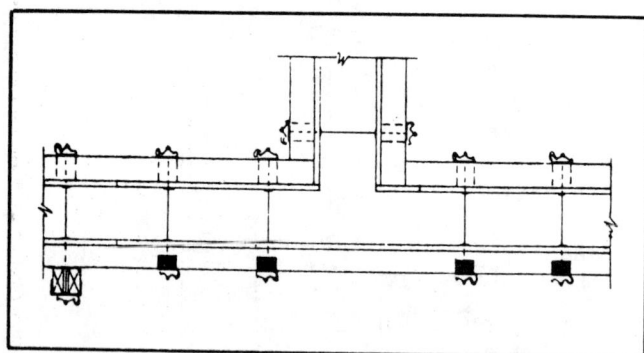

Horizontal walers attached with "C" Brackets
Figure 5-23

Panel Stacking on High Wall Erections

The erection of forming panels for high wall forms is shown at various stages in Figure 5-25.
- Erect strongbacks (1).
- Install Scaffold Jacks (2).
- Install joint cover (3).
- Install second lift of plywood (4).

Fewer ties and walers are required in the top 5 feet of forms (5) when 3/4-inch plywood is used because pressure on the forms begins decreasing at this point.

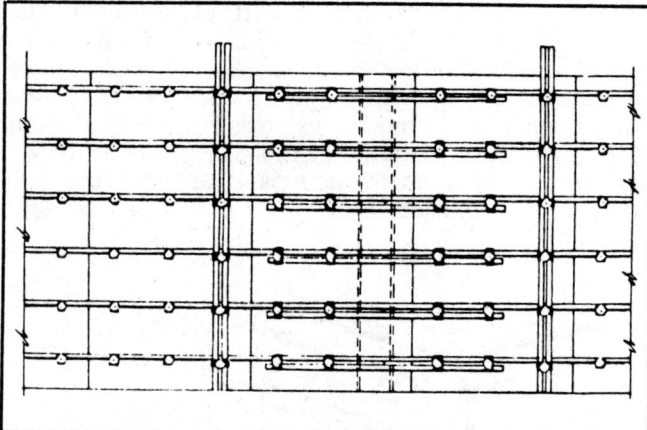

Walers should be doubled opposite the intersecting wall
Figure 5-24

Erect strongbacks These strongbacks serve as liners, and tie panels together on stacked walls. Strongbacks should be spaced every 8 feet for most work (Figure 5-26) using Jahn "C" Brackets with Jahn Tie Extenders (Figure 5-27) or standard 8⅜-inch end 3000 lb. snap ties. Loose 2 by 4's are used for the strongbacks with Jahn "C" Brackets and do not require any nails or spacers (5/8-inch spacing between walers allows for snap ties).

Install Jahn Scaffold Jacks The Scaffold Jacks are installed after strongbacks are in place but prior to installing joint cover walers. The Jahn Scaffold Jack will hold two 2- by 10-inch planks for a comfortable working platform. (See Figure 5-28.)

The horizontal rod slides easily through the "A" Bracket body and is firmly retained in place. The long end of the rod should be slipped in first. The Scaffold Jack has an adjustable support leg for use with either 16-inch or 24-inch tie and waler spacing. (See Figure 5-29.)

Install joint cover Two suggested methods of covering joints when stacking plywood are shown in Figure 5-30. With a single waler, drill holes 1⅛ inch in bottom sheet and put tie, "A" Bracket, and waler in place before adding second plywood sheet. Then tack the top sheet to the waler. For double walers the procedure is basically the same except that "C" Brackets are used instead of "A" Brackets.

Install second lift of plywood The procedure for installing the second lift of plywood is shown in Figure 5-31.
- Lift the plywood sheet into place (A).
- Tack bottom to joint cover waler (B).
- Hold plywood in place with short 2 by 4 spacer and Jahn "C" Bracket (C).
- Set additional panels, tacking to joint cover waler and securing to previously installed panel with Jahn Ply Holders (D).
- Install snap ties, brackets, and walers, working from top to bottom. Brackets can be installed either before or after walers are in place (E).

Concrete Form Construction

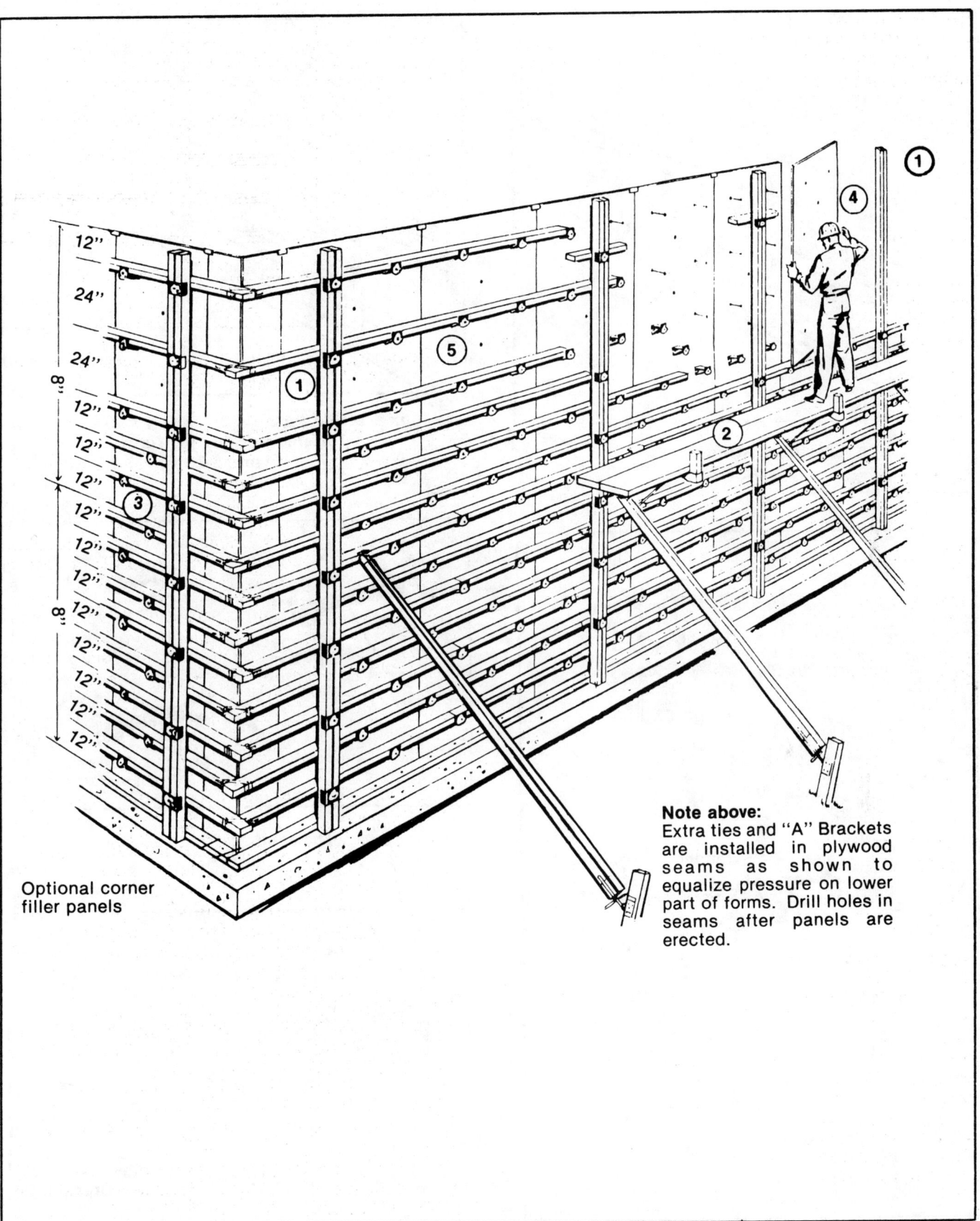

Erection of forming panels for high walls
Figure 5-25

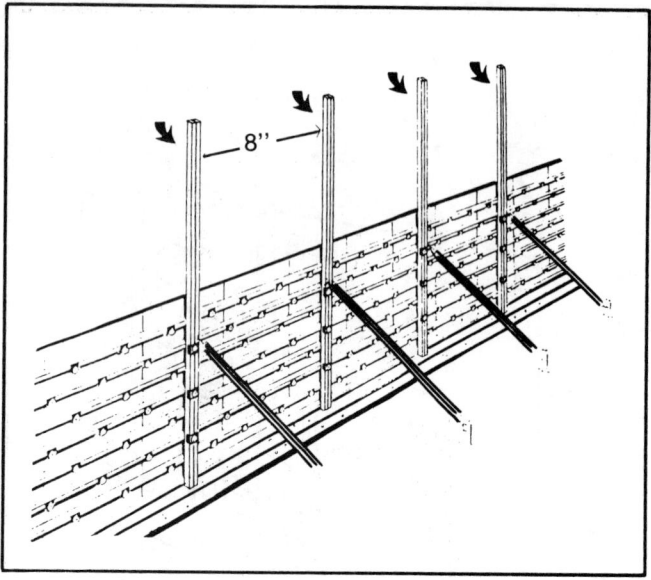

Spacing of strongbacks
Figure 5-26

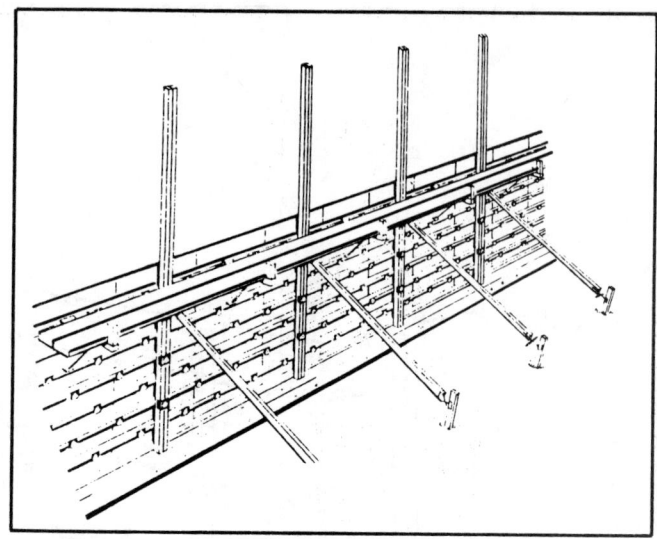

The Jahn Scaffold Jack will hold two 2 x 10's for a comfortable working platform
Figure 5-28

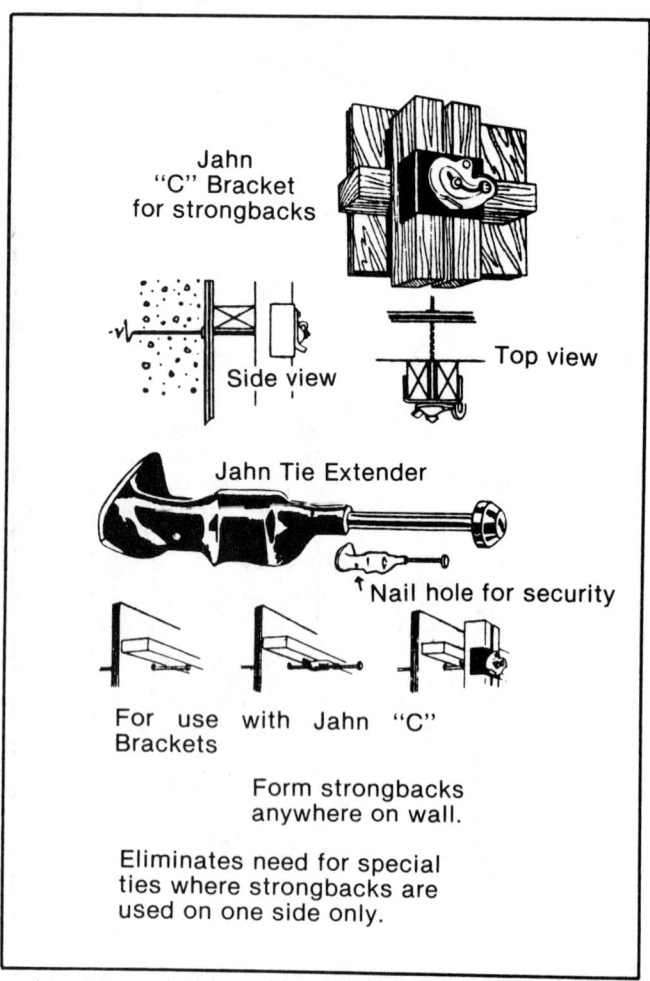

Use of Jahn "C" Bracket with tie extender
Figure 5-27

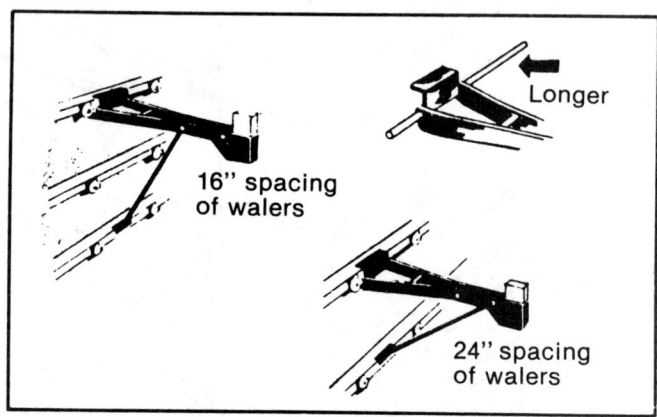

The Jahn Scaffold Jack
Figure 5-29

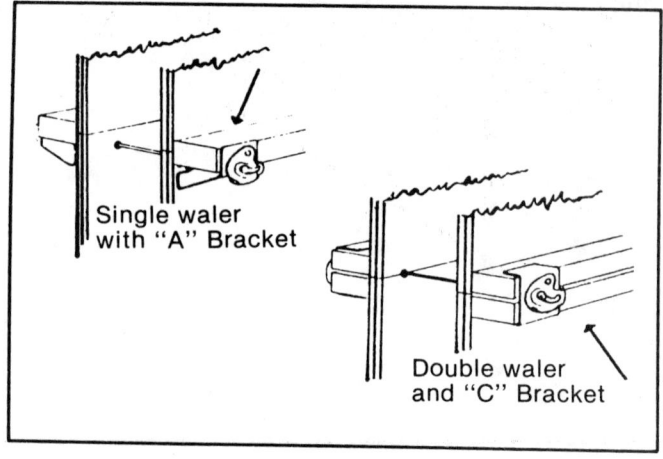

Two methods of covering joints when stacking plywood
Figure 5-30

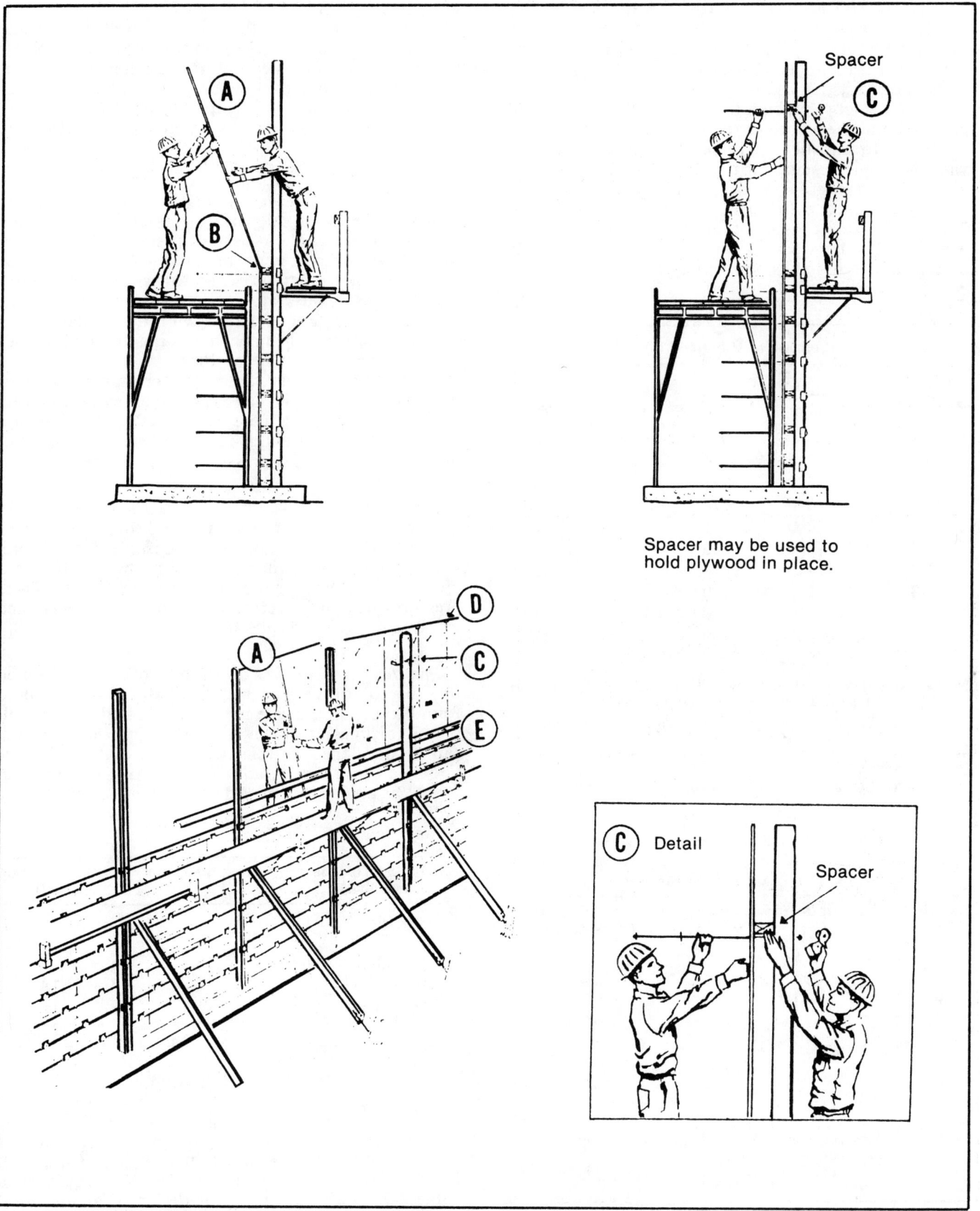

**Installing second lift of plywood
Figure 5-31**

TABLES FOR FORM DESIGN

The tables provided in this section will help you choose the right Plyform thickness for most applications. Also included are tables for choosing the proper size and spacing of joists, studs, and wales.

Though many combinations of frame spacing and plywood thickness meet structural requirements, it is preferable to use only one thickness of plywood and to vary the frame spacing for different pressures.

Plyform is manufactured in various thicknesses, but it is good practice to base designs on 5/8-inch and 3/4-inch Plyform Class I as they are the most commonly available. For large jobs or for those having special requirements, other thicknesses may be preferable, but could require a special order.

Pour Rate (Ft./Hr.)	Pressures of Vibrated Concrete (pcf) (a), (b)			
	50° F		75° F	
	Columns	Walls	Columns	Walls
1	330	330	280	280
2	510	510	410	410
3	690	690	540	540
4	870	870	660	660
5	1050	1050	790	790
6	1230	1230	920	920
7	1410	1410	1050	1050
8	1590	1470	1180	1090
9	1770	1520	1310	1130
10	1950	1580	1440	1170

Notes: (a) Maximum pressure need not exceed 150h, where h is maximum height of pour.
(b) Based on concrete with density of 150 pcf and 4" slump

Concrete pressures for column and wall forms
Table 5-2

Depth of Slab (In.)	Concrete Pressure (psf)	
	Non Motorized Buggies (a)	Motorized Buggies (b)
4	100	125
5	113	138
6	125	150
7	138	163
8	150	175
9	163	188
10	175	200

Notes: (a) includes 50 psf load for workmen, equipment, impact, etc.
(b) includes 75 psf load for workmen, equipment, impact, etc.

Concrete pressures for slabs
Table 5-3

Concrete Pressures

The required plywood thickness, as well as size and spacing of framing, depends on the maximum concrete pressure. So the first step in form design is to determine the maximum concrete pressure. This is based on such things as pour rate, job temperature, concrete slump, cement type, concrete density, method of vibration, and height of form.

Pressures On Column And Wall Forms There are several methods of calculating pressures for wall and column forms. Table 5-2 shows pressures for vibrated concrete at different pour rates and temperatures based on the recommendations of the American Concrete Institute. These values are for internal vibration of the concrete only. For external vibration, double the pressures shown. When concrete is not vibrated, reduce the pressures in the table by 10%.

Concrete pressure is in direct proportion to its density. Pressures shown in Table 5-2 are based on a density of 150 pounds per cubic foot (pcf). They are appropriate for the usual range of concrete poured. For other densities, adjust pressures proportionately.

Pressures On Slab Forms Forms for concrete slabs must be able to support workmen and equipment as well as concrete. Table 5-3 gives some pressures which represent the average when either motorized or non-motorized buggies are used for placing concrete. These pressures include the effects of concrete, buggies, and workmen.

Pressures On Curved Forms Plyform can also be used for curved forms. The radii in Table 5-4 have been found to be appropriate minimums for mill-run panels of the thicknesses shown when bent dry. Shorter radii can be developed by selecting panels that are free of knots and short grain, and/or by wetting or steaming. An occasional panel may develop weaknesses at these radii.

Plywood Thickness (Inches)	Across the Grain (Feet)	Parallel to Grain (Feet)
1/4	2	5
5/16	2	6
3/8	3	8
1/2	6	12
5/8	8	16
3/4	12	20

Minimum bending radii
Table 5-4

Allowable Pressures On Plyform Allowable pressures on Plyform Class I are shown in Table 5-5, Class II in Table 5-6 and pressures for Structural I Plyform in Table 5-7. Use these tables for design of architectural concrete forms where appearance is important. Calculations for the pressures shown in these tables were based on a deflection limitation of 1/360 of the span.

Concrete Form Construction

Table 5-5: Allowable pressures on Plyform Class I for architectural uses (continuous across two or more spans)

Face grain across supports

Support Spacing (Inches)	1/2	5/8	3/4	7/8	1	1-1/8
4	3265	4095	5005	5225	5650	6290
8	970	1300	1650	2005	2175	2420
12	410	575	735	890	1190	1370
16	175	270	370	475	645	750
20	100	160	225	295	410	490
24	--	--	120	160	230	280
32	--	--	--	--	105	130
36	--	--	--	--	--	115

Face grain parallel to supports

Support Spacing (Inches)	1/2	5/8	3/4	7/8	1	1-1/8
4	1860	2350	2910	3450	4615	5455
8	605	905	1120	1325	1775	2100
12	215	360	670	820	1100	1300
16	--	150	300	480	725	895
20	--	105	210	290	400	495
24	--	--	110	180	255	320

Table 5-7: Allowable pressures for Structural I Plyform for architectural uses (continuous across two or more spans)

Face grain across supports

Support Spacing (Inches)	1/2	3/4	5/8	7/8	1	1-1/8
4	3925	5240	6490	6175	6535	7240
8	980	1310	1680	2060	2515	2785
12	415	580	745	915	1335	1540
16	175	270	380	485	725	845
20	100	160	230	305	465	550
24	--	--	120	165	260	315
32	--	--	--	--	115	145
36	--	--	--	--	--	125

Face grain parallel to supports

Support Spacing (Inches)	1/2	3/4	5/8	7/8	1	1-1/8
4	2520	3185	3940	5110	6255	7395
8	830	1225	1515	1965	2405	2845
12	255	425	825	1215	1490	1760
16	105	108	360	570	865	1145
20	--	125	255	400	555	685
24	--	--	130	215	335	435

Table 5-6: Allowable pressures for Plyform Class II for architectural uses (continuous across two or more spans)

Face grain across supports

Support Spacing (Inches)	1/2	3/4	5/8	7/8	1	1-1/8
4	2675	3570	4515	4595	4925	5475
8	670	890	1135	1380	1885	2105
12	295	395	505	615	840	965
16	150	225	285	345	470	545
20	--	135	195	240	325	375
24	--	--	105	140	205	240
32	--	--	--	--	--	115

Face grain parallel to supports

Support Spacing (Inches)	1/2	3/4	5/8	7/8	1	1-1/8
4	1610	2235	2765	3280	4370	5165
8	455	800	1065	1260	1680	1985
12	130	255	485	745	1040	1230
16	--	105	210	335	505	670
20	--	--	150	240	355	490
24	--	--	--	125	195	265

PLATE OR TEMPLATE LAYOUT

The curvature of radius wall forms may be established by many methods, such as saw-cut walers and laminated 1 by 2's. The illustrated methods are simple, practical and economical. Shown in Figure 5-32A is the method to use where exterior forms are to be set first. The method in Figure 5-32B with the interior wall set first, is used more often because it is easier to brace forms to the interior, steel is easier to set, and vertical plywood joints are more easily nailed to studs.

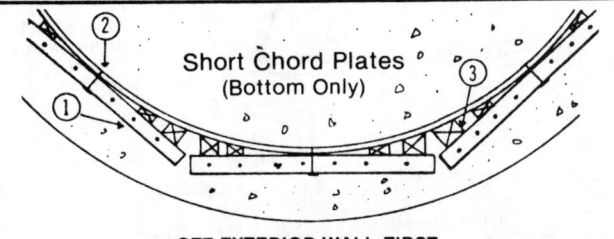

SET EXTERIOR WALL FIRST
1 Nail down series of short 2 x 4 templates as shown setting back the thickness of plywood from chalked radius line. **2** Set up two or more sheets of plywood with temporary bracing and nail to templates where they touch as shown. **3** Drive wedges firmly between plywood and template as shown so plywood conforms to chalkline. Nail as required.

This simple method eliminates costs of sawcutting templates and is generally used with 1 x 4 banding at approximately 4' intervals instead of having a top template.

Setting short chord plates
Figure 5-32A

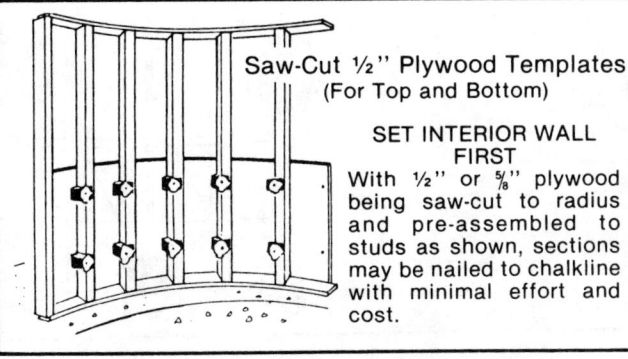

SET INTERIOR WALL FIRST
With ½" or ⅝" plywood being saw-cut to radius and pre-assembled to studs as shown, sections may be nailed to chalkline with minimal effort and cost.

Setting saw-cut ½" plywood templates
Figure 5-32B

Masonry & Concrete Construction

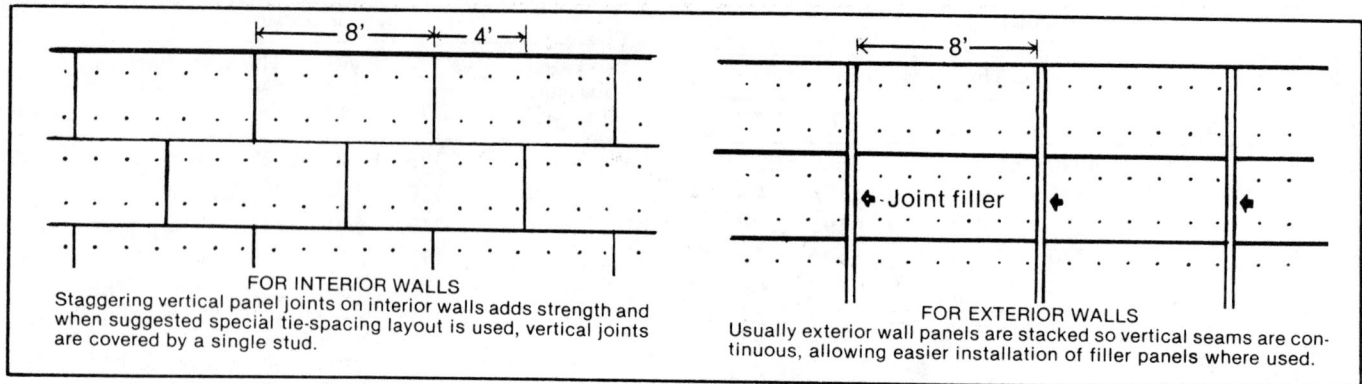

For Interior Walls
Staggering vertical panel joints on interior walls adds strength and when suggested special tie-spacing layout is used, vertical joints are covered by a single stud.

For Exterior Walls
Usually exterior wall panels are stacked so vertical seams are continuous, allowing easier installation of filler panels where used.

Radius wall forming for interior and exterior walls
Figure 5-33

Figure 5-33 shows details for radius wall forming for interior and exterior walls.

To eliminate the need for exterior panel filler strips, trim the two sides of the interior sheets to alleviate the difference in circumference before holes are drilled 1⅛-inch in from panel edge. "A" Brackets and walers may then be used for all joint covers (see Figure 5-34). The trimmed sheets can be cut into 4-by-4-foot pieces at the end of the job.

Always place "A" Brackets on the left side of the studs so the eccentric is properly positioned to be vibrationproof. Install "A" brackets after the studs are in

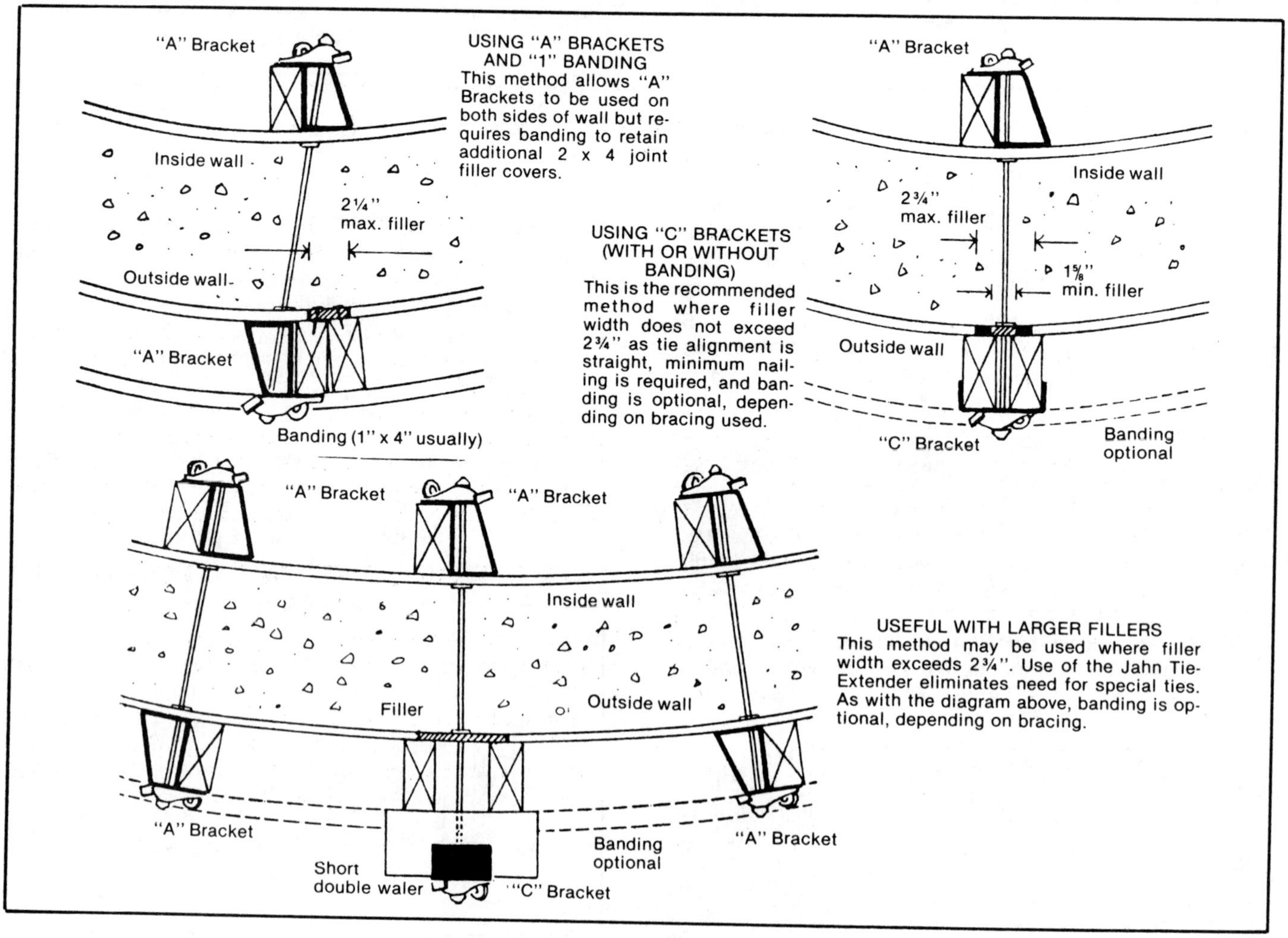

USING "A" BRACKETS AND "1" BANDING
This method allows "A" Brackets to be used on both sides of wall but requires banding to retain additional 2 x 4 joint filler covers.

USING "C" BRACKETS (WITH OR WITHOUT BANDING)
This is the recommended method where filler width does not exceed 2¾" as tie alignment is straight, minimum nailing is required, and banding is optional, depending on bracing used.

USEFUL WITH LARGER FILLERS
This method may be used where filler width exceeds 2¾". Use of the Jahn Tie-Extender eliminates need for special ties. As with the diagram above, banding is optional, depending on bracing.

Vertical joint fillers for exterior walls
Figure 5-34

Concrete Form Construction

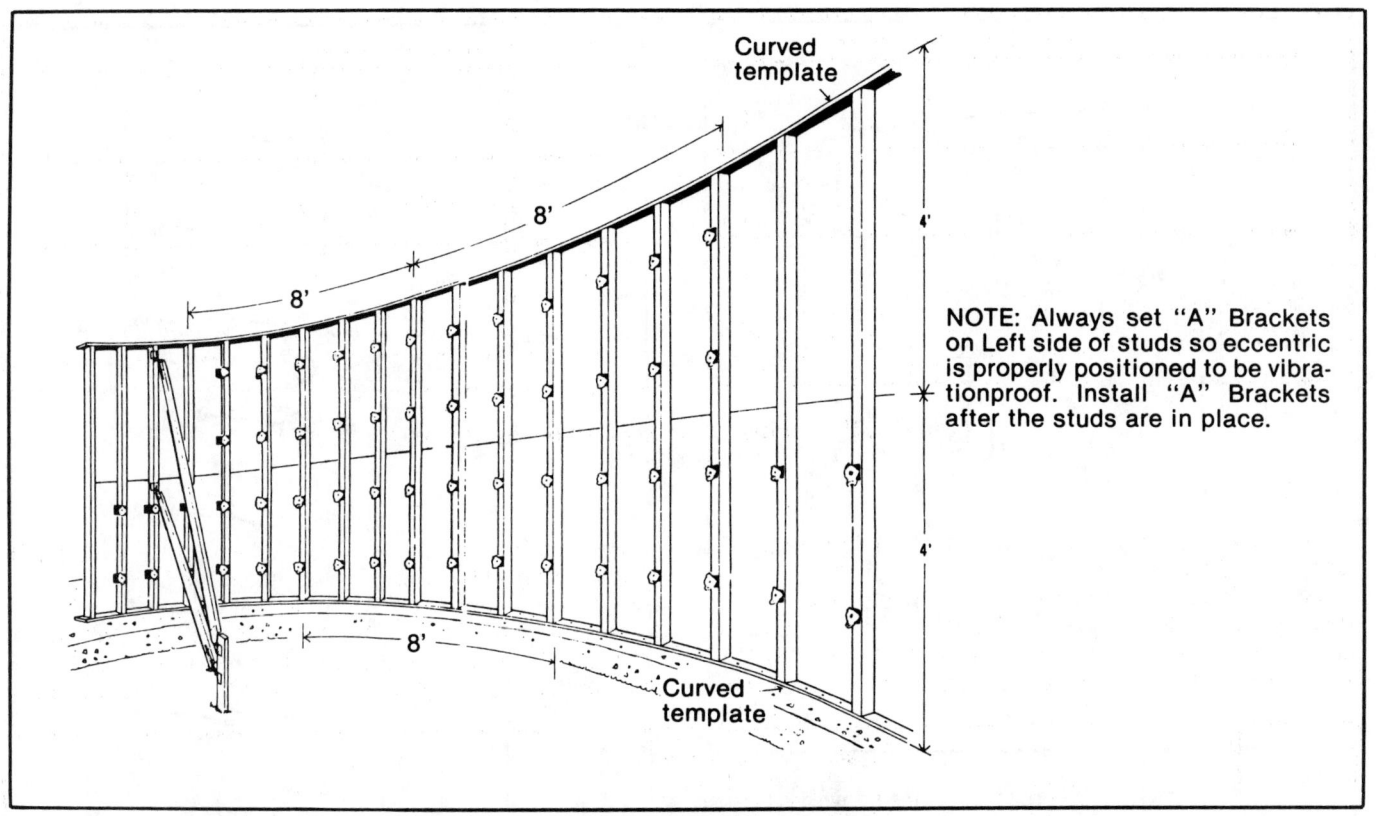

"A" Brackets installed on interior wall
Figure 5-35A

place. Figures 5-35A and B show "A" Brackets installed on interior and exterior walls. Figure 5-36 gives two sectional views of installed "A" Brackets.

OTHER PLYWOOD SUITABLE FOR CONCRETE FORMING

Though not manufactured specifically for concrete forming, grades of plywood other than Plyform have been used for various forming applications. The allowable pressures shown in the following tables give a good estimate of performance for sanded grades, such as A-C Exterior APA and B-C Exterior APA, and unsanded grades, such as C-C Exterior and C-D Interior with exterior glue, provided the face grain is across supports. For Group 1 sanded grades, use Table 5-5 for Class I Plyform. For Group 2 sanded grades, use Table 5-6 for Class II Plyform. For unsanded grades (identification index panels) use the Plyform Class 1 tables assuming 1/2-inch Plyform for 32/16 Identification Index panels, 5/8-inch for 42/20 and 3/4-inch for 48/24.

Textured plywood has recently been used to obtain various patterns for architectural concrete. Many of these panels have some of the face ply removed in the texturing process. Consequently, strength and stiffness are reduced. As textured plywood is available in a variety of patterns and wood species, it is impossible to give exact factors for strength and stiffness reductions. For approximately equivalent strength, specify the desired grade in Group 1 or Group 2 species and determine the thickness assuming Plyform Class II. When 3/8-inch textured plywood is used for a form liner, assume that the plywood backing must carry the entire load.

In some cases, it may be desirable to use two layers of plywood. The allowable pressures shown in Tables 5-5 through 5-7 are addable for more than one layer. These tables are based on the plywood acting as a continuous beam which spans joists or studs. No blocking is assumed at the unsupported panel edge. To minimize differential deflection between adjacent panels, some form designers specify blocking at the unsupported edge, particularly when face grain is parallel to supports.

SELECTION OF FRAMING FOR WALL FORMS

Design the lumber studs and double wales for the Plyform selected. Maximum concrete pressure is 540 psf.

Design Studs: Since the plywood must be supported at 12 inches on center, space studs 12 inches on center. The load carried by each stud equals the concrete pressure multiplied by the stud spacing in feet*.

$$540 \text{ psf} \times \frac{12}{12} \text{ ft.} = 540 \text{ pounds per foot}$$

---

* This method is applicable to most framing systems. It assumes the maximum concrete pressure is constant over the entire form. Actual distribution is more nearly trapezoidal or triangular.

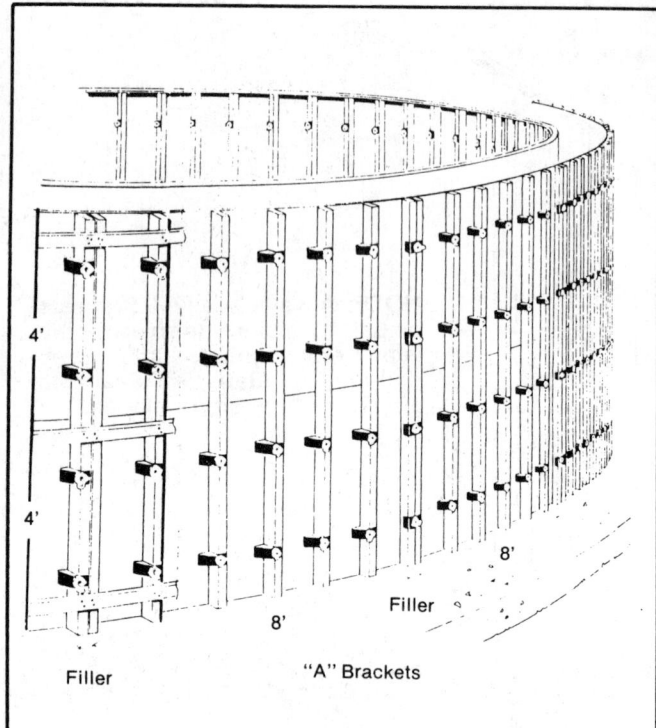

"A" Brackets installed on exterior wall
Figure 5-35B

Assuming 2 by 6 studs continuous over three supports (2 spans), Table 5-8 shows a 53-inch span for 400 pounds per foot and a 43-inch span for 600 pounds per foot. Interpolate between these spans for a load of 540 pounds per foot.

$$\frac{540-400}{600-400} \times (53-43) = \frac{140}{200} \times (10) = 7''$$

for 540 pounds per foot, span = 53"-7" = 46"

The 2 by 6 studs must be supported at 46 inches on center. Assume this support is provided by double 2 by 6 wales spaced 46 inches on center.

Design Double Wales: Load carried by the double wales equals the maximum concrete pressure multiplied by the wale spacing in feet, or:

$$540 \text{ psf} \times \frac{46}{12} \text{ ft.} = 2070 \text{ pounds per foot.}$$

Since the wales are doubled, each wale carries 1035 pounds per foot (2070 divided by 2 = 1035). Assuming 2 by 6 wales continuous over 4 or more supports, the table shows a 14-inch span for 1000 pounds per foot and 30-inch span for 1200 pounds per foot. Interpolation shows that 2 by 6's can span 33 inches for 1035 pounds per foot. Support 2 by 6's at 33 inches on center with form ties. (Place bottom wale about 8 inches from the bottom of the form.)

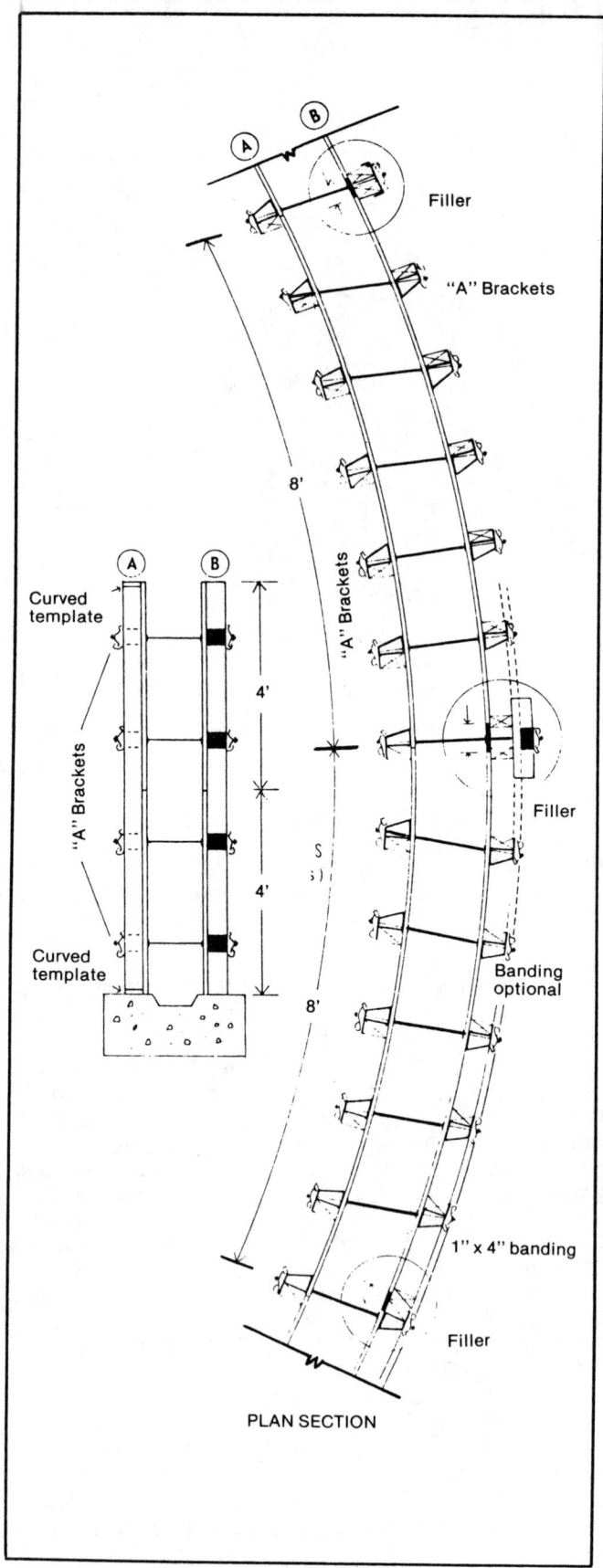

Sectional view of installed "A" Brackets
Figure 5-36

Maximum Spans for Lumber Framing, Inches*

Douglas Fir-Larch No. 2 or Southern Pine No. 2 Kiln dried (KD)

Equivalent Uniform Load (lb/ft)	Continuous Over 2 or 3 Supports (1 or 2 Spans) Nominal Size								Continuous Over 4 or More Supports (3 or More Spans) Nominal Size							
	2x4	2x6	2x8	2x10	2x12	4x4	4x6	4x8	2x4	2x6	2x8	2x10	2x12	4x4	4x6	4x8
400	36	53	70	90	109	50	79	101	41	60	78	100	122	62	91	118
600	30	43	57	73	89	44	66	88	31	49	64	82	99	51	74	98
800	24	38	50	63	77	39	58	76	25	39	52	66	81	44	64	85
1000	21	33	43	55	67	35	51	68	21	34	44	57	69	39	58	76
1200	19	29	38	49	60	32	47	62	19	30	39	50	61	35	53	69
1400	17	27	35	45	54	30	43	57	17	27	36	46	56	31	49	64
1600	16	25	32	41	50	27	41	54	16	25	33	42	52	28	44	58
1800	15	23	30	39	47	25	38	51	15	24	31	40	48	26	40	53
2000	14	22	29	37	45	23	36	48	14	22	29	38	46	24	37	49
2200	13	21	28	35	43	22	34	45	14	21	28	36	44	22	35	46
2400	13	20	26	34	41	20	32	42	13	20	27	34	42	21	33	44
2600	12	19	26	33	40	19	31	40	13	20	26	33	40	20	31	41
2800	12	19	25	32	38	19	29	38	12	19	25	32	39	19	30	39
3000	12	18	24	31	37	18	28	37	12	19	24	31	38	18	29	38
3200	11	18	23	30	36	17	27	35	12	18	24	30	37	18	28	36
3400	11	17	23	29	36	17	26	34	11	18	23	30	36	17	27	35
3600	11	17	22	29	35	16	25	33	11	17	23	29	35	16	26	34
3800	11	17	22	28	34	16	24	32	11	17	22	29	35	16	25	33
4000	10	16	22	28	34	15	24	31	11	17	22	28	34	15	24	32
4500	10	16	21	27	32	14	22	29	10	16	21	27	33	14	23	30
5000	10	15	20	25	31	13	21	28	10	16	20	26	32	14	22	28

Hem-Fir No. 2

Equivalent Uniform Load (lb/ft)	Continuous Over 2 or 3 Supports (1 or 2 Spans) Nominal Size								Continuous Over 4 or More Supports (3 or More Spans) Nominal Size							
	2x4	2x6	2x8	2x10	2x12	4x4	4x6	4x8	2x4	2x6	2x8	2x10	2x12	4x4	4x6	4x8
400	33	48	63	80	97	48	73	96	36	53	70	90	109	56	81	107
600	27	39	51	65	80	41	59	78	27	43	57	72	88	45	66	88
800	22	34	44	57	69	35	51	68	22	35	46	59	72	39	58	76
1000	19	29	39	50	60	31	46	61	19	30	40	51	62	35	51	68
1200	17	26	35	44	54	29	42	55	17	27	36	45	55	31	47	62
1400	15	24	32	41	49	27	39	51	16	25	33	42	51	27	43	57
1600	14	23	30	38	46	24	36	48	15	23	30	39	47	25	39	52
1800	14	21	28	36	43	22	34	45	14	22	29	36	44	23	36	47
2000	13	20	27	34	41	21	33	43	13	21	27	35	42	21	33	44
2200	12	19	26	33	40	19	31	40	13	20	26	33	40	20	31	41
2400	12	19	25	31	38	18	29	38	12	19	25	32	39	19	30	39
2600	12	18	24	30	37	18	28	36	12	18	24	31	38	18	28	37
2800	11	18	23	30	36	17	26	35	11	18	24	30	37	17	27	36
3000	11	17	23	29	35	16	25	33	11	17	23	29	36	17	26	34
3200	11	17	22	28	34	16	24	32	11	17	22	29	35	16	25	33
3400	10	16	22	27	33	15	24	31	11	17	22	28	34	15	24	32
3600	10	16	21	27	32	15	23	30	10	16	22	27	33	15	23	31
3800	10	15	20	26	32	14	22	29	10	16	21	27	33	15	23	30
4000	10	15	20	25	31	14	22	29	10	16	21	27	32	14	22	29
4500	10	14	19	24	29	13	21	27	10	15	20	26	31	13	21	28
5000	9	13	18	23	28	12	20	26	9	15	20	25	30	13	20	26

*Spans are based on PS-20 lumber sizes. Single member stresses were multiplied by a 1.25 duration-of-load factor for 7-day loads. Deflection limited to 1/360th of the span with ¼" maximum. Spans are center-to-center of the supports.

Maximum spans for lumber framing
Table 5-8

Load on Ties: The load on each tie equals the load on the double wales multiplied by the tie spacing in feet or, 2070 pounds per foot x $^{33}/_{12}$ = 5690.

If allowable load on the tie is less than 5690 pounds, decrease tie spacing accordingly. For instance, a tie with 5000 pounds allowable load should be spaced no more than:

$$\frac{5000}{2070} \times 12 \text{ inches} = 29 \text{ inches}.$$

Other Loads On Forms
Concrete forms must also be braced against lateral loads due to wind or any other construction loads. Lateral loads should be at least 10 pounds per square foot for wind load, or greater if prescribed by local building codes. In all cases, forms over 8 feet high should be designed to carry at least 100 pounds per linear foot applied at the top of the form.

Wall forms should be designed to withstand wind pressure applied from either side. Inclined wood braces can be designed to take both tension and compression, so braces on only one side may be required. Good bracing must then be designed that will not buckle under load. Guy-wire bracing, on the other hand, can take only tensile loads. If used, it is required on both sides of the forms.

In general, wind bracing also resists uplift forces on the forms, provided the forms are vertical. If forms are inclined, uplift forces may be significant. Special tiedowns and anchorages may be required in some cases.

CONCRETE SURFACE CHARACTERISTICS RESULTING FROM USING PLYWOOD FORMS
Dusting
Surface dusting of concrete has been observed in concrete poured against a variety of forming materials, including plywood. Although there appears to be no single reason for this, dusting has been traced to many possible causes, including excess oil, dirt, dew, smog, and unusually hot, dry weather, as well as chemical reactions between the form surface and the concrete.

Several methods have been successful in dealing with this problem. Some of these include proper form storage (cool, dry conditions) and cleanliness (avoiding needless exposure to dust, oil, and weathering). If dusting does occur, a fine water spray helps speed surface hardening. The State of California Division of Highways reports that "...rather than attempt to employ inconvenient methods of preventing dusting, final results will be satisfactory if affected areas are subsequently cured for a few days with water in a spray fine enough not to erode the soft surface." Other concrete specialists have recommended surface treatment solutions such as magnesium fluosilicate or sodium silicate.

Staining
Occasionally, a reddish or pinkish discoloration may appear when the concrete has been poured against High-Density Overlaid plywood forms. The stain, a fugitive dye, is temporary and usually disappears with exposure to sunlight and air. Where sunlight cannot reach the stain, natural bleaching takes longer. Household bleaching agents such as Clorox or Purex (5% solutions of sodium hypochlorite), followed by clear-water flushing, have been effective in hastening stain removal.

On rare occasions, other discolorations have been observed in new concrete. For example, iron salts resulting from iron sulfides and ferrous oxides in slag cement have been found to stain concrete a greenish-blue color.

Both occurrence and intensity of color seem to be related to the length of time between oiling of forms and pouring of concrete, as well as to the length of time before the forms are stripped. Loosening or opening the forms at the earliest possible time after placing the concrete sometimes prevents discoloration in slag concrete. The discoloration usually fades and disappears with time. Hydrogen peroxide solutions are useful in removing the color, particularly when applied to the concrete immediately after form removal.

Turkey-red staining with Plyform used in factory-casting of concrete products has occurred when vegetable oil (i.e. castor oil) is applied to the form. Using mineral oil instead is the most direct solution. If castor oil must be used, test the concrete mix to determine whether staining will occur.

FORM MAINTENANCE
Do not use metal stripping bars or pries on plywood. Use wood wedges, tapping gradually when necessary. The strength, light weight and large panel size of the plywood helps reduce stripping time. Cross-laminated construction resists edge splitting.

Cleaning
Soon after removal, inspect plywood forms for wear. Clean, repair, spot-prime, refinish and lightly oil before reusing. Use a hardwood wedge and a stiff fiber brush for cleaning (a metal brush may cause wood fibers to "wood"). Light tapping with a hammer will generally remove a hard scale of concrete. On prefabricated forms, plywood panel faces (when the grade is suitable) may be reversed if damaged, and tie holes may be patched with metal plates, plugs, or plastic materials. Remove nails and fill the holes with patching plaster, plastic wood, or other suitable materials.

Coatings And Parting Agents
Protective sealant coatings and parting agents for plywood increase form life and aid in stripping. Mill-oiled Plyform panels may require only a light coating of oil or parting agent between uses (Figure 5-37). For regular plywood, apply a liberal amount of oil or parting agent a few days before the plywood is used, then wipe to leave a thin film. Check specifications before using any oil or compound on the forms.

Specially coated panels with long-lasting finishes are available that make stripping easier and reduce maintenance costs. Care in handling will provide a maximum number of reuses.

Concrete Form Construction

Oil forms after each use to increase their life and to make stripping easier
Figure 5-37

Hairline cracks or splits may occur in the face ply. These "checks" may be more pronounced after repeated use of the form. Checks do not mean the plywood is delaminating. A thorough program of form maintenance including careful storage to assure slow drying will minimize face checking.

Plywood Form Coatings Lacquers, resin, or plastic-base compounds and similar field coatings sometimes are used to form a hard dry waterproof film on the plywood surface. The performance level of the coating is generally rated somewhere between B-B Plyform and High Density Overlaid plywood. In most cases the need for oiling between pours is reduced by the field-applied coatings, and many contractors report obtaining significantly greater reuse than with the B-B Plyform, but generally fewer than with HDO plywood.

Mill-coated products of various kinds are available in addition to mill-oiled Plyform. Some plywood manufacturers suggest no oiling with their proprietary concrete forming products, and claim exceptional concrete finishes and a large number of reuses. In any event, make the selection of a release agent with an awareness of the product's influence on the finished surface of the concrete. For example, some release agents that include waxes or silicones should not be used when the concrete is to be painted.

Handling and Storage

Exercise care to prevent panel chipping, denting and corner damage during handling. Panels should never be dropped. Carefully lay the forms flat, face-to-face and back-to-back, for hauling. You can solid-stack them or stack them in small packages with their faces together. This slows the drying rate and minimizes face checking.

Using plywood stack handling equipment and small trailers for hauling and storing panels between jobs will reduce handling time and minimize damage. If possible, store the plywood panels inside a building or a shed, or cover them loosely to allow air circulation without heat build-up. Panels no longer suitable for formwork can be used for subflooring or wall and roof sheathing if their condition permits.

PREFABRICATED FORMS

For years concrete forms were made strictly of wood and were built on the job. In most construction today, prefabricated forms built of wood, steel and wood, and

Masonry & Concrete Construction

Symons Company Steel-Ply system of panels and fillers
Figure 5-38

aluminum are used. These forms come in many sizes and are assembled on the job site.

Formwork labor and material often average 30 to 50 percent of the total concrete wall costs, with labor averaging 2 to 3 times the material cost. It's vital to select the least expensive forming system.

Higher labor productivity means lower labor costs and higher profit for contractors. It also means faster construction cycles, a better chance of meeting schedules, lower overhead per job, less chance of delaying other trades, and more jobs handled per year.

Steel-Ply System
The Steel-Ply system consists of panels and fillers (Figure 5-38). The Symons Company makes over 75 different panels and fillers. With such a wide variety to choose from, you can form practically any dimension by combining panel and filler sizes and erecting them vertically or horizontally. A 2-foot by 8-foot panel weighs 75 pounds. When starting, use an outside corner piece and two fillers as shown in Figure 5-39. The outside corner pieces come in 3-, 4-, 5-, 6-, and 8-foot lengths and are 2½ inches by 2½ inches. There are slots

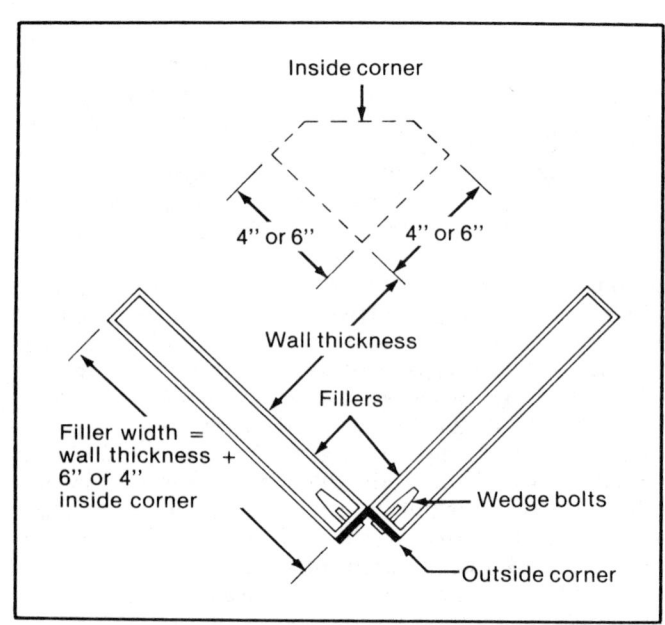

Plan of corner with Steel-Ply system
Figure 5-39

Concrete Form Construction

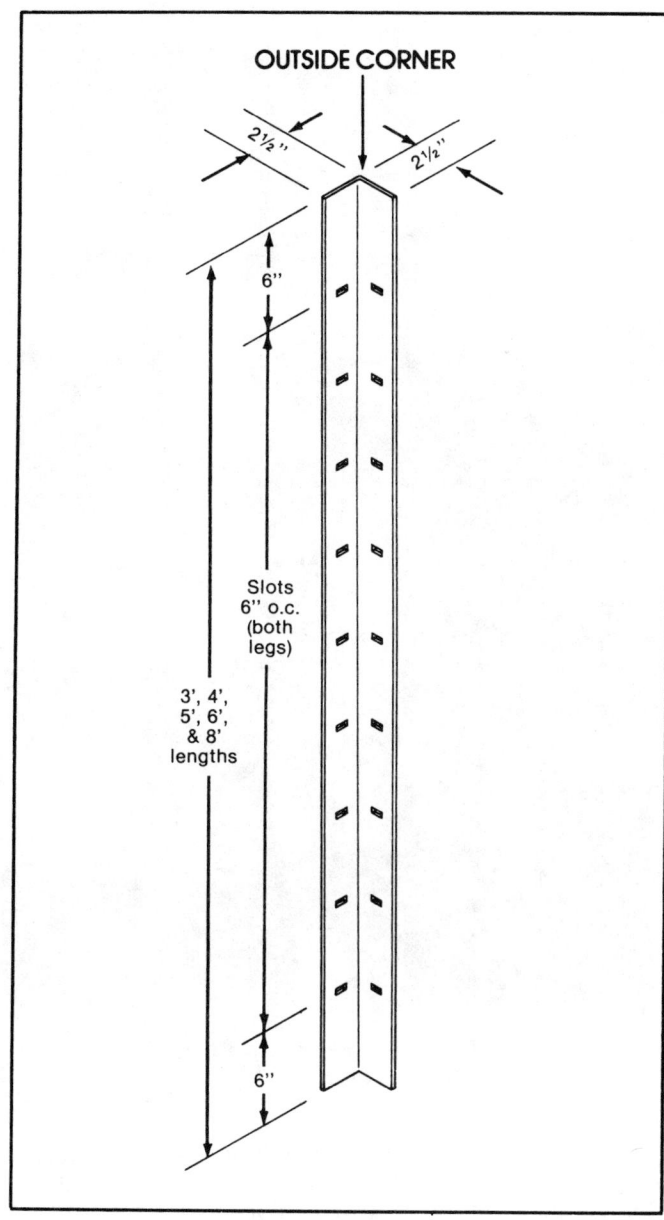

Corner piece of Steel-Ply system
Figure 5-40

hammer to secure the filler, tie, and panel. As an alternative method, set both sides of the wall, wedge-bolting adjacent panels at the top and bottom. Then insert and wedge-bolt the ties using other slots.

Proper tie selection is determined by several factors: spacing, safety, lateral concrete pressure, type of formwork (vertical two-sided, battered wall or unopposed single side), and concrete finish desired. Various ties are shown in Figure 5-44. Table 5-9 lists safe load ratings of wire and flat ties.

Handsetting: Aligning After setting 8 feet to 10 feet of panels, install walers using a one-piece waler clamp, or by the methods shown in Figure 5-45. Position single or double 2 by 4's and tap the locking wedge with a hammer.

With this type of prefabricated form, less material and labor are used for erection. Also, there is no measuring, sawing, or nailing. Use walers only on one side of the form. Waling is for horizontal panel alignment only and is not a structural part of the formwork. Job-built plywood forms use up to five rows of walers on each side of the wall.

Handsetting: Bracing Plumb and brace to vertical alignment as shown in Figure 5-46. Use the materials shown in Figure 5-47 for quick and efficient operation.

Special configuration of wedge bolts allowing their use in a wide range of form construction
Figure 5-41

in both sides for the wedge bolts. The slots are 6 inches on center. (See Figure 5-40.) The special configuration of the wedge bolts (Figure 5-41) allows them to be used in many different types of form construction.

Set the corner piece and two fillers on a lumber plate or concrete footing (Figure 5-42), working along a chalk line. The plates help to ensure a level work surface. Connect the fillers to the outside corner by wedge-bolting through the corner piece and the filler side rails adjacent to the cross members. Tap the vertical wedge bolt with a hammer to secure connection.

Place a panel next to the filler. (See Figure 5-43.) Insert a tie through the dado slot. Now insert a wedge bolt horizontally through the filler slot, tie loop, and panel slot. Insert a second wedge bolt vertically and tap with a

Masonry & Concrete Construction

Setting the corner piece and fillers
Figure 5-42

Placing panel next to filler
Figure 5-43

	Ultimate Load (lb)	Rating According to Factor of Safety	
		1.5 (lb)	2.0 (lb)
Standard Duty Wire Tie	4,500	3,000	2,250
Standard Duty Threaded Tie(1)(2)	4,200	2,800	2,100
Standard Duty S-Base Tie	3,000	2,000	1,500
Heavy-Duty Wire Tie	6,000	4,000	3,000
Standard Duty Flat Tie	6,000	4,000	3,000
Heavy-Duty Flat Tie	7,000	4,500	3,500
Heavy-Duty Adjustable Flat Tie	7,000	4,500	3,500
Toggle Tie(1)	4,200	2,800	2,100

(1) Tie capacity is dependent on adequate anchorage.
(2) When anchored with threaded inserts in 3500 psi concrete, an ultimate load of 4000 lb and a safety factor of 4:1 is recommended.

Safe load rating of Symons wire and flat ties
Table 5-9

Only one side of the formwork needs to be braced, thus saving time. By using the steel brace plate, turnbuckle and steel stakes, you eliminate the need to cut, fit and nail together scrap lumber each time. This type of bracing can be used repeatedly.

Handsetting: Second Side Start with an inside corner and set opposing panels of identical width so the ties are aligned. (See Figure 5-48.) Slightly deflect the ties for clearance and rotate the panel into position. Secure the corner, ties and panel with wedge bolts. (See Figure 5-49.)

Install the scaffold brackets after the first tier is set. Starting at a corner, set the next tier using the same hardware and techniques as the lower tier. Wedge-bolt the upper tier to the lower through panel end rails, using two connections through corner fillers and one in each other panel as shown in Figure 5-50. Align and wale as shown in Figure 5-51. For vertical alignment, strongback using one of the methods shown in Figure 5-50.

Stripping Tap the wedge bolts loose to remove braces, strongbacks, walers and panel connections. If a form-

WIRE TIES—clean and deep (1") breakback. Standard and heavy-duty. Special lengths and Stainless Steel available.

S-PANEL TIE—For handset general use

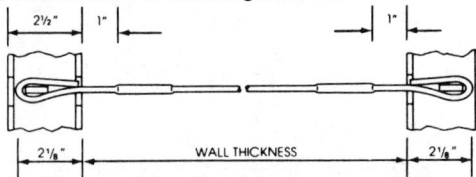

THREADED TIE—For battered or unopposed single side forming. Standard duty only

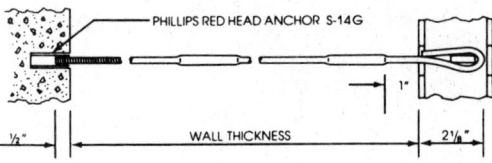

S-PILASTER TIE—Permits 1" breakback where needed

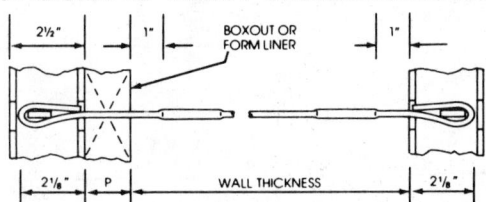

PRE-BENT TIE—For non-aligned tie slots, and where opposite forms are at angles to each other.

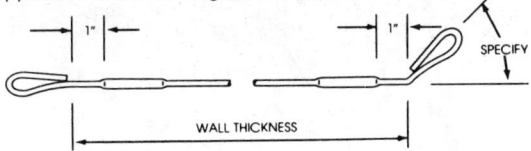

S-SPANDREL TIE—Ties outside form to inside deck form.

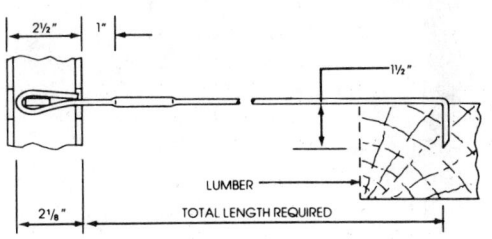

S-BASE TIE

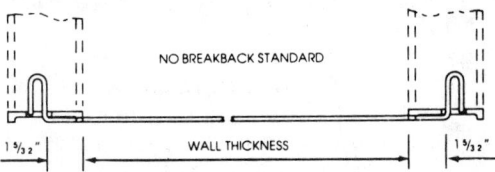

CONE TIES—For use where an architect specifies a smoother, cleaner appearance than standard wire tie breakback.

S-NO. 2 CONE TIE—For use with 2" x 4" waler and strongback   U.S. Patent No. 3,785,610

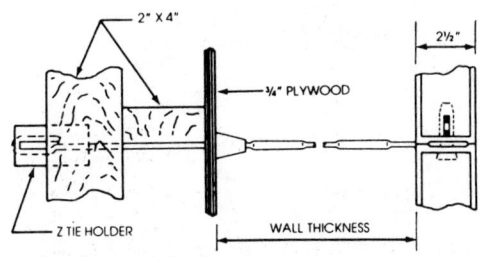

TOGGLE TIES—3" threaded end permits approximately 2" of adjustment.

SINGLE END THREADED TOGGLE

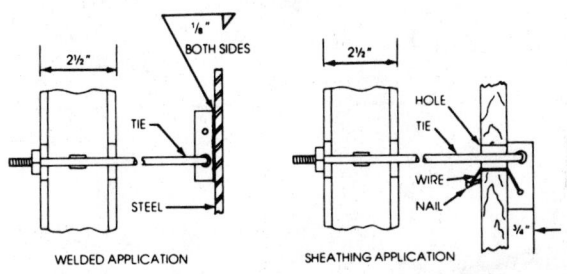

S-GANG FORM TIE—For crane handled forms. Allows tie breakback and stripping gang without disassembly.

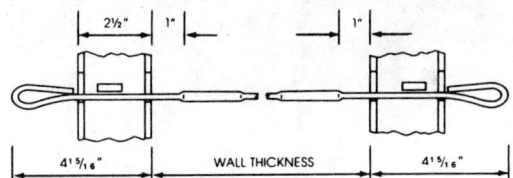

FLAT TIES—economy and speedy erection for walls. Wall appearance less important. Up-down action and shallow ¼" breakback. Standard and heavy-duty. Special lengths available.

X-FLAT TIE & HEAVY-DUTY FLAT TIE

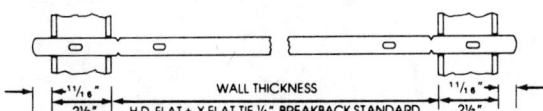

ADJUSTABLE FLAT TIE—For flexibility to handle unexpected situations U.S. Patent No. 3,362,678

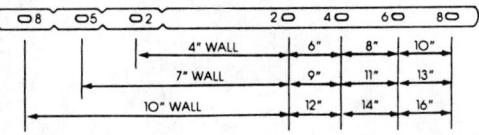

Selection of ties
Figure 5-44

Masonry & Concrete Construction

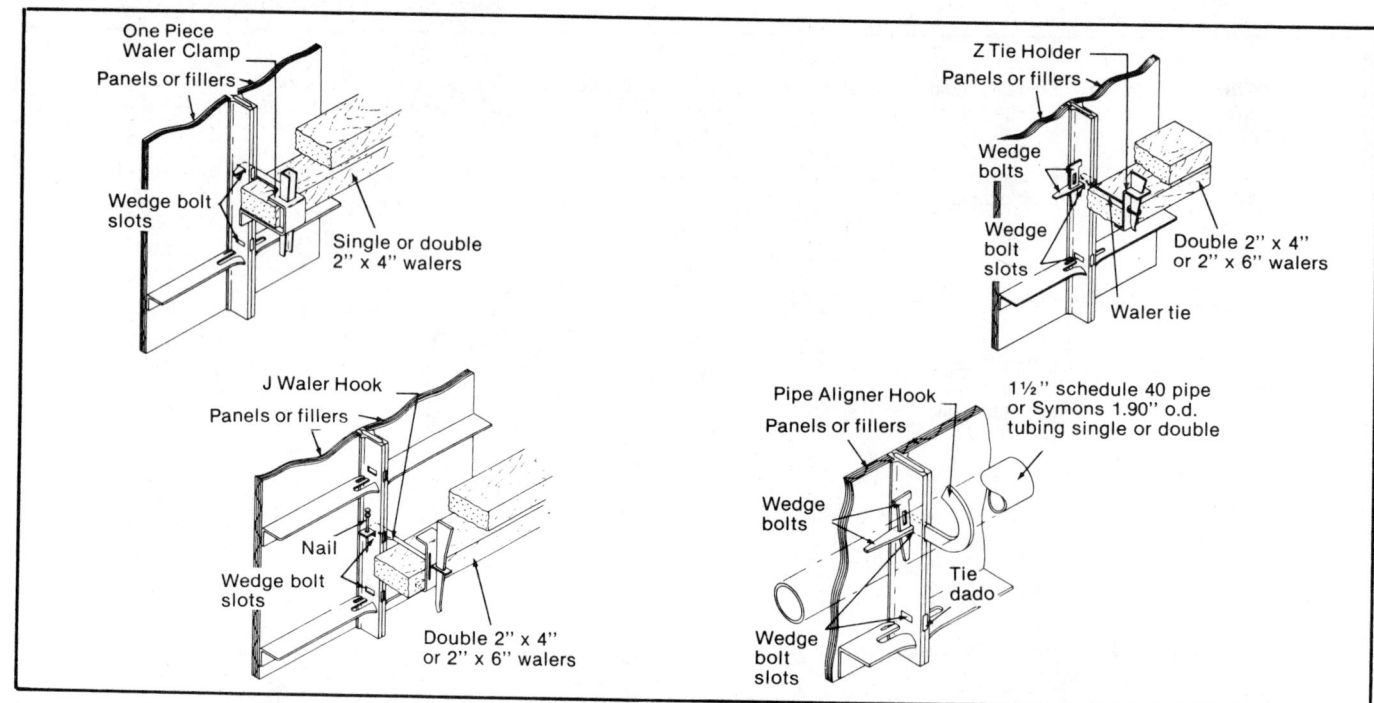

Installation of walers
Figure 5-45

Plumbing and bracing to vertical alignment
Figure 5-46

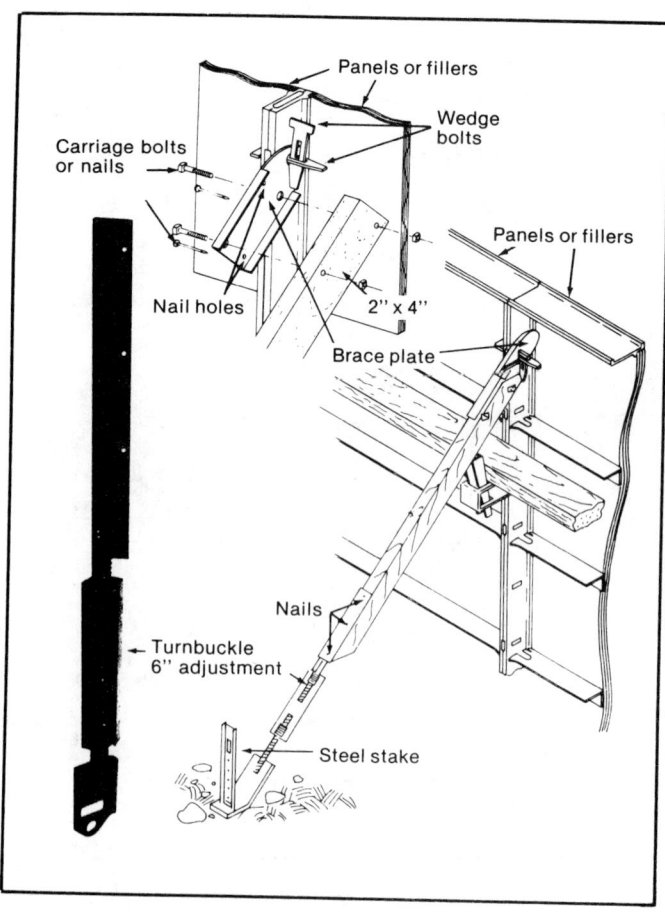

Materials for quick and efficient bracing
Figure 5-47

Concrete Form Construction

Handsetting the second side
Figure 5-48

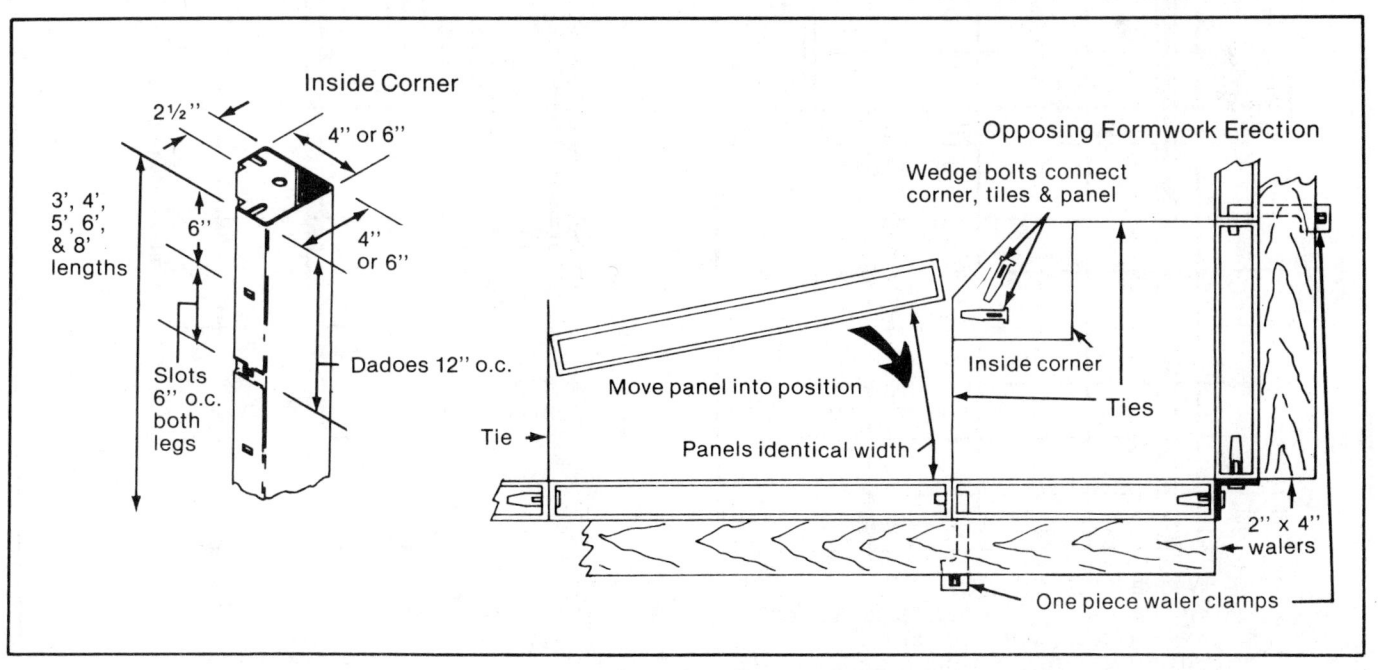

Method of securing corners
Figure 5-49

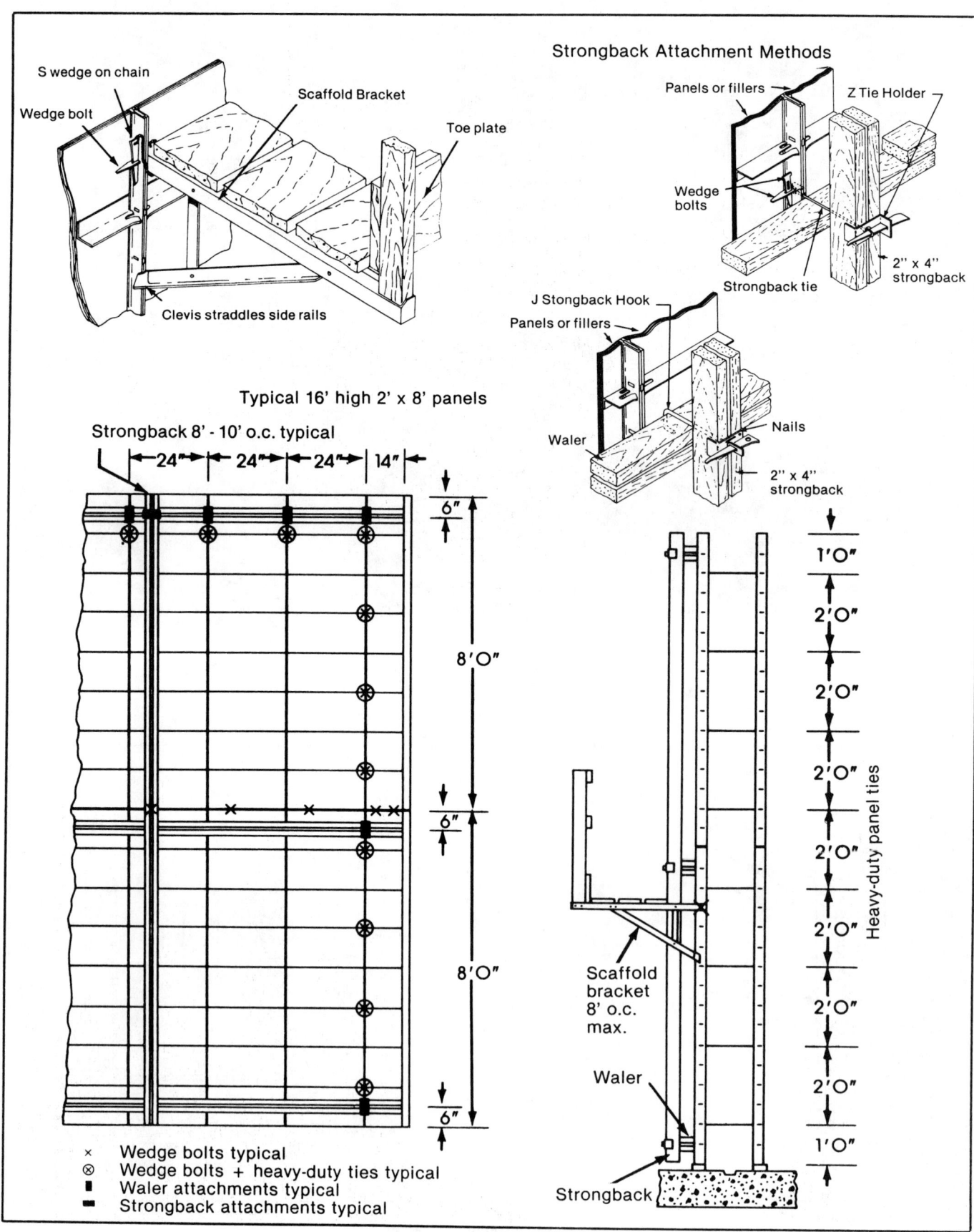

Strongback attachment methods
Figure 5-50

Correct installation of scaffold after first tier is set
Figure 5-51

Stripping forms
Figure 5-52

Masonry & Concrete Construction

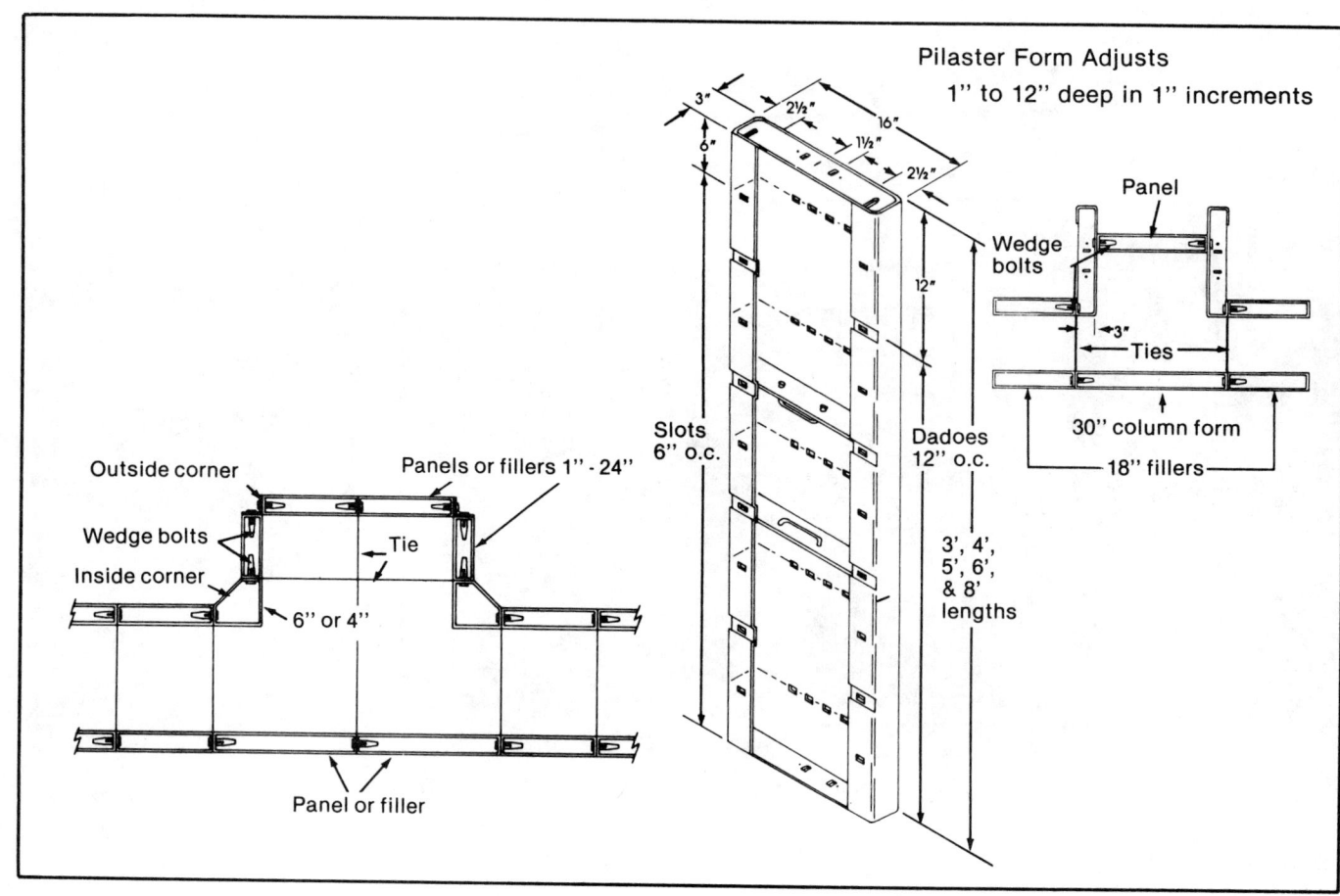

Pilaster form adjusts
Figure 5-53

coating was used, the forms will remove easily, leaving a smooth clean finish. (See Figure 5-52.)

Pilasters With this type of prefabricated form, walers, braces or ties are not needed on most jobs.

Combine inside and outside corners, panels and fillers to form just about any pilaster. (See Figure 5-53.)

Pilaster forms are non-symmetrical. Wedge-bolt one edge to the wall form for 2-, 4-, 6-, 8-, 10- or 12-inch depths. Simply turn the pilaster form end for end for 1-, 3-, 5-, 7-, 9-, or 11-inch depths. One or more panels or fillers easily form any pilaster width.

A 24-inch wide pilaster needs no ties if a 30-inch column form is used opposite the pilaster form.

Chapter 6
Reinforcing Masonry in Seismic Zones

Seismic zones are areas that are subject to damage from earthquakes. The map and chart in Figure 6-1 show the zones and the seismic probability of damage in the zones.

An earthquake produces random erratic vibratory ground motions to which a building actively responds. Thus, seismic design involves two steps: (1) the prediction of ground motion at the base of a structure according to the seismic nature of the site, and (2) the selection of a structure which will deform but not collapse when responding to these ground motions. Either of the following two approaches to the problems in design of earthquake-resistant structures may be used: (1) dynamic analysis (or response spectrums), or (2) empirical analysis. Empirical methods will be considered here.

FOUNDATIONS

Path of Seismic Forces
The base shear of the building is a measure of the horizontal force transmitted from the ground to the building. The medium used for this transmission of horizontal force may be friction between floor slab and ground, friction between bottom of footing and ground, or passive resistance of earth against vertical surfaces of the footing, grade beams or basement walls. The exact path that this transmission of forces will follow is not readily subject to a mathematical solution.

Footing and Raft Foundations
For ordinary nonseismic conditions a suitable footing (spread or wall) or raft (mat) foundation must: (1) be placed to a suitable depth, and (2) be safe with respect to bearing capacity of the structure. Primary damage to structures can be caused by differential settlements. An estimate of total and differential settlements which the structure can safely tolerate is made by a structural engineer. The foundation engineer can provide criteria for proportioning footings to equalize settlements and size so as not to exceed allowable bearing pressures.

In case of an earthquake, the overturning effect must be carried into the foundations; therefore, a careful analysis of permissible overloads from the combined effect of vertical and lateral loads will be made as part of the foundation design. The unbalance tensile force from the supported structure, if any, must be resisted by anchorage to the foundation. The problem is to provide stability against overturning for the short-time loading effects produced during an earthquake (or wind) without imposing such restrictions as to create wide disparity in foundation settlements under normal loading. This disparity could create more damage to the structure than that which might occur in an earthquake under highly increased soil pressures. The soil pressure for resisting combined static and prescribed seismic loads can generally exceed the normal allowable pressure for static loads by one third. However, the various types of soils react differently to short-time seismic loading, and any increase over normal allowable static loading will be confirmed by a soil analysis. Under no circumstances should the footing size be less than that required for static loads alone.

Pile Foundations
Earthquake vibrations may cause consolidation or liquidizing of loose soils; consequently, the resulting settlement of building foundations is rarely uniform. In the case of rigid structures supported on individual spread footings bearing on such material, differential settlements can result in damage to the superstructure.

Masonry & Concrete Construction

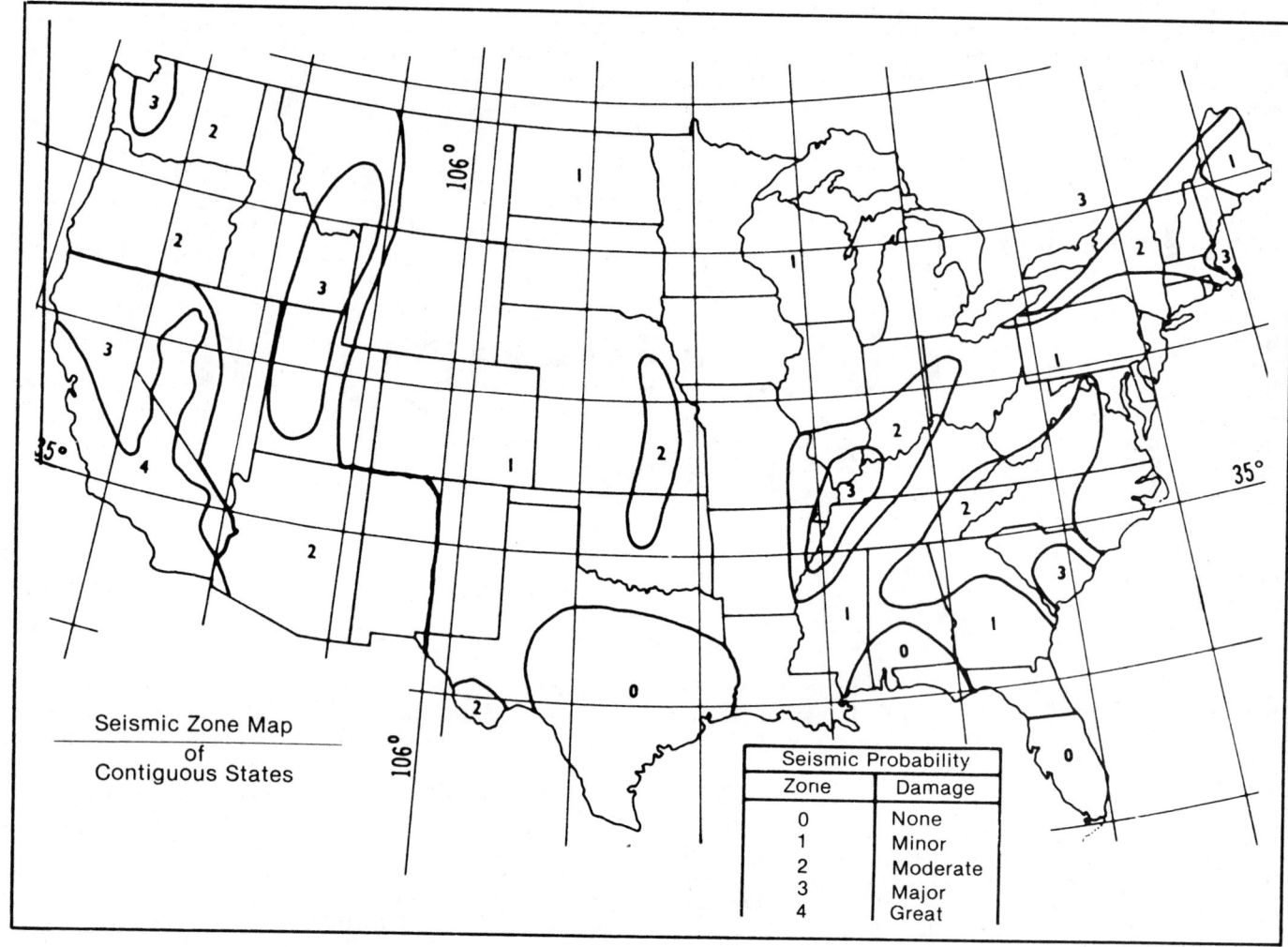

Seismic Zones
Figure 6-1

Either the stabilization of the soil prior to construction or the use of piles, caissons, or deep piers bearing on a firm stratum is a solution to this problem. For pile-supported structures subjected to horizontal loads, it must be decided whether the lateral load-carrying capacity of the vertical piles is adequate or whether batter piles should be used. The lateral load-carrying capacity of vertical piles depends on the properties of the soil; the size, length, and material of the pile; the pile grouping and spacing; and the duration and frequency of loading. These factors should be taken into consideration in estimating the ability of vertical piles to withstand the horizontal loads. Values which have been used for the allowable horizontal loads at the tops of piles vary from 1000 to 1500 pounds per pile. When the horizontal load is greater than the value suitable for vertical piles, batter piles may be used. If both the batter piles and vertical piles beneath a structure are point bearing and all are driven into the same stratum, the bearing capacity of the batter piles may be considered to be the same as that of the vertical piles. In analyzing groups of batter piles, it may be assumed that the pile caps are rigid, that the piles are end bearing and pinned at the top, and that analysis of any arrangement is possible by simple statics. However, the pile caps and their interconnecting members must be able to resist any eccentric pile loads. For piles subject to both vertical and horizontal loads, the applied horizontal thrust is equally shared among the horizontal components of the batter piles.

SPECIAL SEISMIC DETAILING

Separation of Structures

Buildings placed in close proximity to one another can suffer severe damage from the hammering which occurs during an earthquake. This motion is produced partly by the deflections of the structures themselves and partly by the rocking or settling of foundations. The simplest way to prevent damage is to provide adequate clearance between buildings to allow sufficient movement. Proper clearance is determined by calculating the total number of deflections. This determination is based on a summation of the story drifts plus the building flexural deflections (column lengthening and shortening) of the level involved. In case of a normal building less than 80 feet in height using concrete or unit-masonry shear walls, the gap shall not be less than the arbitrary rule of 2 inches for the first 20 feet of height above the ground plus 1/2 inch for each 10 feet of additional height. For higher or more flexible buildings, the gap or seismic joint between the structures should be equal to at least four (4) times the design deflections as determined from required (prescribed) lateral forces plus the resulting tipping distance of each structure towards the other at lower roof level due to racking or to settlement of the foundations.

Seismic Joints

Junctures between distinct parts of buildings, such as the intersection of a wing of a building with the main portion, are often designed with flexible joints to allow movement. With such a design, each part of the building must be considered as a separate structure with its own independent bracing system. A large one-story industrial building with a relatively flexible frame presents similar considerations. At one end of the building is a small office section with stiff exterior or interior walls. The office unit is generally more rigid than the rest of the building. If these two units are tied together, the horizontal force of the entire structure will be delivered to the small rigid office unit which may be incapable of resisting such large forces (or excessive torsion may be developed in the larger structure). Earthquakes have caused extensive damage to buildings of this design which were not properly separated.

Flexible Couplings. Certain types of structures commonly found in industrial installations are often connected at or near their tops by such parts as piping, conveyors and ducts. It may be desirable to connect two buildings by a covered bridge or passageway. In most cases it would not be economically feasible to make such a bridge sufficiently rigid to allow both buildings to vibrate together. Instead, a sliding joint can usually be installed at one or both ends of the bridge.

Stairways. Stairways may be considered as inclined extensions of horizontal diaphragms. Since the stairway has a vertical component, it must be considered as a vertical shear wall and designed as such, or be cut loose so it will not act as a shear wall in an earthquake. If the stairway acting as an inclined shear wall is relatively flexible when compared to other vertical resisting elements in the building, the problem is reduced. Thus, the use of concrete stairs in a stiff building with masonry or concrete walls may be satisfactory. However, more flexible steel stairs should usually be used in buildings having a more flexible frame. In most cases, interior stairs create a hole in the diaphragm which should be treated as an opening in the web of a plate girder.

Short-Column Effects. Whenever the lateral deflection of any column is restrained, the column will generally carry a larger portion of the lateral forces than assumed in the structural analysis. During earthquakes, column failures have frequently been caused by the stiffening (shortening) effect of deep spandrels, stairways, partial-height filler walls, or intermediate bracing members. Unless considered in the analysis, such stiffening effect can be eliminated by providing the adequate isolation at the juncture of the column and the resisting element.

BASIC SEISMIC SYSTEMS

Actually, there are only two basic approaches to providing a seismic structural system capable of resisting lateral forces. All types of buildings may be divided into two categories: (1) frame action by column and beam bending, and (2) shear walls as vertical cantilevers or by x-bracing (Figure 6-2). A combination of these may be used (termed a dual system). Additionally, there are variants, such as the tube concept for distribution of forces, but all rely on either space frames or shear walls to provide the necessary rigidity.

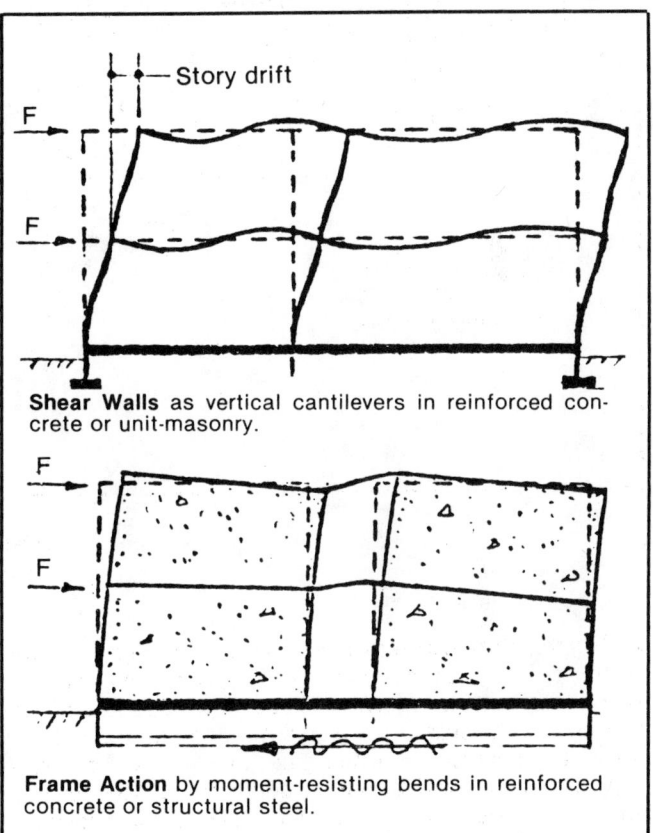

Shear Walls as vertical cantilevers in reinforced concrete or unit-masonry.

Frame Action by moment-resisting bends in reinforced concrete or structural steel.

The two basic seismic systems
Figure 6-2

Space Frames

In a building with a steel or reinforced concrete frame capable of resisting earthquake forces, the columns and beams must be designed to bend. The story-to-story deflection (story drift) may be in terms of inches and still not cause failure of columns or beams. However, the story drift which can be expected to occur during an earthquake will shatter brittle partitions, stairways, plumbing, exterior walls and other items extending between floors. Therefore, buildings can sustain substantial interior and exterior damage, possibly approaching 50% of the building value, and still be considered safe because of the soundness of the frame, columns and beams.

Shear Walls

A shear-wall building is normally quite rigid compared to a framed structure. With low-design stress limits in shear walls, deflection due to shearing forces (for low buildings) is negligible. Shear-wall construction is an excellent method of bracing buildings to limit damage, and is normally economically feasible up to at least eight stories.

Shear walls are usually constructed of reinforced unit masonry or concrete, but wood or plywood in wood frames may be used in buildings up to and including three stories. Using shear walls (box system concept) for an earthquake resistant design is quite valid, but their effectiveness depends primarily on the connections between the structural elements. Excessive openings in any shear wall will alter the effectiveness of the wall. And if the height-to-width ratio becomes great enough, overturning can occur. Also, if the soils beneath its footings are relatively soft, the entire shear wall may rotate, causing localized damage around the wall. Interior columns do not resist any significant amount of horizontal forces in a box system except in tall buildings.

Damage which is directly or indirectly due to permanent ground distortion cannot be overcome by existing earthquake design and construction practices, nor is it practical to do so. However, careful site planning can reduce the amount of damage caused by earthquakes and can alleviate some of the many hazards they pose.

Chapter 7
Reinforced Masonry Construction

This chapter describes structural unit masonry construction. Layout and details of construction are similar to modular masonry.

There are four types of reinforced masonry walls: (1) reinforced grouted masonry, (2) reinforced hollow masonry, (3) reinforced filled-cell masonry, (4) reinforced faced masonry. For any specific job, the type of construction, use of the bases and wainscots, and selection of materials, including contractor's options, are governed by the specifications and plans.

BASIS OF DESIGN
Lintel Beams
Form lintels by placing beam units over openings and reinforcing with a minimum of two #4 bars embedded in concrete corefill. Extend reinforcement 40 bar diameters or 24 inches, whichever is greater, beyond the face of each opening; support reinforcement by wire chairs to ensure proper coverage of steel. Provide steel stirrups as required. Supply bond beams serving as lintels with supplemental steel as required.

Bond Beams
Lap reinforcement bars in bond beams 40 bar diameters or 24 inches, whichever is greater, at splices, intersections and corners. Stagger bar splices. Provide bond beams at the top of masonry foundation wall stems, below and at the top of openings or immediately above lintels, at floor and roof levels, and at the top of parapet walls. (See Figure 7-1.) Provide intermediate bond beams as required to conform to Table 7-1. Whenever the height is not a multiple of this normal spacing, increase the spacing to a maximum of 24 inches provided the bond beams are supplemented with joint reinforcement. Provide one line of joint reinforcement for each 8 inches of spacing. No additional bond beam will be required between window openings which do not exceed 6 feet in height provided you install the prescribed supplemental joint reinforcement. To facilitate placement of steel or concrete corefill, place the top bond beam for filler walls or partitions in the next-to-top course. The area of bond beam reinforcement will be included as part of the minimum horizontal steel.

Walls And Partitions
Design masonry walls for applicable vertical loads and horizontal forces, both parallel and normal to face, with due allowance for the effect of any eccentric loadings.

(A) Height and Thickness Limitations. The minimum nominal thickness of a wall is controlled by the type (structural role) of the wall and the height between supporting diaphragms.

(B) Minimum Reinforcement. Unit masonry needs to be reinforced not only for structural strength but also to provide ductile properties and to enable it to "hold together" in the event of a severe seismic disturbance. Reinforce all walls and partitions as required by the structural calculations, but in no case should there be less reinforcement than the minimum area of steel and the maximum spacing of bars prescribed in Table 7-2. The minimum percentage reinforcement and the maximum spacing of bars is controlled by the type of wall and the seismic zone. In computing the minimum area of reinforcement, consider only reinforcement which is continuous in any wall panel. Joint reinforcement used for crack control or mechanical bonding may be considered as part of the total minimum horizontal reinforcement, but do not use it to resist computed stresses. Further, provide additional bars around openings, at corners, at anchored intersections, in wall piers and at the ends of wall panels. (See Figures 7-2 through 7-8.)

Masonry & Concrete Construction

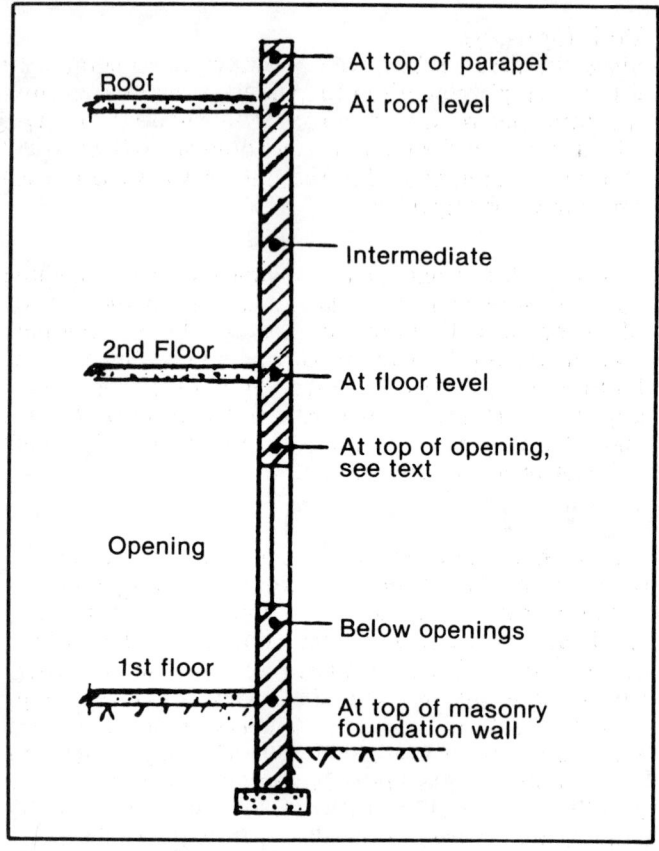

Location of bond beams
Figure 7-1

Type of Wall	Nominal Wall Thickness (Inch)	Max. Height Between Diaphragms (Feet)
Structural	6	12
	8	16
(Load-bearing or shear)	10	20
	12	24
	14	28
	16	32
Non structural	6	18
	8	24
	10	30
	12	36
	14	36
	16	36

Maximum wall height
Table 7-1

Columns and Pilasters

Construct masonry walls and pilasters of reinforced masonry. The design must be able to withstand all horizontal and vertical loads. Do not use masonry columns or pilasters to qualify a structure as a complete vertical load-carrying space frame so as to reduce the factor of a box system (Figure 7-9), and do not use them in rigid frame construction.

Limiting Dimensions. The smallest nominal dimension of every masonry column or wall pilaster should

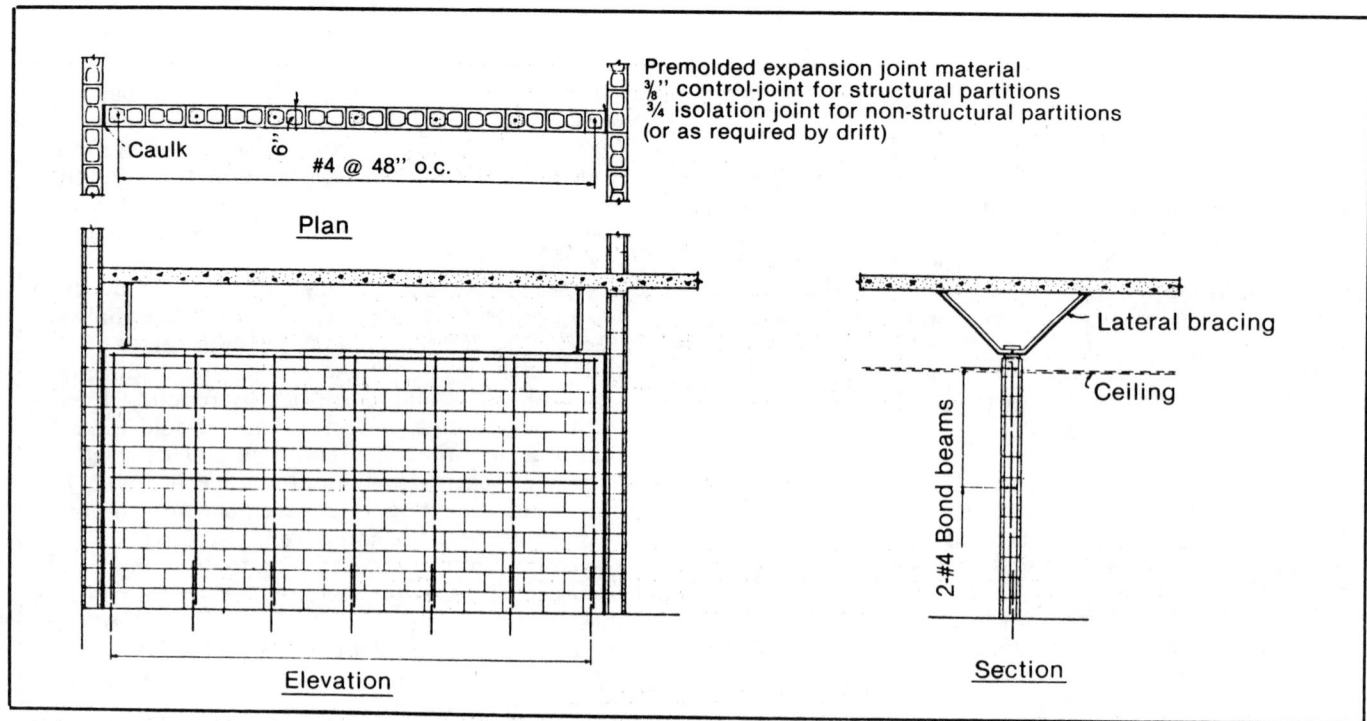

Non-structural partition with minimum reinforcement
Figure 7-2

Type of Wall	Total Minimum Reinforcement (Percent)			Maximum Spacing of Bars (Inches)					
				Vertical Bars			Horizontal Bars		
	Seismic Zone			Seismic Zone			Seismic Zone		
	4 & 3	2	1	4 & 3	2	1	4 & 3	2	1
Load-bearing	0.24	0.20	0.15	24	36	48	48	60	72
Shear (non-load-bearing)	0.20	0.15	0.15	24	48	60	48	72	84
Non-structural	0.15	0.15	0.15	48	60	72	84	84	96

Notes: 1. The total minimum reinforcement is the sum of the vertical and horizontal reinforcement; not less than 1/3 of the prescribed total minimum reinforcement will be used in each direction.

2. Only reinforcement which is continuous in the wall panel will be considered in computing the minimum area of horizontal reinforcement.

Minimum wall reinforcement
Table 7-2

not be less than 12 inches. No masonry column should have an unsupported length greater than 18 times its nominal dimension.

Wall Piers

Design masonry piers to withstand all horizontal and vertical loads. Construct every pier or wall section whose width is less than three times its thickness or 24 inches, whichever is greater, as required for columns. Use horizontal steel for ties in every pier or wall section whose width is between three times its thickness and less than one half the height of adjacent openings. (See Figure 7-10 and Table 7-3.)

Wall Openings

Since the area around wall openings is vulnerable to failure, supplemental reinforcement is required around the perimeter of the openings. (See Figure 7-11.) Use "jamb bars" of the same size and number as the vertical stud reinforcement used in the wall. Always use at least one bar #4 or larger.

A. Case I. Provide jamb bars on each side of opening and at least one bar, #4 or larger, at the top and bottom of the opening. The lintel bars above the opening may serve as the top horizontal bar and a bond beam at the bottom may serve as the bottom horizontal bar. Case I applies to: (1) all openings in non-structural partitions, and (2) any opening in structural or exterior walls which is 2 feet or less in both directions.

B. Case II. The perimeter reinforcement will be the same as in Case I plus additional reinforcement as follows: Provide at least 1 bar, #4 or larger, on the cavity adjacent to the bars as required in Case I. Extend the bar or bars not less than 40 bar diameters or 24 inches, whichever is larger, beyond corners of the opening. However, only two horizontal layers of reinforcement are required at the top and bottom of the openings, and only two reinforced cells are required at each side of the opening. For example, if there is both a bond beam and a lintel bar at the top of the opening, no additional horizontal bars are required above the opening. Case II applies to exterior walls and structural partitions for any opening exceeding 2 feet but not over 4 feet in any direction.

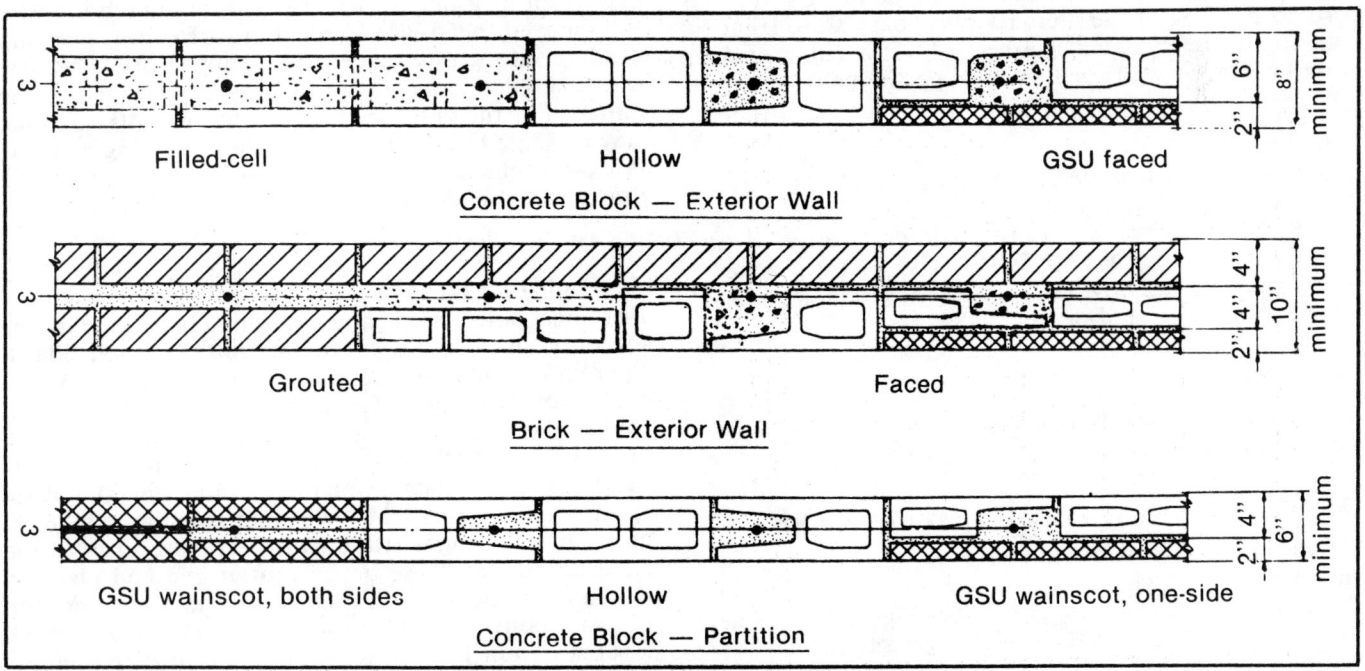

Typical reinforced masonry wall sections
Figure 7-3

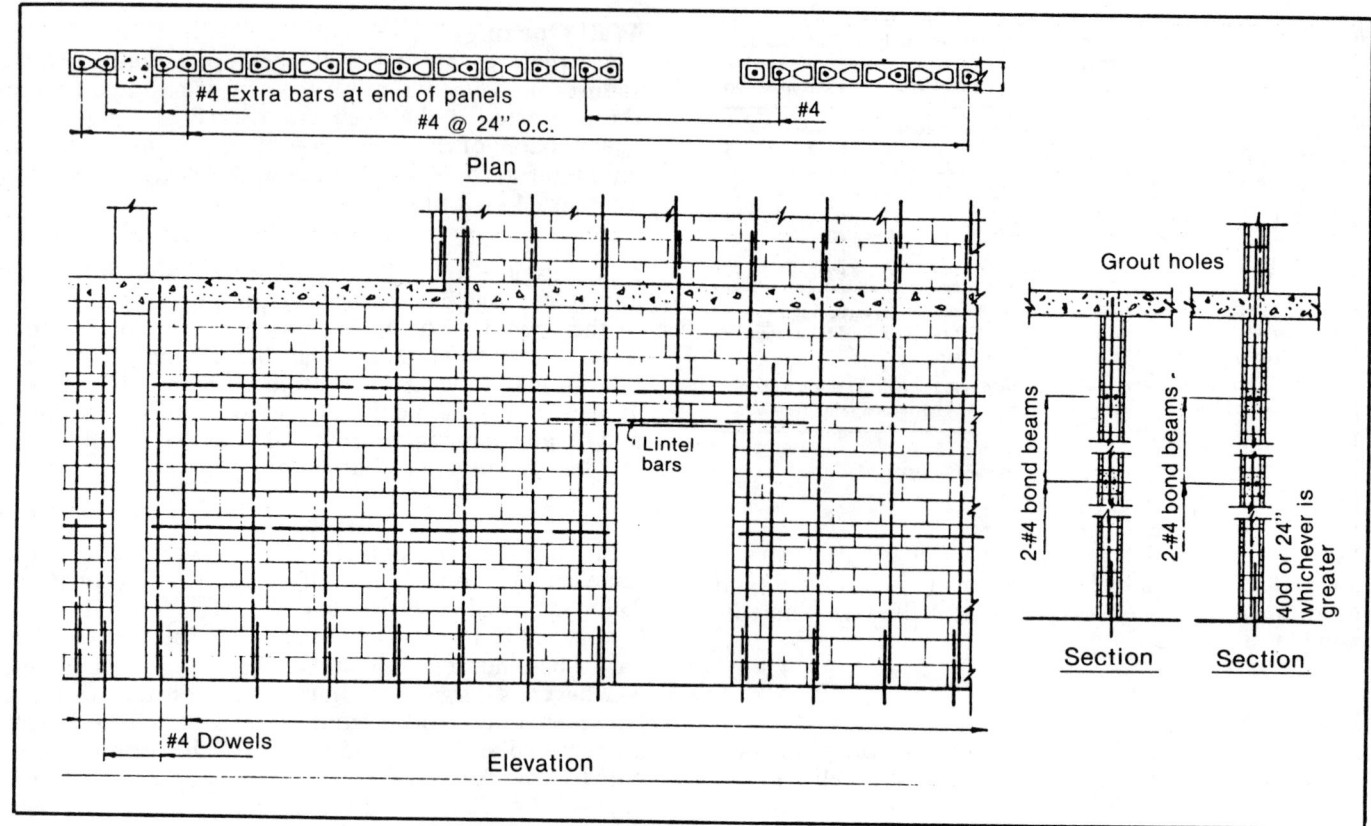

Structural partition with minimum reinforcement
Figure 7-4

C. Case III. The perimeter reinforcement will be the same as in Case II, except that vertical jamb bars will be provided in lieu of the shorter vertical bars. Case III applies to any opening which exceeds 4 feet in either direction in exterior walls or structural partitions.

Stacked Bond

Running bond (staggered head joints for all stretchers) is the strongest and most economical of the bonding patterns. Consequently, it is used more often than stacked bond for masonry walls. Use of the stacked bond pattern is usually restricted to walls essential to architectural layout and design.

Cavity Walls

Do not use cavity walls unless each wythe is individually designed as an independent structural wall.

REINFORCED GROUTED MASONRY

This is a type of construction made with two wythes of masonry units in which the collar joint is reinforced and filled solidly with concrete grout. (See Figure 7-12.) The grout may be placed as the work progresses or after the masonry units are laid. Reinforce collar joints with deformed bars, both vertically and horizontally. Position reinforcement and embedded items such as structural connections and electrical conduit to allow proper placement of grout. Lay all units in running bond with full shoved head and bed joints. Project masonry headers into grout spaces. Use clipped-brick headers where a good appearance of masonry headers is required. For the outer wythes of exterior walls, use only clay or shale brick, unless the option of using concrete brick or split-faced blocks is shown or noted on the drawings. (See also Figure 7-17.)

A. High-Lift Grouting

First erect both vertical and horizontal bars; then lay the masonry units, one wythe of masonry on each side of the reinforcement with space between for grout. When the masonry reaches a full story in height, fill the collar joint solidly with concrete grout. As the work progresses, keep both wythes at approximately the same height to accommodate wall ties (or ladder bars) that are spaced not to exceed 24 inches horizontally and 16 inches vertically to resist the hydrostatic pressures of the fluid grout. Lay these ties in mortar bed and place all ties in the same line vertically to facilitate the vibration of grout pours. Width of the grout space should not be less than 3½ inches. Construct the wall so as to preserve an unobstructed vertical alignment of the grout space.

Provide cleanout openings at the bottom of each pour. The openings should be of sufficient size and

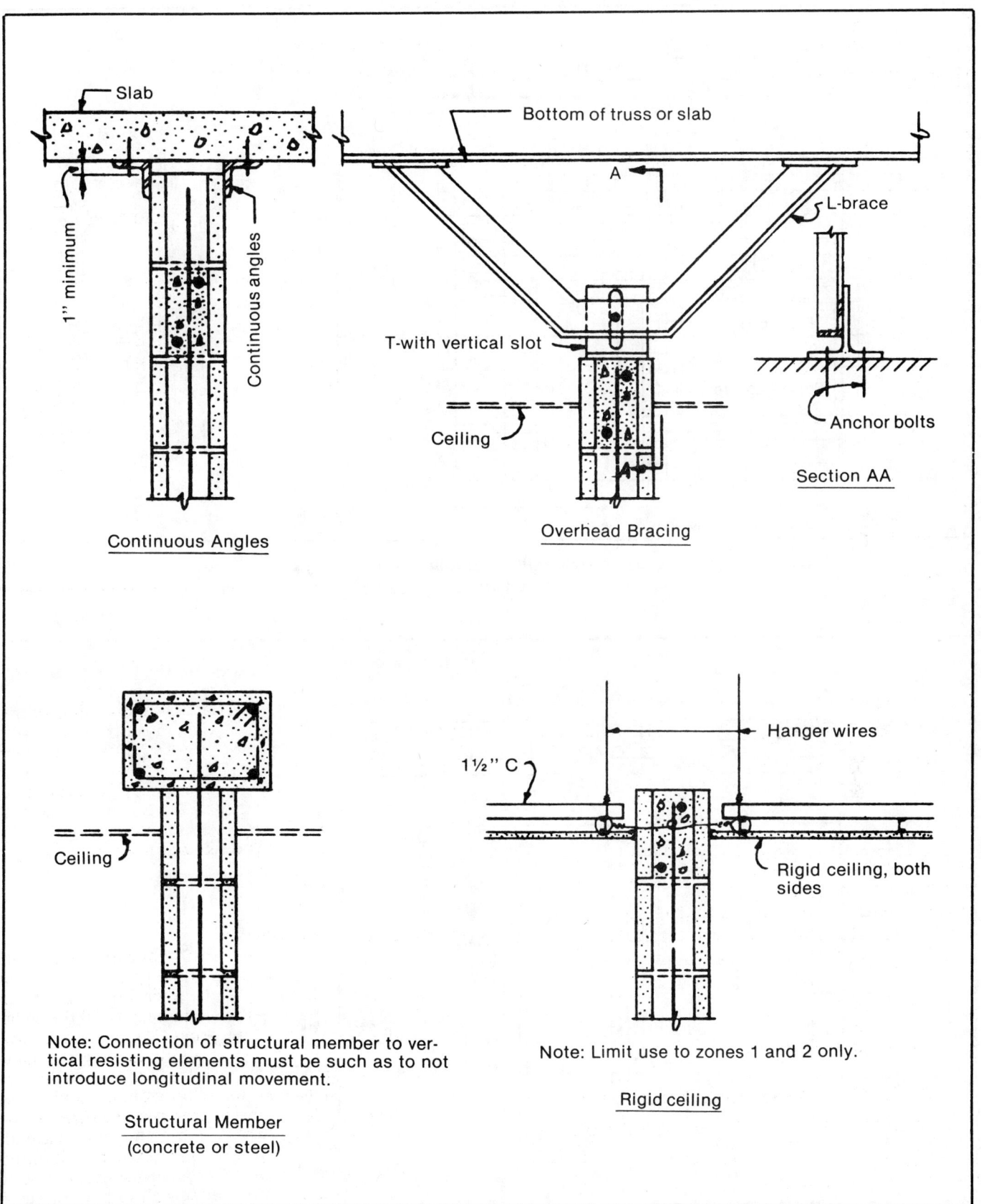

Lateral supports for non-structural partitions
Figure 7-5

Masonry & Concrete Construction

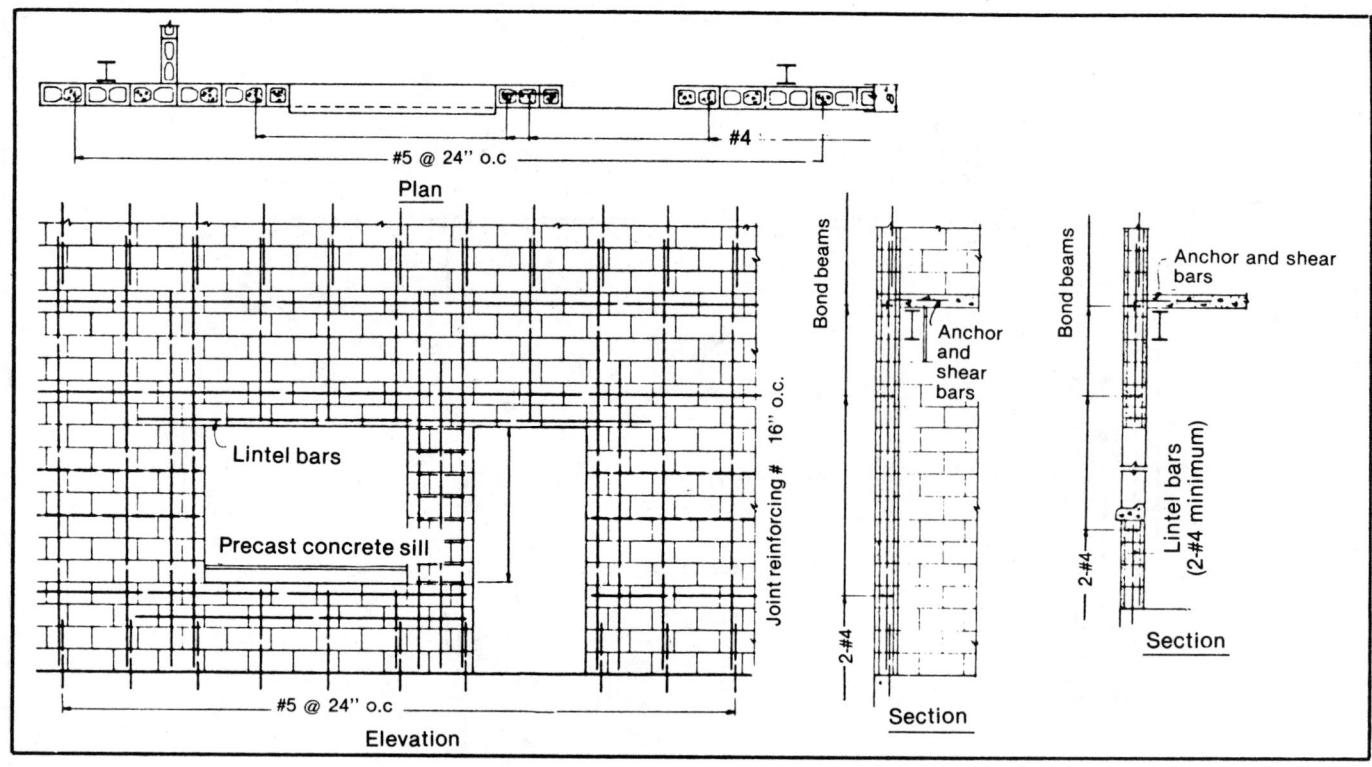

Curtain wall with minimum reinforcement
Figure 7-6

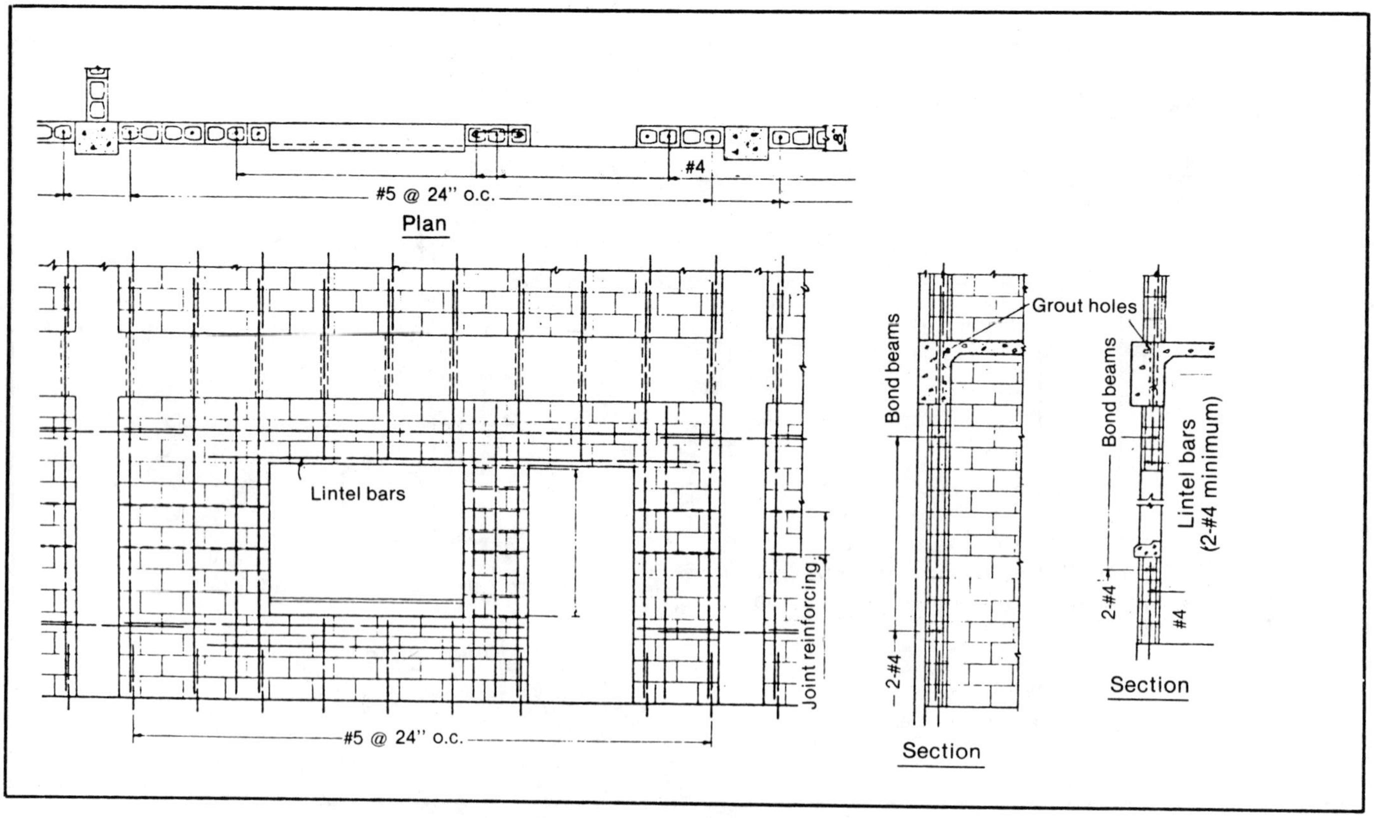

Filler wall with minimum reinforcement
Figure 7-7

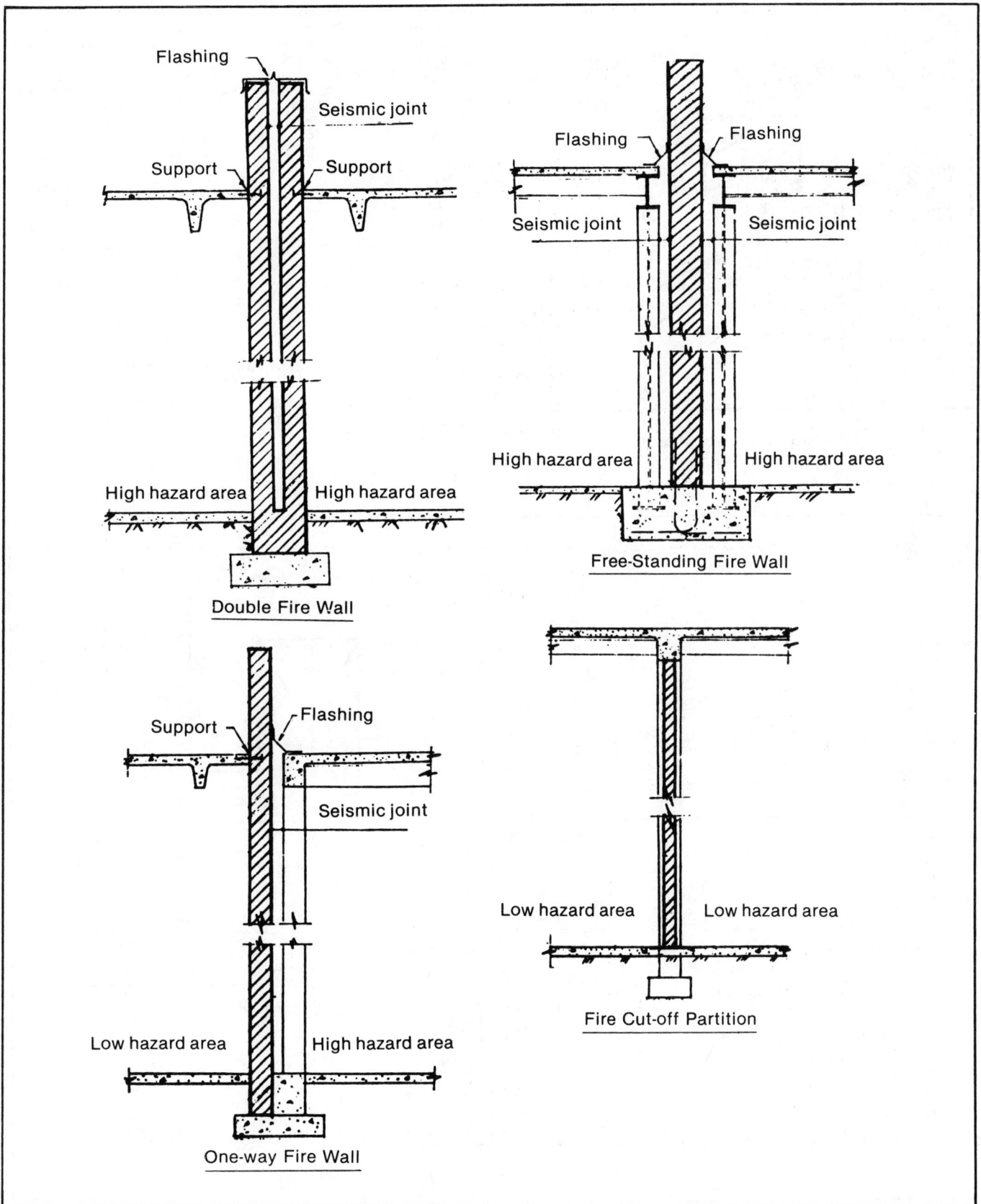

Typical fire walls
Figure 7-8

Masonry & Concrete Construction

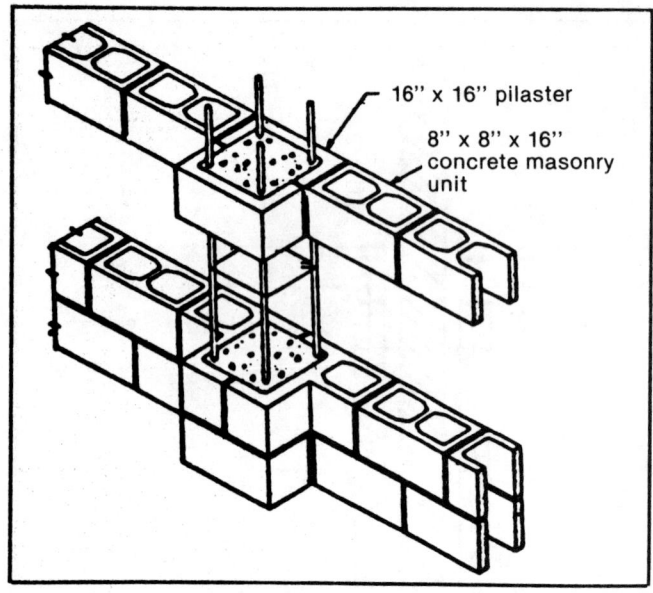

Hollow unit masonry pilaster
Figure 7-9

Nominal Wall Thickness (Inches) T	Design as Column if 2W H or W less than (Inches)	Design as Wall Pier if 2W H and V between, incl. (Inches)	Design as Wall if 2W H and W exceeds (Inches)
6	24	24 - 32	32
8	24	24 - 40	40
10	32	32 - 48	48
12	40	40 - 64	64
16	48	48 - 80	80
Design Criteria	Paragraph 4 - 14	Paragraph 4 - 15	Paragraph 4 - 13

Dimensions of wall piers
Table 7-3

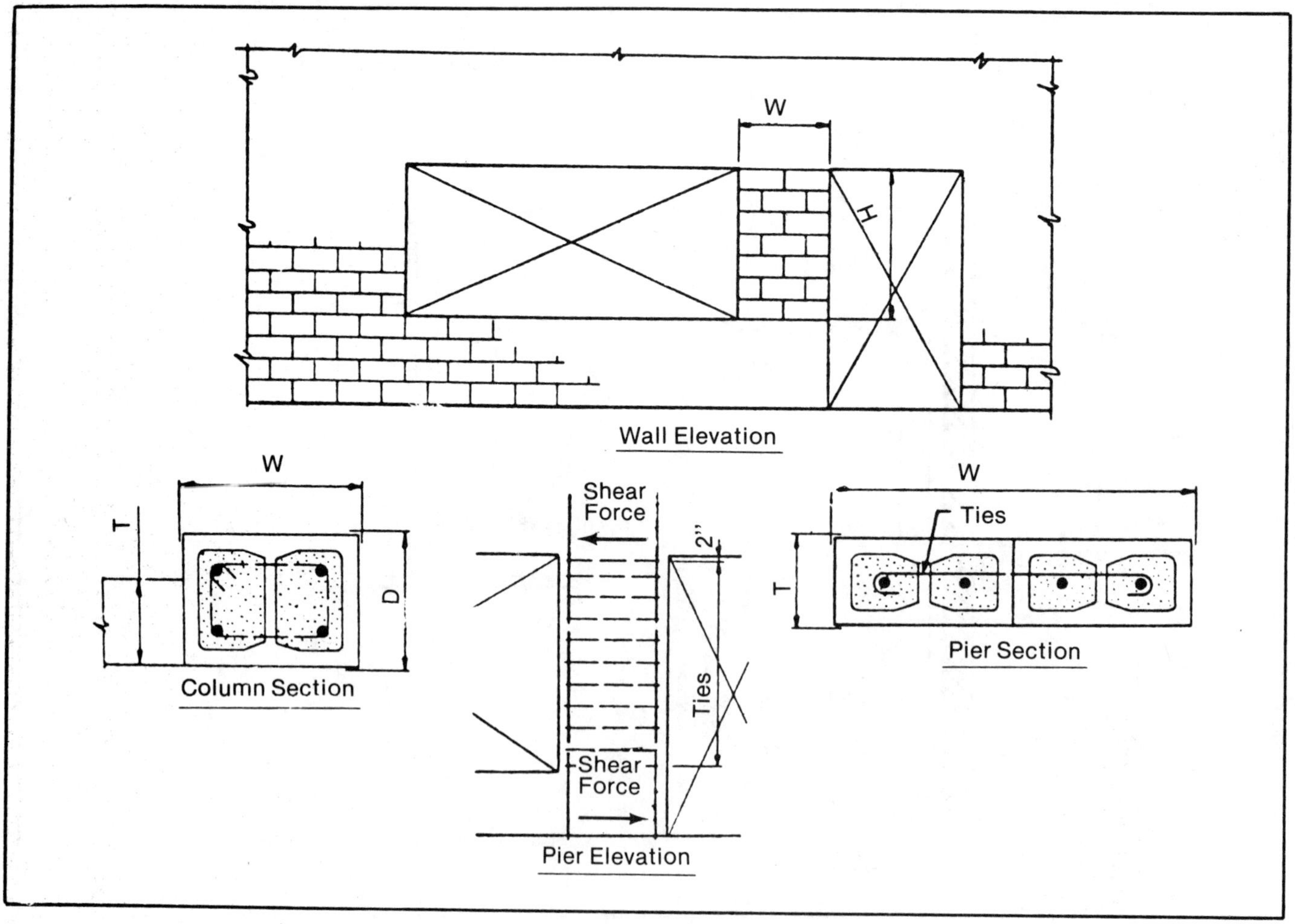

Masonry wall piers
Figure 7-10

Reinforced Masonry Construction

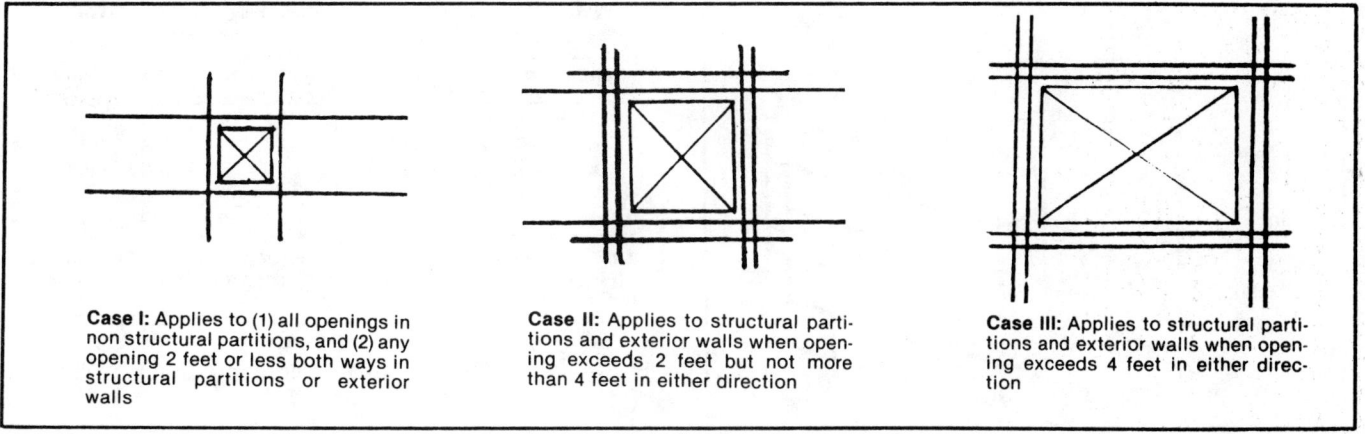

Case I: Applies to (1) all openings in non structural partitions, and (2) any opening 2 feet or less both ways in structural partitions or exterior walls

Case II: Applies to structural partitions and exterior walls when opening exceeds 2 feet but not more than 4 feet in either direction

Case III: Applies to structural partitions and exterior walls when opening exceeds 4 feet in either direction

**Reinforcement around wall openings
Figure 7-11**

location to allow flushing away of mortar droppings and debris. Remove all mortar droppings and overhangs from the foundation or bearing surface and from the reinforcing. There are two ways to do this. One is by hosing at least twice a day (at midday and at quitting time). The other is by spreading a 2- or 3-inch layer of dry sand over the exposed surface of the foundation and then dislodging any hardened mortar from the collar-joint wall surfaces. The mortar debris can then be swept up with the sand prior to cleanup and grouting.

Place in position all cleanout closures, reinforcing, bolts and embedded connection items before you start grouting. Handle the grout from the mixer to the point of deposit in the grout space as rapidly as practical by pumping and placing methods which will prevent segregation of the mix and cause a minimum of grout splatter on reinforcing and masonry-unit surfaces not being immediately encased in the grout lift.

Restrict use of the high-lift grouting methods to walls where wall openings, arrangement of piers, special reinforcing details, or embedded items do not prevent the free flow of grout or inhibit the use of mechanical vibration to properly consolidate the grout.

B. Low-Lift Grouting

First, erect the vertical bars. Then place and grout the horizontal bars as laying of the masonry work progresses. Clean and roughen the contact surface of all foundations and floors that are to receive masonry work to ensure a good bond between the grout fill and the concrete surfaces. Set the width of collar joints to provide at least a 3/4-inch grout coverage around all reinforcement bars.

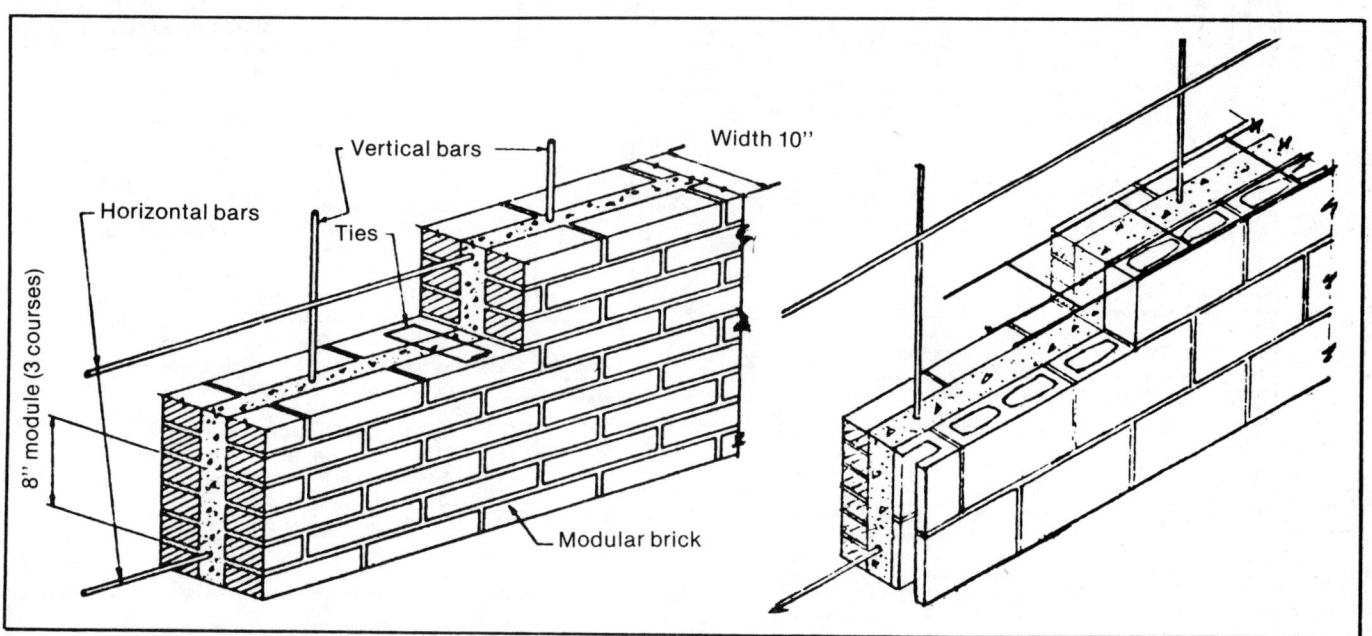

**Reinforced grouted masonry
Figure 7-12**

Masonry & Concrete Construction

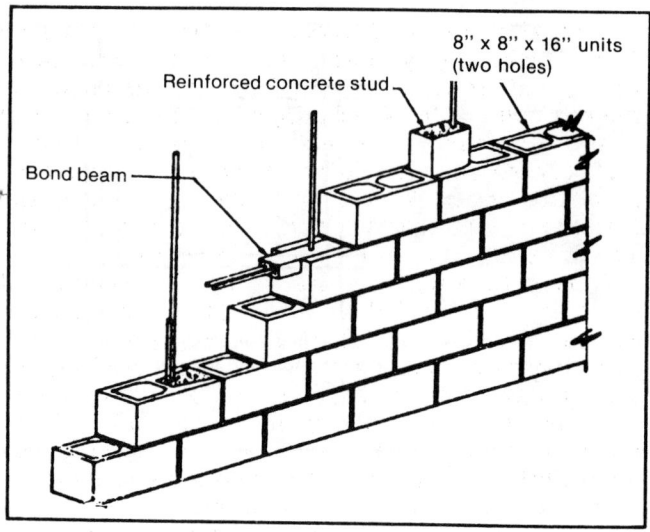

Reinforced hollow masonry
Figure 7-13

Width CMU (Inches)		Assumed Dimension For Design (Inches)					
Nominal	Design	t_s	t_w	x	d_1	d_2	d_f
4*	3-5/8	1	1	7½	--	--	2.81
6	5-5/8	1	1	7½	2.81	--	3.81
8	7-5/8	1-1/4	1	7½	3.81	5.31	4.81
10	9-5/8	1-3/8	1-1/8	7½	4.81	7.06	5.81
12	11-5/8	1-1/2	1-1/8	7½	5.81	8.81	--

*In order to provide adequate space for placement of grout, the use of 4-inch units is limited to faced construction where one face shell is eliminated from reinforced cells

Effective area of concrete block
Table 7-4

REINFORCED HOLLOW MASONRY

Reinforced hollow masonry is made with a single wythe of hollow masonry units (usually concrete blocks) reinforced vertically and horizontally with steel bars. Cores and voids containing reinforcing bars or embedding items are filled with grout as the work progresses. First, erect the vertical bars; then place and grout the horizontal bars and joint reinforcement as laying of the hollow masonry work progresses (Figure 7-13). See Figure 7-14 and Table 7-4 for the effective area of hollow masonry units. Figure 7-17 also shows reinforced hollow masonry in detail.

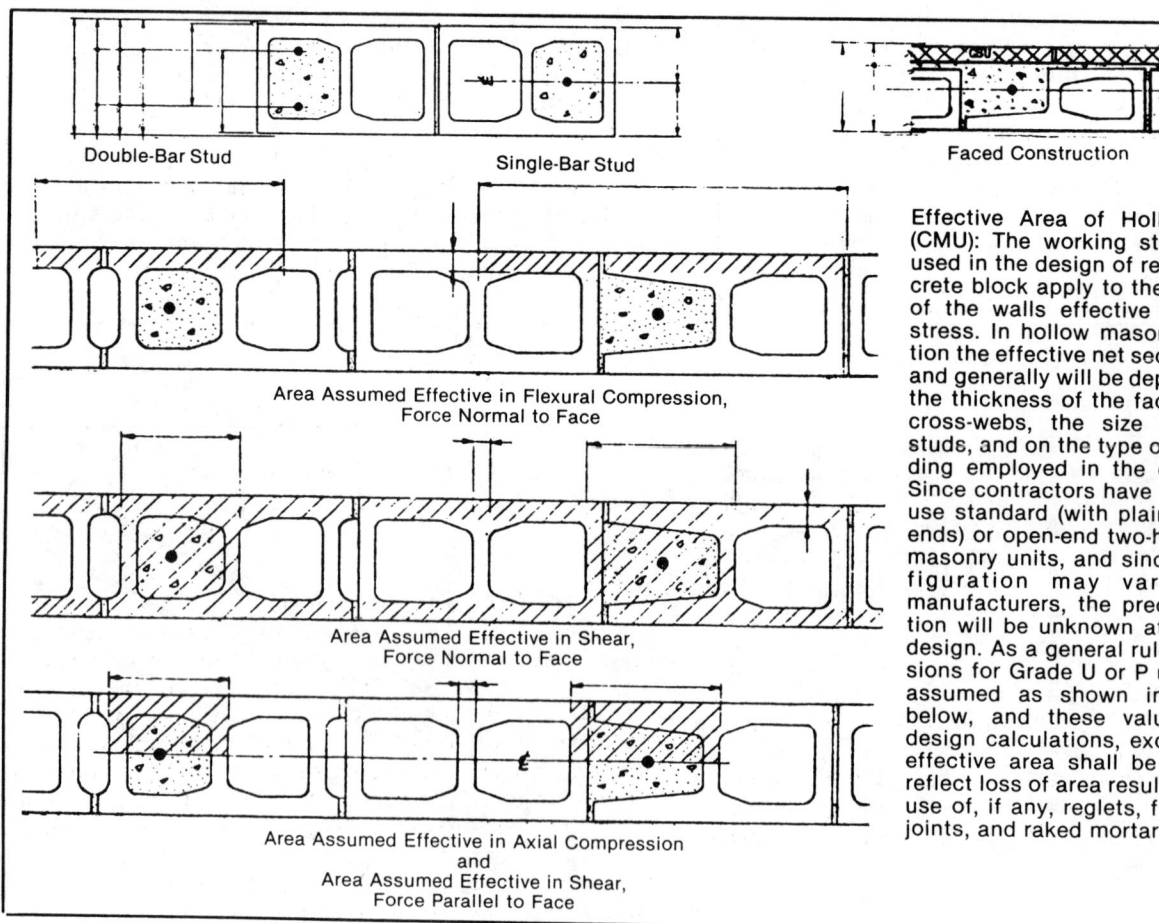

Effective Area of Hollow Masonry (CMU): The working stresses to be used in the design of reinforced concrete block apply to the net section of the walls effective for resisting stress. In hollow masonry construction the effective net section will vary and generally will be dependent upon the thickness of the face shells and cross-webs, the size of concrete-studs, and on the type of mortar bedding employed in the construction. Since contractors have the option to use standard (with plain or concave ends) or open-end two-hole concrete masonry units, and since exact configuration may vary between manufacturers, the precise net section will be unknown at the time of design. As a general rule, the dimensions for Grade U or P units may be assumed as shown in Table "X" below, and these values used in design calculations, except that the effective area shall be adjusted to reflect loss of area resulting from the use of, if any, reglets, flashing, slip-joints, and raked mortar joints.

Effective area of hollow masonry units
Figure 7-14

REINFORCED FILLED-CELL MASONRY

Reinforced filled-cell masonry is a type of construction made with a single wythe of hollow masonry units, reinforced vertically and horizontally with deformed steel bars. All cores and voids are filled solidly with grout after the wall is laid. (See Figure 7-15.) First, lay the hollow masonry units to the full height of the wall, placing horizontal bars and joint reinforcement as the work progresses. Then place the vertical bars and grout all cores and voids using the high-lift method. You may use open-end units, but bond-beam units are preferred to facilitate the horizontal flow of concrete grout and are required at all horizontal bar locations. Secure both horizontal and vertical reinforcement with wire ties or spacing devices near the ends and at intervals not exceeding 160 bar diameters.

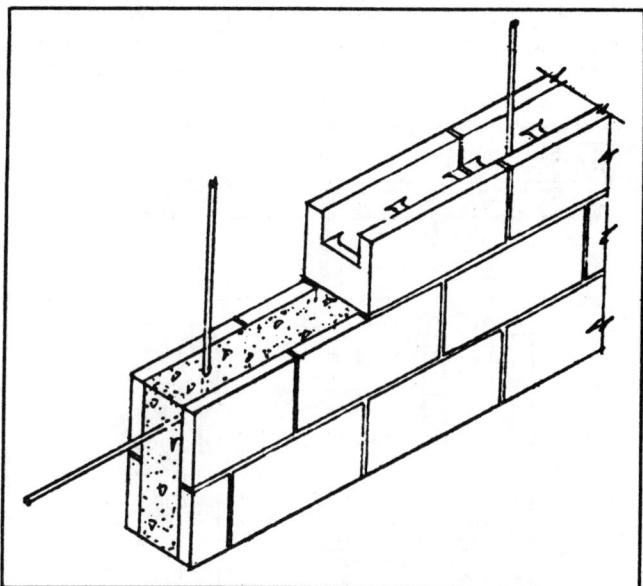

Reinforced filled-cell masonry
Figure 7-15

Before laying, clean and roughen the contact surface of all foundations and floors that are to receive masonry work. These surfaces should be protected during construction to ensure good bonding. Provide cleanout openings of sufficient size and location through the block faces at the bottom of each pour to allow flushing away of mortar and debris. After you have laid the masonry units, cleaned the cells, positioned reinforcing and closed the cleanouts, place the high-lift grout in one continuous pour by lifts which allow time for consolidation and loss of water. Proceed at such a rate as not to form intermediate construction joints or blowouts. Limit the maximum height of any pour to 12 feet for 8-inch walls and 16 feet for 12-inch walls. (See also Figure 7-17.)

REINFORCED FACED MASONRY

Reinforced faced masonry is made with two widths of masonry units in which the structural-bonded facing and backing are of different materials. (See Figure 7-16.) Use hollow concrete masonry units (CMU) faced with glazed structural units (GSU). Lay the facings in running bond as the work progresses and anchor them to the reinforced hollow masonry backup by 16-inch vertically spaced joint reinforcement to form composite action. Completely fill the joint between the facing and the hollow unit masonry with mortar. Set faces of bases and wainscots flush with the CMU wall above.

To permit placing of vertical bars on the centerline of the composite wall, the backup units are b-shaped without an interior face shell on those cells with vertical bars. Partitions constructed with GSU exclusively are limited to pipe spaces, dwarf or stub partitions, and to areas where partition requires two GSU faces. Whenever both faces are exposed to view, the GSU partition thickness is composed of two units to maintain flush alignment of each face. Reinforce all GSU partitions with deformed bars, both vertically and horizontally. Figure 7-17 also shows detail of reinforced faced masonry.

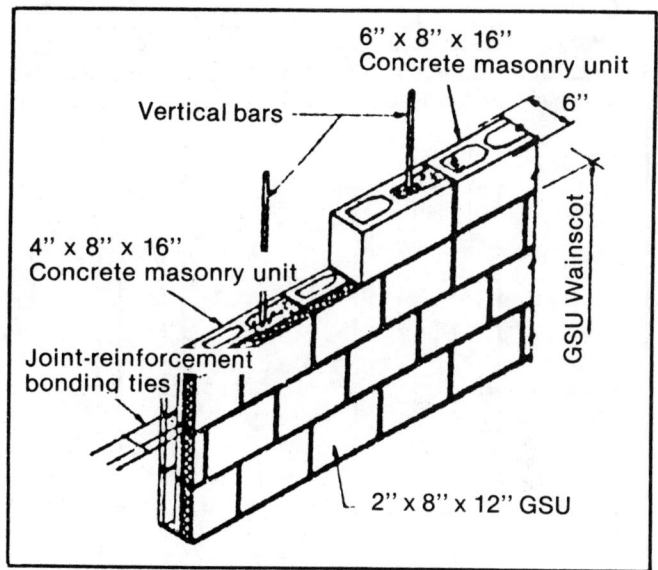

Reinforced faced masonry
Figure 7-16

CRACK-CONTROL JOINTS

Cracking of walls constructed with concrete masonry units is caused by the development of tensile stresses within the wall assembly which exceed the tensile strength of the materials comprising the assembly. Generally this occurs when temperature and moisture changes cause wall movement, or when concrete masonry places restraint on the movement of adjoining elements. Moisture loss depends on the shrinkage potential of the masonry units and the drying conditions at the building site, expressed in relative humidity. Common methods employed to control cracking in masonry structures are (1) materials designed to limit the drying-shrinkage potential, (2) reinforcement to increase crack resistance, and (3) control joints and other details to ac-

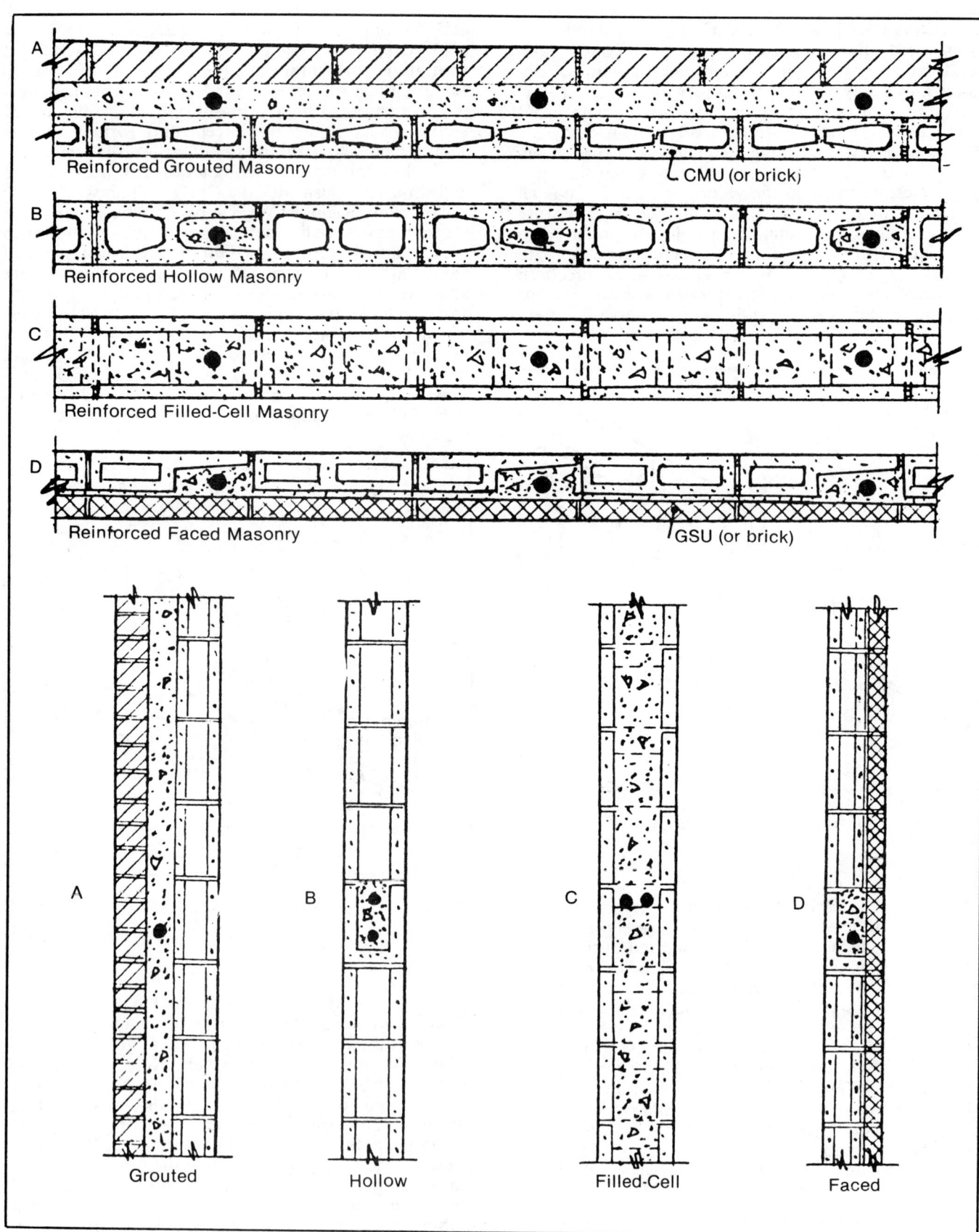

Basic types of reinforced masonry
Figure 7-17

commodate movement. Control joints provide a complete separation of the masonry. Hence, location of control joints fixes the length of wall panels and, in turn, fixes the rigidity of the walls, the distribution of seismic forces and the resulting unit stresses. Therefore, adding, eliminating or relocating control joints will not be permitted once the structural design is complete. Control joints do not transfer bending or diagonal tension across the joint. Joint reinforcement and bars in nonstructural bond beams are continuous for the length of the diaphragm. Using quality-controlled concrete masonry units and the prescribed minimum reinforcement of seismic design, cracking is not a problem when maximum horizontal spacing of control joints is limited to 4 times the diaphragm-to-diaphragm height or 100 feet on center, whichever is less. (See Figure 7-18.)

DEFINITIONS

Collar Joint
The continuous vertical, longitudinal joint between two widths of masonry.

Control Joint
A continuous vertical joint in a wall designed to accommodate movements resulting from temperature and moisture changes.

Grout
A mixture of portland cement, aggregates and water which is proportioned to produce pouring or pumping consistency without segregation of the constituents. Used to fill voids and cells, or collar joints in masonry walls so as to encase steel and bond units together for composite action.

High-Lift Grouting Method
Indicates that grout will be pumped into all wall voids after the masonry units, reinforcing steel and embedded items are built to full story height. High-lift grout is placed in one continuous pour by lifts which allow time for consolidation and loss of water, but placed at such a rate as not to form intermediate construction joints or blowouts.

Lateral Support
These are members such as cross walls, columns, pilasters, buttresses, floors, roofs, or spandrel beams which have sufficient strength and stability to resist the horizontal forces transmitted to them.

Low-Lift Grouting Method
Indicates that grout will be poured in small increments as the masonry work progresses.

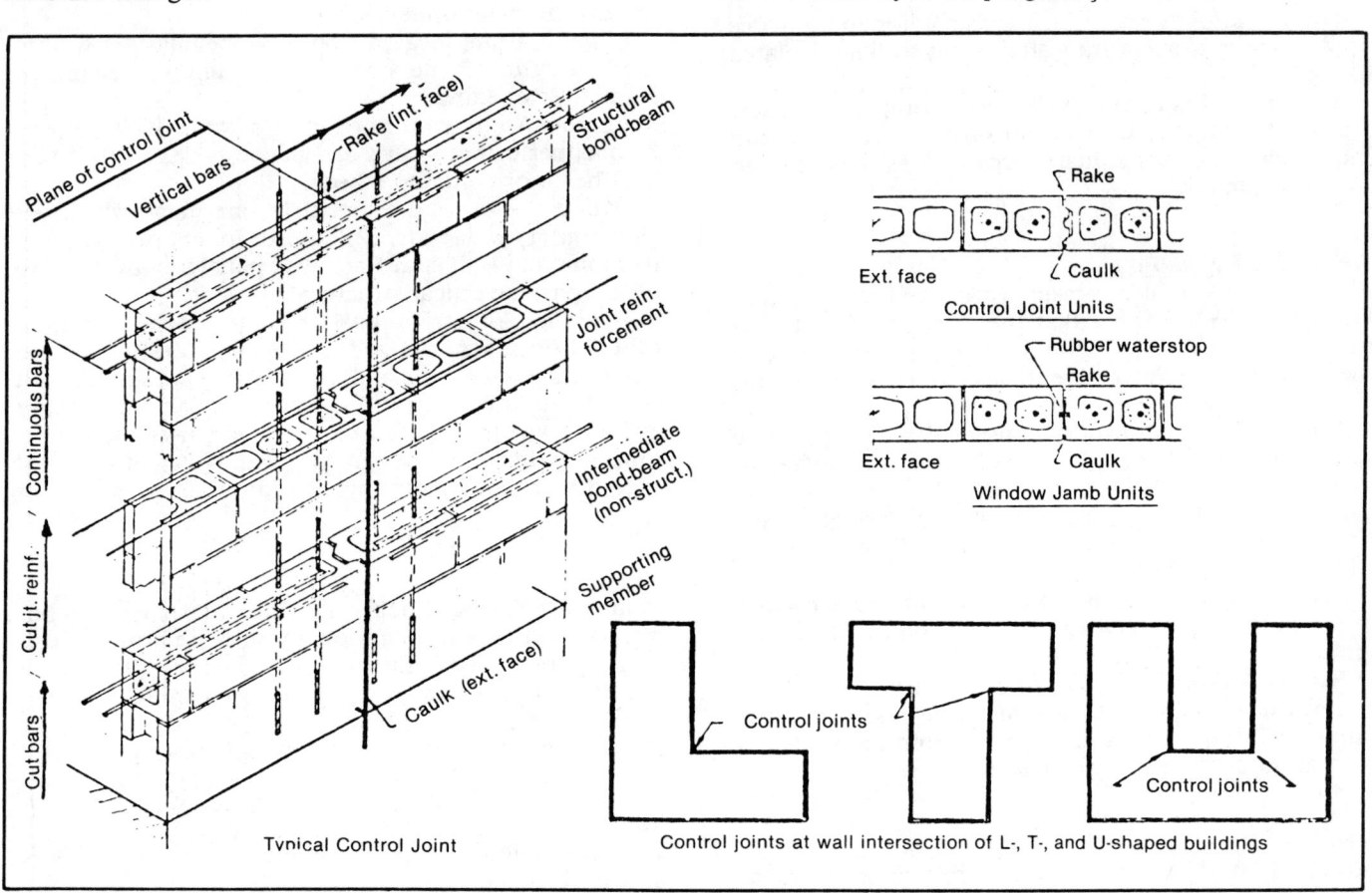

Crack control joints
Figure 7-18

Masonry Wall
A vertical, plate-like element (whose horizontal dimension exceeds five times its thickness) constructed of stone, brick, concrete masonry units, glazed structural units or other suitable masonry materials.

Cavity Wall A hollow wall built of masonry units so arranged as to provide a continuous air space within the wall (with or without insulating material), and in which both the inner and the outer wythes of the wall are reinforced so as to separately resist seismic forces in proportion to their rigidities.

Curtain Wall An exterior, nonbearing wall built outside the building frame, generally with vertical support at ground level only, but may be (and generally is) laterally supported at each story level by anchoring to floors, roof or spandrel beams.

Exterior Wall Any outer wall serving as a vertical enclosure of the building.

Filler Wall A nonbearing wall in skeleton frame construction, built between steel or concrete columns and wholly supported at each story.

Load-Bearing Wall Any wall which in addition to supporting its own weight supports the structure above it without benefit of a complete load-carrying space frame in structural steel or reinforced concrete.

Non Load-Bearing Wall Any wall which does not intentionally support the structure above it.

Partition Any interior wall.

Shear Wall Any wall which resists a horizontal force applied in the plane of the wall (i.e. any wall not isolated along 3 edges)

Structural Wall Any wall which supports vertical loads other than its own weight or which resists lateral movement from horizontal forces such as those caused by an earthquake.

Reinforced Masonry
Masonry units, reinforcement, grout, and mortar combined in such a manner that the component materials act together in resisting seismic forces, and with at least the minimum reinforcement as prescribed by this chapter.

Faced Masonry Masonry construction in which the structural-bonded facing and backing are of different materials.

Filled-Cell Masonry Single-wythe masonry construction composed of hollow units in which all voids are filled with grout after the wall is laid.

Grouted Masonry Multi-wythe masonry construction in which the space between wythes is solidly filled with grout.

Hollow Masonry Single-wythe masonry construction composed of hollow units in which cells and voids containing reinforcing bars or embedded items are filled in with grout as the work progresses.

Reinforcement
Structural steel shapes, deformed reinforcing bars or joint reinforcement embedded or encased in unit masonry in such a manner that it works with the masonry in resisting stress.

Joint Reinforcement An assemblage of steel reinforcing wires designed for use in masonry bed joints, serving to distribute stresses and to tie separate wythes together.

Ladder Bar Prefabricated joint reinforcing wires to which parallel deformed side rods are connected by perpendicular cross wires, forming a ladder design.

Structural Members
Bond Beam A horizontal reinforced masonry beam, serving as an integral part of the wall. Its principle purpose is to provide structural integrity and in-turn crack control. It may also serve as a chord (flange) member of a horizontal diaphragm provided reinforcement steel is made continuous for full length of the diaphragm.

Column A compression member, vertical or nearly vertical, the width of which does not exceed three times its thickness and the height of which exceeds four times its least lateral dimension. Any portion of a bearing wall not bonded at the sides into associated masonry shall be considered a column when its horizontal dimension does not exceed three times its thickness. The least nominal dimension of every masonry column or wall pilaster shall not be less than 12 inches. No masonry column will have an unsupported length greater than eighteen times its least nominal dimension.

Lintel A beam located over any opening in a wall to carry weight of the construction and superimposed loads over opening.

Pilaster An integral portion of a wall which projects from either or both faces and may serve as either a vertical beam or column or both.

Wall Panel A wall segment in one plane which lies between: (1) wall ends, (2) control joints, or (3) a control joint and wall end. Each wall panel is considered to be a separate vertical structural element.

Wall Pier An upright part of a wall between (or adjacent to) openings, the width of which does not exceed five times its thickness and the height of which does not exceed two times its width. Design as a column if height exceeds two times width or if width is less than three times the thickness; design as a wall if height does not exceed two times width and if width exceeds five times the thickness.

Veneer
A masonry facing which is attached to the backup but not so bonded as to intentionally act with it under load or movement. To limit potential damage, the use of veneer construction will be restricted to masonry veneers of less than 1½ inches nominal thickness, such as ceramic tile. Otherwise, use faced construction.

Wythe
Each continuous vertical section of a wall, one masonry unit in thickness.

Chapter 8
Concrete Slabs and Sidewalks

Sidewalks and driveways are usually the last masonry jobs to be completed on a project. Many contractors schedule this work last because the final grade usually has been completed and the materials have been used up or removed from the site by this time.

A sidewalk can be just about any design or shape and can be poured in various colors or finished in different textures. Generally sidewalks are made four feet wide except for service walks, which are generally three feet wide. Always use air-entrained concrete.

Pouring and Tamping

Pour sidewalks directly on the ground unless there is a problem with dampness. Before pouring clear the ground of any roots, large stones or other debris. Then compact the ground under the slab by tamping. This helps prevent settling of the slab and subsequent cracking. Power-driven tampers such as the one shown in Figure 8-1 help compact the base of large slabs or long sidewalks. Tampers can be rented, as can most of the larger power tools used in concrete work. If the tools are not needed very often it is usually cheaper to rent than to buy.

In areas where there is suspected dampness, excavate the area where the walk is to be poured a few inches below the bottom level of the slab and place a fill of course gravel or crushed stone to help drain excess water. (See Figure 8-2.) Make the forms for the sidewalk of 2 by 4 lumber, and stake the outside at least every three feet using duplex-headed nails.

Expansion Joints

A sidewalk generally doesn't need reinforcement because of the placement of control joints or expansion joints. These joints are usually placed about every four feet in the length of the sidewalk. They can be placed at three or five feet also, but they should be uniform.

Put an isolation joint against any building that might come into contact with concrete using asphaltum-impregnated felt 1/2 inch thick. Expansion joints are made in walks by cutting into the slab with a groover such as the one shown in Figure 8-3. These groovers cut a joint into the slab that controls the cracking which may occur when the concrete shrinks. Some groovers are made with a deeper bit than others. Some masons use their finishing trowel or brick trowel to cut the joint to the bottom of the slab before using the groover, making a full joint. Thin deep-bit groovers are made that cut a thin, deep groove to prevent spike heels from catching in the walks. They are available with up to 2-inch deep bits.

Sloping and Edging

Slope sidewalks 1/8 to 1/4 inch to one side or the other to drain off water. Round the edges of the walk to dress it up and to prevent spalling of the concrete.

Concrete edgers of stainless steel and bronze are made in many sizes. Sidewalk edgers are usually 6 inches long and 2¾ inches wide. There are combination edgers such as the edger-groover shown in Figure 8-4. This type of edger cuts costs and working time when the walk and the curb are poured together. The heavy-gauge edger forms the top edge of the curb and cuts a groove about 1/4 inch deep and 3/8 inch wide at the top in one pass along the slab. (See Figure 8-5.)

Cure concrete at least six days before using it for heavy traffic.

Before pouring a concrete sidewalk, prepare the base. If the area is damp and you suspect a problem with moisture, place gravel or crushed stone to a depth of 4 inches. (See Figure 8-2.) If there is no moisture problem,

Masonry & Concrete Construction

Power-driven tamper
Figure 8-1

Groover
Figure 8-3

A base of crushed stone helps drain excess water
Figure 8-2

Concrete Slabs and Sidewalks

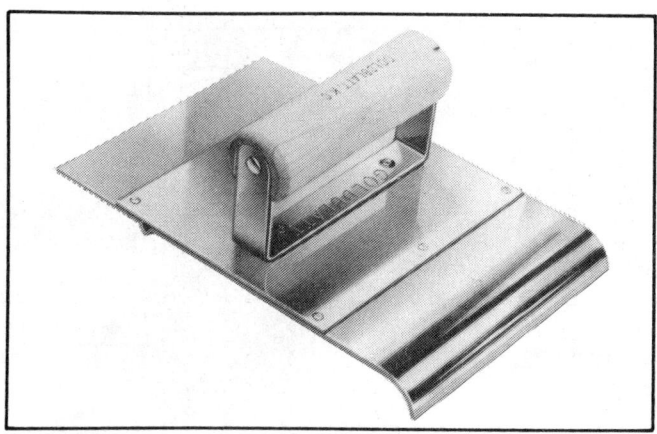

Edger-groover
Figure 8-4

compact the surface and pour the concrete on the ground.

Do the actual pouring of the concrete as near as possible to the place it is needed. Moving the concrete long distances tends to separate the aggregates, causing weak concrete. Screed the concrete as it is poured into the forms. (See Figure 8-7.) Be careful not to leave any concrete on the top edge of the forms. When it dries it will disrupt the smooth movement of the screed by causing it to rise. This rising motion will produce humps in the concrete.

After the concrete is screeded, it must be bullfloated. Using the bullfloat on the surface smooths out the screed marks and also makes the finishing of the surface easier by causing the aggregate to come to the surface of the concrete. Run the bullfloat at a 90-degree angle to the direction of the screed. Push the float across the slab with its front slightly raised, and pull it back across with its back raised to keep it from digging into the wet concrete. (See Figure 8-8.)

When the bullfloating is finished and the concrete has had a chance to start to set (about 20 minutes under normal conditions), use a hand float to further smooth the surface. Move the float in a 180-degree arc across the slab, raising the edge slightly to prevent digging in. Note in Figure 8-9 the expansion joint material that has been set into the wet concrete. It will be pushed down into the concrete until it is just below the surface.

After the surface is floated, give the edges a once-over with an edger. (See Figure 8-5.) Finish the edge soon after pouring the slab since the form material causes the concrete to dry out faster than the rest of the slab. The edging is done two or three times before the job is com-

After the surface of the concrete is floated, finish the edges with an edger.
Figure 8-5

Cover the finished walk to promote curing.
Figure 8-6

pleted, but the first time is the most important because it smooths the aggregate while it is still soft.

Once the walk is finished and the area cleaned up, cover the slab to promote curing of the concrete. A minimum of six days should be given to the curing process. There are various ways to cover concrete, using burlap, wet sand (after the concrete has set), straw, or plastic. Plastic is the most economical and the quickest way. However, if put on too soon, plastic will make marks in concrete as well as discolor it slightly. Place rocks, boards or other heavy objects on the edges of the plastic to keep it from blowing away. (See Figure 8-6.)

Concrete Slab Sidewalks of Precast Concrete

There are companies that make concrete in a precast slab for use in spanning openings, permitting thicker slabs and for sub slabs.

In Figure 8-10, a precast slab is being placed over an opening where a sidewalk will be located. The precast slabs are very heavy and require a machine to remove them and to place them in the correct spot.

In Figure 8-11, the precast slabs are shown in place. It is very important to get the slabs level with each other. They may have a pitch away from a building for runoff, but the edges should be level with each other.

After the slabs are in place, grout them with a mixture of portland cement and a grouting aid such as Interplast-N, which is an expanding grouting aid. This type of grouting aid increases fluidity and produces a slow, controlled expansion prior to hardening. A grout of this type doesn't contain calcium chloride, nitrates, or other chemicals that might contribute to corrosion of steel.

The precast panels with grout added to the joints between the slabs are shown in Figure 8-12. Place the grout in the joints within one hour after the grout has been mixed.

Where areas to be grouted require form work, forms should be tight and well fitted. When using Interplast-N grout, restrain the expansion of the grout to produce the highest possible density, bond and strength. Use top forms where there are open areas. Unformed exposed surfaces will have a substantially lower strength for a limited depth.

After the precast slabs are in place and grouted, coat them with Hydrocide liquid membrane to a thickness of about 55 mils. This is a product of the Contech Inc. Company of Minneapolis, Minnesota.

On top of the liquid membrane, place a material called "Elastomeric membrane." It has the appearance of building paper. Next, place the top or finished slab of concrete. (See Figure 8-13.)

The finish over the precast slabs could be either concrete or some other material such as paving brick. Both are shown in Figure 8-14.

Concrete Slabs and Sidewalks

Screed the concrete as it is poured into the forms.
Figure 8-7

Bullfloating concrete
Figure 8-8

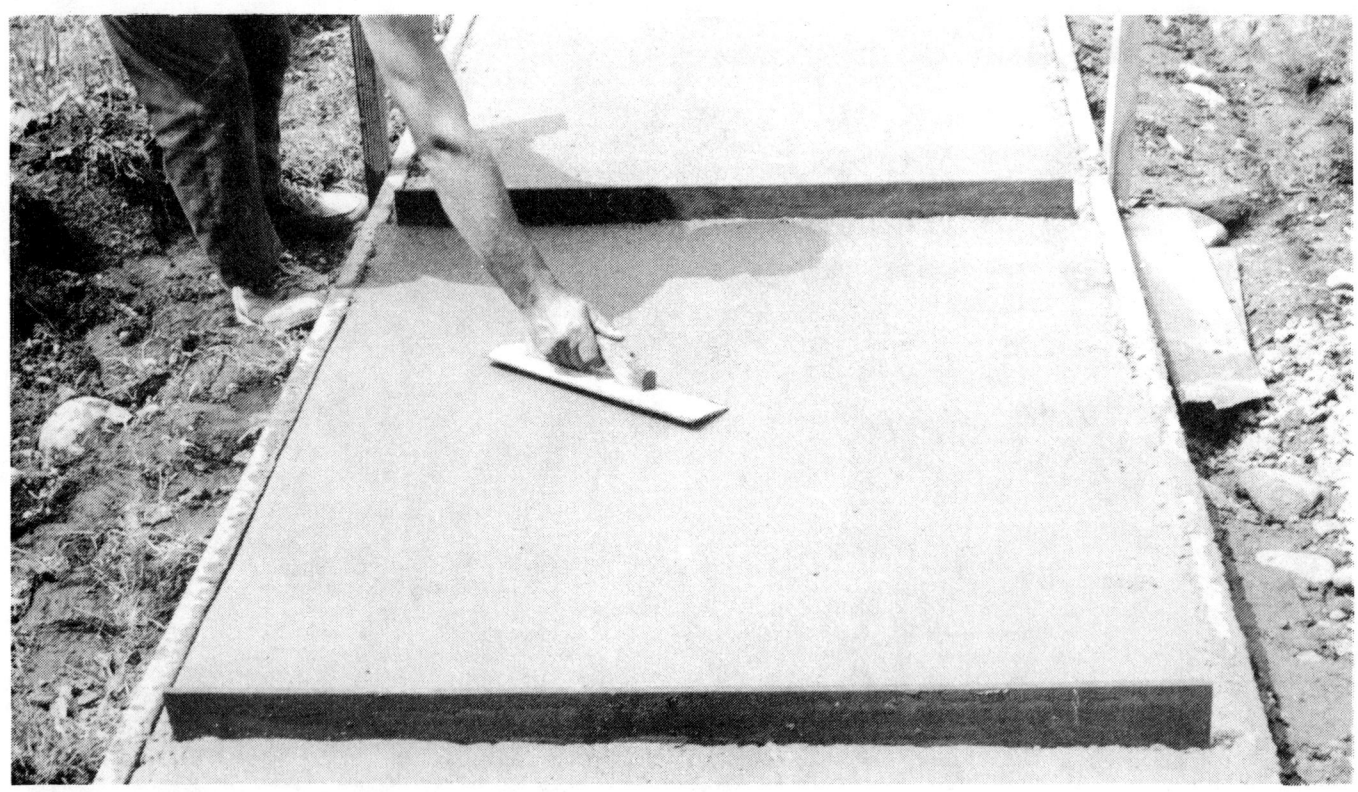
Use a hand float to smooth the surface of the concrete.
Figure 8-9

Placing precast slabs
Figure 8-10

Concrete Slabs and Sidewalks

**Precast slabs in place
Figure 8-11**

**Grouted precast panels
Figure 8-12**

Precast slab coated with liquid membrane and elastomeric membrane
Figure 8-13

Precast slabs finished with concrete and paving brick
Figure 8-14

CONCRETE SLABS

Because of the occasional use of heavy vehicles, make the concrete for driveways about 6 inches thick by using thicker forms than are used for sidewalks. Reinforce concrete for driveways with wire mesh or with rods crossing each other in the slab. Drives are usually at least 10 feet wide and double drives are usually 20 feet wide.

Much of the information about sidewalks is true about driveways. If the subgrade is damp or if a problem with dampness arises, then, as with sidewalks, excavate the subgrade at least 4 inches and fill with crushed stone or gravel. The driveway should have a slope from one side to the other of about 1/8 inch to a foot for drainage of surface water. The driveway should also have expansion joints no more than ten feet apart. Install isolation joint material between the house foundation and the drive if it is near the foundation. Make concrete for driveways with air-entrained cement.

Slab-On-Ground Construction

Slab-on-ground construction is shown in Figures 3-23 through 3-28 in Chapter 3. For the most part, there are two kinds of concrete slabs. One is the slab-on-ground which is constructed in warmer climates. It is also used to make interior floors such as basements or slab floors. Slab-on-ground construction as shown in Figures 3-23 through 3-28 is only the perimeter construction. The remainder of the slab is poured in a manner similar to other slabs.

Two #4 bars should be used for a wood-bearing partition if one story; two #5 bars should be used for a wood-bearing partition if two stories and for masonry nonbearing partitions. Masonry bearing partitions are to be supported on separate foundations. (See Figure 8-15.)

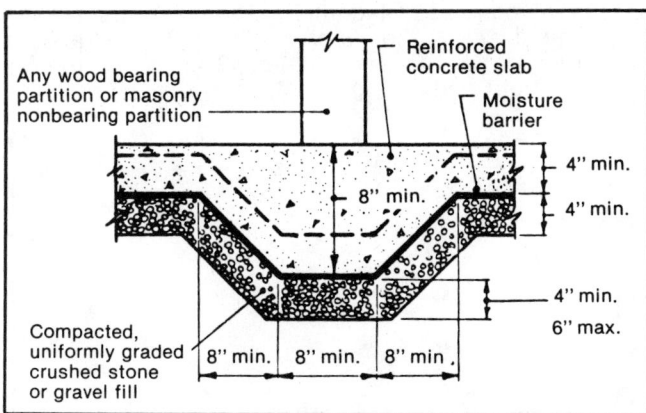

Masonry bearing partition on separate foundation
Figure 8-15

The Superior Screed Joint. A product has been developed for making concrete pouring easier and for saving time and money. It was developed by the Superior Concrete Accessories Company. The Superior screed key joint shown in Figure 8-16 is a 24-gauge galvanized metal contraction joint that acts as a screed and is left in the slab; no form removal, joint repair or filling is necessary. It is available in 10-foot lengths for 4-, 5-, and 6-inch slabs. The top edge is designed to act as a screed and gives a "fine-line" terrazzo appearance. A built-in engineered key gives best load transfer. This key has round edges to eliminate sharp-angle fracture points in the slab. The dowel knockouts are on 6-inch centers. Installation procedure is shown in Figure 8-17.

Reinforcing is generally placed in the top half of concrete slabs. (See Figure 8-18.) Some contractors pour a layer of concrete and drop the mesh in the first half of the pour and then pour the rest over the first layer immediately. Others prefer to lay the mesh on the ground and, while pouring the concrete, pull the mesh up into the slab using special hooks. Consult local codes to see if they specify the procedure or the placement of the wire mesh reinforcement.

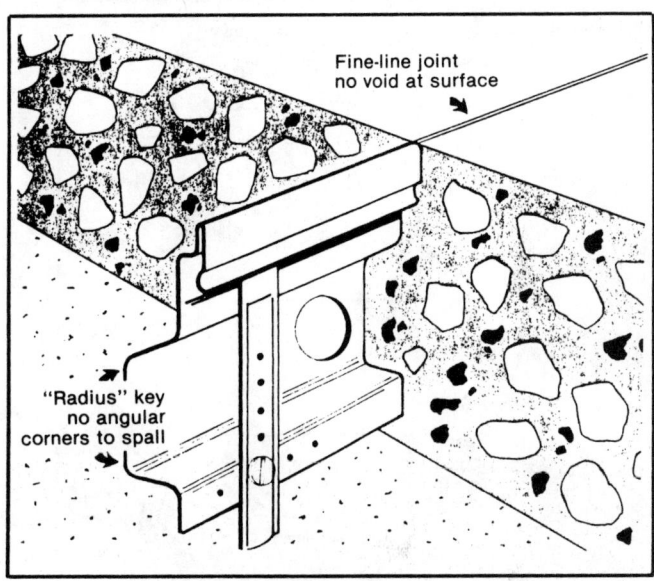

Superior screed key joint
Figure 8-16

On large concrete pours a power screed can be used. (See Figure 8-19.) It is possible to screed up to 10,000 square feet per hour with a power screed. Screed boards clamp onto the unit about 15 inches apart. This power trowel will strike off, compact and float the slab in one pass to leave a semifinished surface. For slabs 16 feet to 24 feet wide, two equally spaced units can be used. This type of equipment can be rented if needed.

The Rollerbug is used for floating and finishing. (See Figure 8-20.) Contractors who have used the rollerbug are delighted with the surface it leaves. It is ideal for all finishes: exposed aggregate, float, broom, or trowel. You can use it as a key for two-course work, and it is excellent for depressing exposed aggregate. The rollers are the only moving parts. Maintenance is simple: just clean the rollers the same way you clean other cement tools.

The Jitterbug tamper is used to bring fine material and fat to the surface for fast, easy finishing and to settle larger aggregate to consolidate the slab for maximum strength. (See Figure 8-21.)

Masonry & Concrete Construction

Stretch line over entire length. Drive stakes at approximately 20' centers. Set stakes ⅜" below finished floor elevation. Secure line to top of the stakes as in inset. The intermediate stakes are driven to bottom of line on 2' maximum centers.

Stake Driving Tool

This groove in top cap prevents top of stake from being flattened. A 2 x 4 is still used to guide the stake vertically.

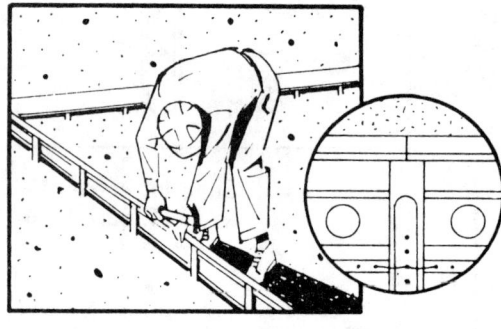

Install Screed Key Joint on stakes as shown. It is preferable to have Radius Formed Key facing initial concrete placement. Butted joints are aligned over stake (as shown in inset), and are wired to nail holes in stakes. This keeps the joint tight and in alignment.

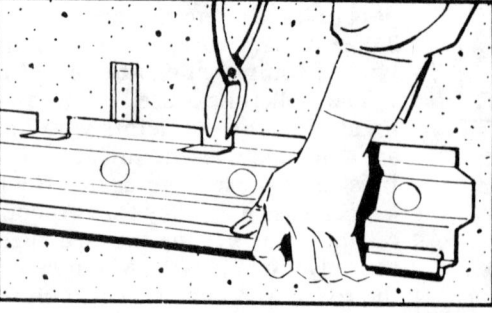

Tinsnips or a metal cutting saw may be used where cutting to length, or trimming around conduit, pipe, etc. is necessary.

Installation of the Superior screed key joint
Figure 8-17

Concrete Slabs and Sidewalks

5 Where joints meet at right angles, Screed Key Joint may be trimmed to fit as shown. Note stake support near all corners and that nails, wire or metal screws may be used as a tie to stake.

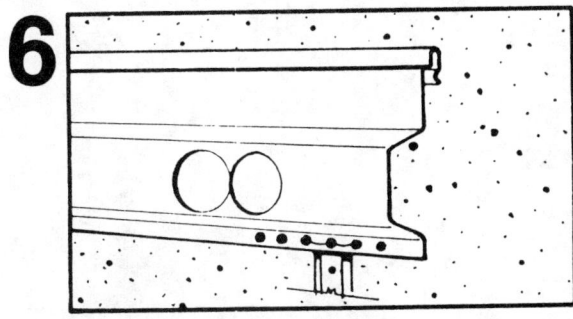

6 1⅛" knockouts are supplied as shown on 6" centers where dowelling is specified. When Screed Key Joint is used as a bulkhead for a construction joint, the knockout tab would be bent back into the pour at 45° angle as shown. Additional bracing may be required for bulkheaded pours.

7 Entire slabs may be poured at once, by leaving out 10' sections of Screed Key Joint temporarily until the concrete trucks can pull ahead.

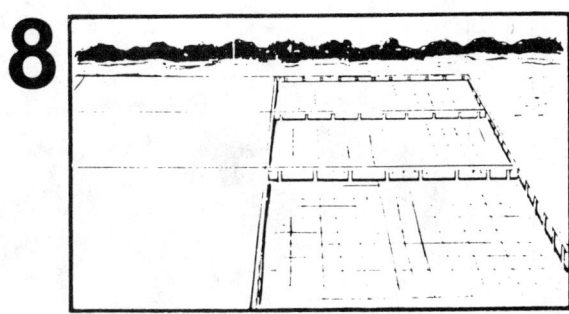

8 The more common method is to pour concrete in strip fashion as shown. When a strip is poured and finished, there are no added and costly steps of cutting, joint treatment, form stripping, edge-damage or concrete surface damage.

Installation of the Superior screed key joint
Figure 8-17 (continued)

Masonry & Concrete Construction

Placement of wire mesh reinforcement
Figure 8-18

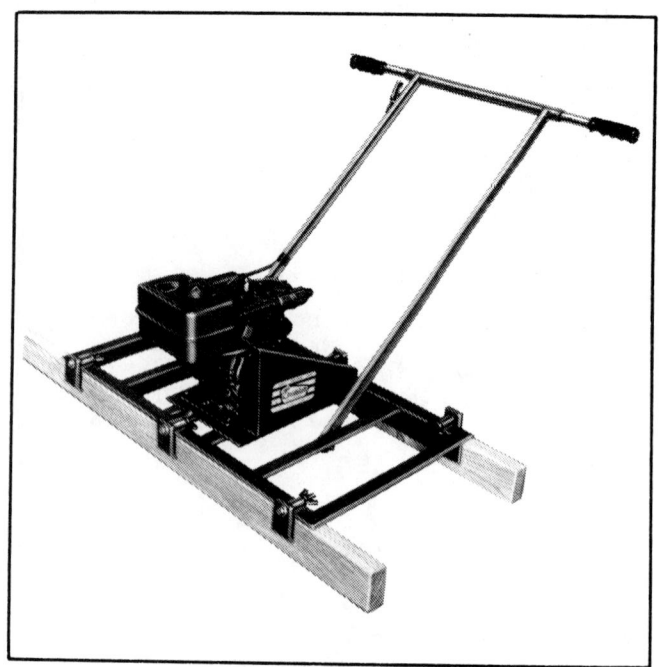

Power-driven screed
Figure 8-19

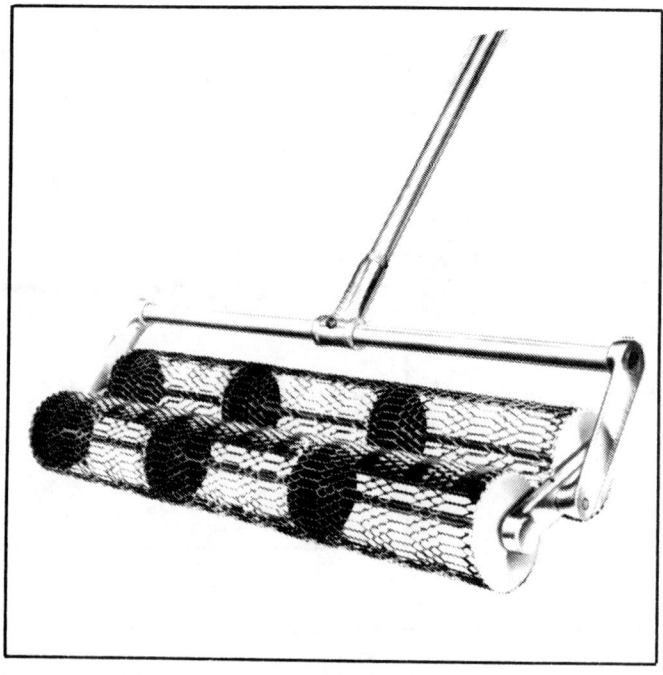

Rollerbug
Figure 8-20

Concrete Slabs and Sidewalks

Jitterbug tamper
Figure 8-21

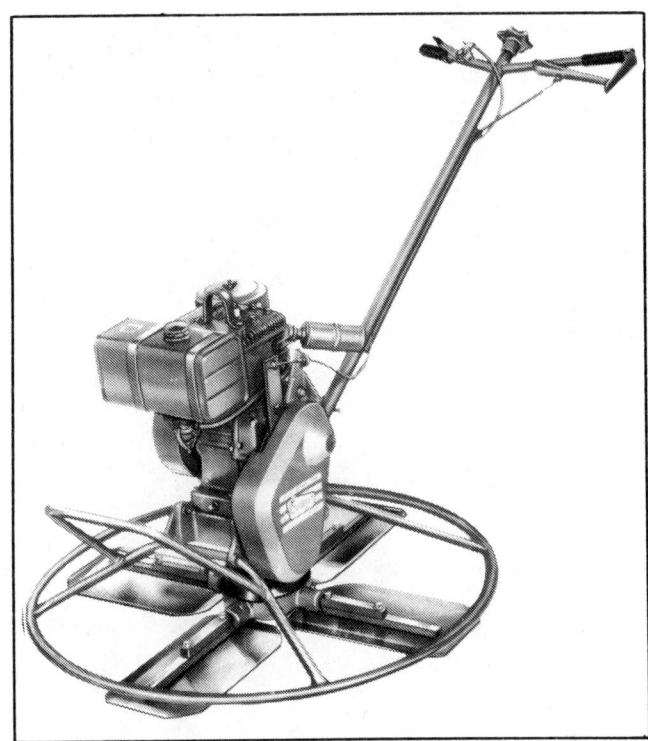

Power-driven trowel
Figure 8-22

The rugged power trowel shown in Figure 8-22 is designed for the roughest concrete work you have to do. It helps achieve a perfect finish even on the largest of projects. This machine has a deadman's control lever that stops the blades the moment the lever is released. The pony trowel shown in Figure 8-23 is a low-cost easy-to-use trowel that can float up to 2,000 square feet per hour. You have to hook a slow-speed heavy-duty drill to the trowel to finish with it. The drill should be 1/2 inch or larger with a chuck speed of 500 rpms or less.

OUTSIDE CONCRETE WORK IN COLD WEATHER

Do not place concrete over frozen ground. Cure concrete at least 48 hours before allowing it to freeze; it is best to prevent freezing for four to five days.

Use Type III portland cement, or Type I with calcium chloride dissolved in the mixing water at the rate of 2 pounds per bag of cement.

When air temperatures are below 40 degrees:
1. Heat the sand, gravel, and water to just below 150 degrees. Heat the sand and gravel in separate piles over culvert pipe or an improvised firebox. Stir and rake frequently for even heating.
2. Remove snow and ice from the forms.
3. Place the concrete immediately after mixing.
4. Cover the concrete and try to retain as much heat as possible with canvas, straw, or hay for four or five days.

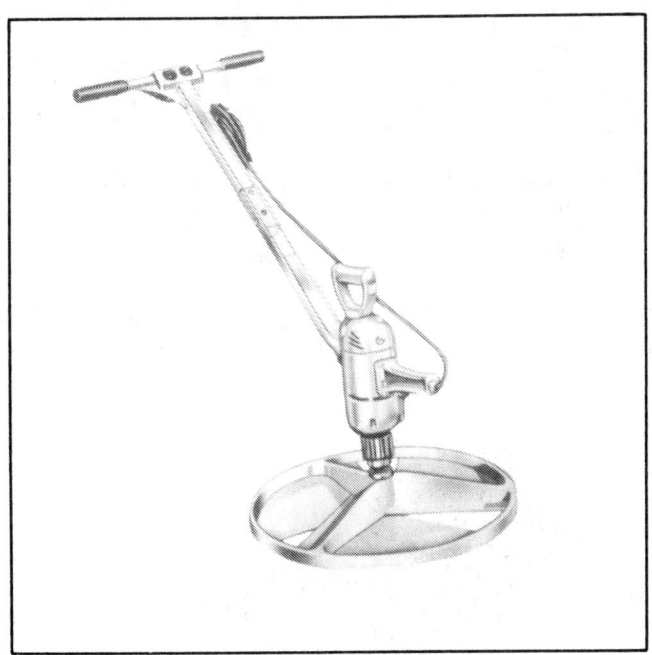

Pony trowel
Figure 8-23

5. Remove forms only after sufficient curing. Pour hot water on the concrete. If the concrete has properly cured, there will be no effect, but if the concrete has frozen when setting it will now soften up on the surface.

Chapter 9
Coloring Concrete

The art of coloring concrete has many practical applications. Care must be taken, however, in selecting coloring materials. Materials should meet the following specifications:

Alkali Resistant
Coloring materials such as certain organic pigments react chemically with alkalies and lose their color value. To test the effect of alkali on a pigment, mix 20 parts of cement with one part of coloring, using sufficient water to form a buttery paste. Keep the samples moist and observe them for several days. If considerable fading occurs, the coloring materials are unsuitable. Under these test conditions, it is possible to develop efflorescence, which must be removed before the true color can be judged.

Lightfast
The coloring material must be able to withstand the bleaching effects of strong sunlight. There is no quick way to determine color retention characteristics. Unless you have access to a fadeometer or a weatherometer, you will have to depend upon published technical data and the experience of the color manufacturer.

Chemically Inert
Coloring materials must not react chemically with the cement or admixtures to cause a reduction in strength of the concrete products when used within the recommended concentrations.

Calcium Sulfate Content
The color additive should not contain more than 15 percent by weight of calcium sulfate because of its effect on setting time.

Insoluble in Water
The colors must not dissolve in water.

Fineness
At least 98 percent of the color should pass a 325 mesh screen.

Uniformity
The color must be manufactured to a given standard so it will be uniform from shipment to shipment.

IRON-OXIDE PIGMENTS
The coloring agents which meet these specifications are selected natural and synthetic iron-oxide pigments. The iron oxides are available in numerous shades of red, yellow, brown, and black. Your use of the selected natural iron oxide, a synthetic iron oxide, or a combination of the two will depend upon the color you wish to produce in your concrete. There are times when it is more economical to use a synthetic oxide with its greater coloring power, but in the majority of applications, the color desired is the determining factor. Both types of iron oxide possess excellent color durability.

Natural Iron-Oxide Pigments
The natural iron-oxide pigments are produced from mined ores which are selected for their color value and chemical composition. They are calcined, pulverized and blended to obtain the desired colors and shades. Their iron-oxide content ranges from 40 to 90 percent. However, in order to obtain a reasonable coloring value per unit of weight, do not use a natural iron oxide containing less than 70 percent iron oxide.

Synthetic Iron-Oxide Pigments
Synthetic iron oxide, available from *E.I. du Pont de Nemours and Company*, is sometimes referred to as pure iron oxide. The synthetic iron oxides are developed by precipitating ferric hydroxide and oxidizing it in the presence of steam and iron to form a hydrated ferric ox-

ide, and by calcining metallic salts. The color range of the precipitated oxides is produced by controlling their particle size during oxidation. The color of the calcined synthetic oxides is determined by the temperature of calcination.

The synthetic oxides have an iron-oxide content of from 96 to 98 percent. Therefore, they have fewer impurities, such as silica, than the natural oxides. This fact in addition to their smaller particle size gives them greater coloring power per unit of weight.

COLOR SELECTION AND PROPORTIONING

Insofar as the coloring characteristics of a pigment are concerned, two factors are of interest. One is the masstone or the top color, and the other is the tint. The masstone refers to the color of the pigment before it is mixed with cement or any other material.

The tint is the characteristic with which you are more concerned. It is the shade resulting when a small quantity of pigment is added to a white material. For a given weight of pigment, the deeper the shade produced, the greater the tinting power of the pigment. For example, three pounds of a natural yellow ochre will barely change the color of a bag of cement, but three pounds of a synthetic yellow oxide will produce a very definite yellow shade. The shade of the tint and whether it is light, dark, clean or dirty, is affected by the color of the cement. The lighter the cement, the brighter and cleaner the resulting tint.

The color of the resulting aggregate also has a direct bearing on the final color of the concrete product. You may have seen colored concrete products which derive their color primarily from the aggregate with no coloring pigment added. The coloring power of the cement and the aggregate in addition to the pigment determines the final color of the end product. A white cement plus a light-colored aggregate will produce the cleaner and clearer tints. The darker cements and aggregates will produce grayer tints.

The determination of the exact shade and quantity of pigment to be used to produce a particular color in a concrete mix can be determined only by experiment. In general, for white cement and light-colored aggregates, one half to one pound of pigment per bag of cement will produce delicate light pastel shades. With gray cement and dark-colored aggregates, three to seven pounds per bag will produce deeper shades.

There are over 150 shades of red, yellow, and brown oxide. Terms such as tan, buff, sun yellow, and adobe tan express shades of color which have not been standardized in the pigment industry. A color designated as adobe tan, for instance, made by two or three pigment manufacturers, will vary in shade from manufacturer to manufacturer. Each manufacturer assigns his own name to his pigments. Therefore, the most practical starting point for selecting a color is to obtain color cards or samples of colors, which are readily available from all manufacturers of cement colors.

These manufacturers can usually make suggestions concerning the quantity and shade of pigment required to produce a specific color. However, if you desire to do some experimenting to determine the effect of color when used with your particular cement and aggregates, the procedure is relatively easy. Use your regular mix, weigh the cement and aggregate in a quantity small enough to be mixed in a steel pan or bucket, add the pigment, and mix thoroughly. Add the water and mix again. The mixing must be thorough to eliminate any streaks. Start with 5 percent by weight of cement if gray is used, and one percent if white cement is used. Pour the mix into a steel pan or mold and allow it to dry. Make three or four such mixes, increasing or decreasing the quantity of pigment by 25 percent for each mix. It will require about two days for your cement to dry to the point where you can determine what the color will be.

The great many shades available and the lack of standardization of the color descriptions among manufacturers makes it impossible, without using proprietary names, to state specifically the exact type and amount of pigment to be used to produce a given shade. Table 9-1 gives the percentage range by weight of cement for various pigments when intermixed with gray portland cement. If white portland cement is used, one or two percent of pigment generally is sufficient.

Desired Color		Pigment to use	Amount (1) Required Percent	
			Patio Stones	Block and Split Units
Yellow	(P)	Synthetic yellow oxide (pure yellow oxide) (ferrite yellow oxide)	5-10	4- 6
	(N)	Yellow ochre (natural yellow oxide)	8-10	6- 8
Red	(P)	Pure red oxide (synthetic red oxide) (precipitated red oxide)	5-10	3- 6
	(N)	Natural red oxide	6-10	4- 8
Brown	(P)	Concentrated brown oxide (pure brown oxide)	6-10	4- 6
	(N) (N) (N)	Burnt umber Raw umber Burnt sienna	5-10	3- 5
Gray	(P)	Synthetic black iron oxide	3- 5	3- 5
Black	(P)	Synthetic black iron oxide	8-10	8-10
		Lamp black	1- 2	1- 2

(1) Percent of pigment by weight with gray cement
(P) Pure or synthetic pigments
(N) Natural pigments

Coloring pigments for use in concrete made with gray portland cement
Table 9-1

Colors

Yellows The synthetic yellow iron oxide, available in shades from light lemon to dark orange, provides the greatest coloring power per unit of weight. To produce light shades, two percent of synthetic yellow iron oxide by weight of cement is adequate. For deeper shades use five or six percent of the deeper orange shades with gray cement. This will produce a buff shade.

The natural yellow pigment is classified as ochre or natural yellow oxide. Its coloring power is considerably less than that of the synthetic yellow oxide and its tint is not as clean and bright.

Red Synthetic red iron oxides range in shade from yellow-red, resembling common red-face brick, to a blue-red or purple. For coloring split masonry units, use three to six percent pigment with gray cement. The pigment concentration depends upon the intensity of the color desired. When used with white cement and light-colored aggregates for such items as patio stones and decorative molds, one half to two percent of color is sufficient.

The color range of natural red iron oxides is not as extensive as with the synthetic red oxides. In general, the colors are not as bright, and this characteristic is admirably suited to the production of colored concrete units which resemble natural stone. For work of this nature, the quantity of color used depends upon the intensity of the color desired in the final product. Some very pleasing shades of pink can be obtained by using white cement with three percent of a natural red oxide.

Browns Natural brown oxides are known as burnt umber, raw umber and burnt sienna. The burnt umbers produce medium to dark shades of brown. Raw umber develops a brown with a yellowish cast. Burnt sienna is reddish brown. In split masonry units, which are generally made with gray cement and dark colored aggregates, three to five percent of these pigments is sufficient. If more intense colors are desired, such as for patio stones, increase the pigment concentration up to 10 percent by weight of cement. The burnt sienna in particular develops some very interesting warm hues.

Synthetic iron oxide browns are produced by direct chemical precipitation or by blending black with pure yellow iron oxide and pure red iron oxide. Either type of brown is equally satisfactory in all respects. They range in color from a light tan to a deep chocolate. Four to six percent of pigment is used for the lighter shades and up to 10 percent for the deeper shades.

Grays and Blacks Three to five percent of iron oxide black produces shades of gray, and up to 10 percent produces black.

A specially treated lamp black, which is readily dispersible in water, produces a jet black at concentrations of two percent of pigment.

Green Dupont's special phthalocyanine green is available in one shade of green for use in concrete. For specific recommendations, consult the manufacturer.

Blues Use blue pigments with caution. They are entirely different pigments from the iron oxides that we have been discussing.

Phthalocyanine blue looks promising, but no long-time exposure data are available. The results with ultramarine blue have been erratic; color failures have occurred without apparent reason. It is possible that the pigment is attacked by the lime during the setting period of the cement. A stabilizer has been developed by the Reichard-Coulston Company of New York to prevent the ultramarine blue from being affected by the lime and results are encouraging. The stabilizer is added at the ratio of one part stabilizer to six parts of cement by weight.

Mixing Colors in Concrete

The addition of pigments to the process of producing concrete products is no more difficult than the application of any other admixture. Add the pigment to the mix prior to adding the water. Three or four minutes mixing time for the cement, aggregate and color should be sufficient to obtain the proper pigment dispersion. Then add the water and mix for an additional four minutes.

There are many variations of this procedure which have been used successfully. If lightweight aggregate material is used, add the aggregate to the mixer prewetted with a small quantity of water before charging the other materials and mixing as mentioned previously. Some producers successfully use a method whereby they disperse the pigment in the mixing water prior to its addition to the mix.

In order to obtain a uniform color from batch to batch, add the pigment by weight and not by volume.

Color Permanence

Iron oxides are noted for their ability to retain their color over long periods of exposure to the elements. Wilson has reported on a nine-year exposure test of colored concrete mortars:

"Mineral pigments of the types comprising the greater part of the pigments tested are not affected by exposure to weather in the presence of portland cement. The more extensively used of these pigments are the various pure or impure oxides of iron, chromium oxide, and carbon black."*

The apparent loss of color in a concrete surface is due to causes other than the fading of the pigment. Weathering of the pigmented cement paste exposes more of the aggregate to view. If the color of the aggregate is in contrast to that of the pigment, a change in the mass color of the surface may be noted. Efflorescence, or blooming, is another cause of apparent color failure. It is usually caused by calcium hydroxide or other hydroxides being leached to the surface where they react with the carbon dioxide in the atmosphere to form a calcium carbonate. This compound is deposited in the surface as a white film which masks the color. The white calcium carbonate crystals are more noticeable on colored concrete than on the natural gray. This has caused the phenomenon to be attributed to the color pigment. Actually this is not so. Properly selected pigments do not contain soluble materials in quantities sufficient to cause efflorescence.

The use of selected iron-oxide pigments in concentrations up to 10 percent by weight does not adversely affect the compressive strength of concrete. In fact, there is evidence that the compressive strength is increased.†

---

*Raymond Wilson, "Tests of Color for Portland Cement Mortars," *Proceedings of the American Concrete Institute,* Vol. 32 (1936), p. 228.

†*Color Your Concrete — It's Easy,* Technical Report C-1 (New York: Reichard-Coulston, Inc.)

Chapter 10

Crack Control in Masonry Walls

Cracks in masonry walls may result from stresses induced by live loads; dead loads; movement of foundations; earth, water and wind pressures; deflection of beams or slabs supporting the walls; differential or restrained movements of building components either connected to or abutting the walls; and restrained volume change of the wall materials. When such cracking occurs in concrete masonry walls, it generally is due to tensile stresses which develop when wall movements accompanying temperature and moisture change are restrained by the other elements of the building, or when concrete masonry places restraint on the movements of adjoining elements.

Controlling cracking of masonry walls is accomplished by three divisions:

(1) Product specifications which limit moisture movement
(2) Control joints which accommodate movement
(3) Joint reinforcement which increases tensile strength and crack resistance between control joints

With proper design, the combination of these three controls will eliminate cracking of masonry walls.

CONTROL JOINTS AND JOINT REINFORCEMENT

It is generally agreed that a combination of joint reinforcement and control joints provides the best design means of controlling cracking. Typical recommendations pertaining to concrete masonry walls using both joint reinforcement and control joints are shown in Table 10-1.

Typical Control-Joint Locations

- At major changes in wall height or thickness
- At construction joints in foundation, roof, and in floors
- At chases and recesses for piping, columns, fixtures, etc.
- At abutment of walls and columns
- At return angles in L-, T-, and U-shaped structures
- At one or both sides of wall openings

Generally a control joint is placed at one side of an opening less than six feet in width and at both jambs of openings over six feet wide. When joints are omitted, extra joint reinforcement should be placed above and below wall openings. Since the control joint is a weakened plane in the wall, it must be detailed to accommodate lateral load across the joint but at the same time be free to move longitudinally. Control-joint materials which provide these features are shown later in this chapter. Unless specifically called for in the plans, joint reinforcement should not extend through the control joints.

Where concrete masonry is used as a backup for clay brick or stone:

(1) Extend control joints through facing if it is rigidly bonded (masonry bonded or with full-collar joints).
(2) Do not extend through facing or veneer when bond is flexibly tied (wire ties) to the backup and the collar joint is not filled with mortar.

Extend control joints through plaster applied directly to masonry units. Plaster applied directly on lath which is furred out from masonry may not require vertical separation at every control joint.

Figures 10-1 through 10-3 show typical installation, placement and spacing suggestions for truss- and ladur-type reinforcing as recommended by the Dur-O-Wal Company.

Masonry & Concrete Construction

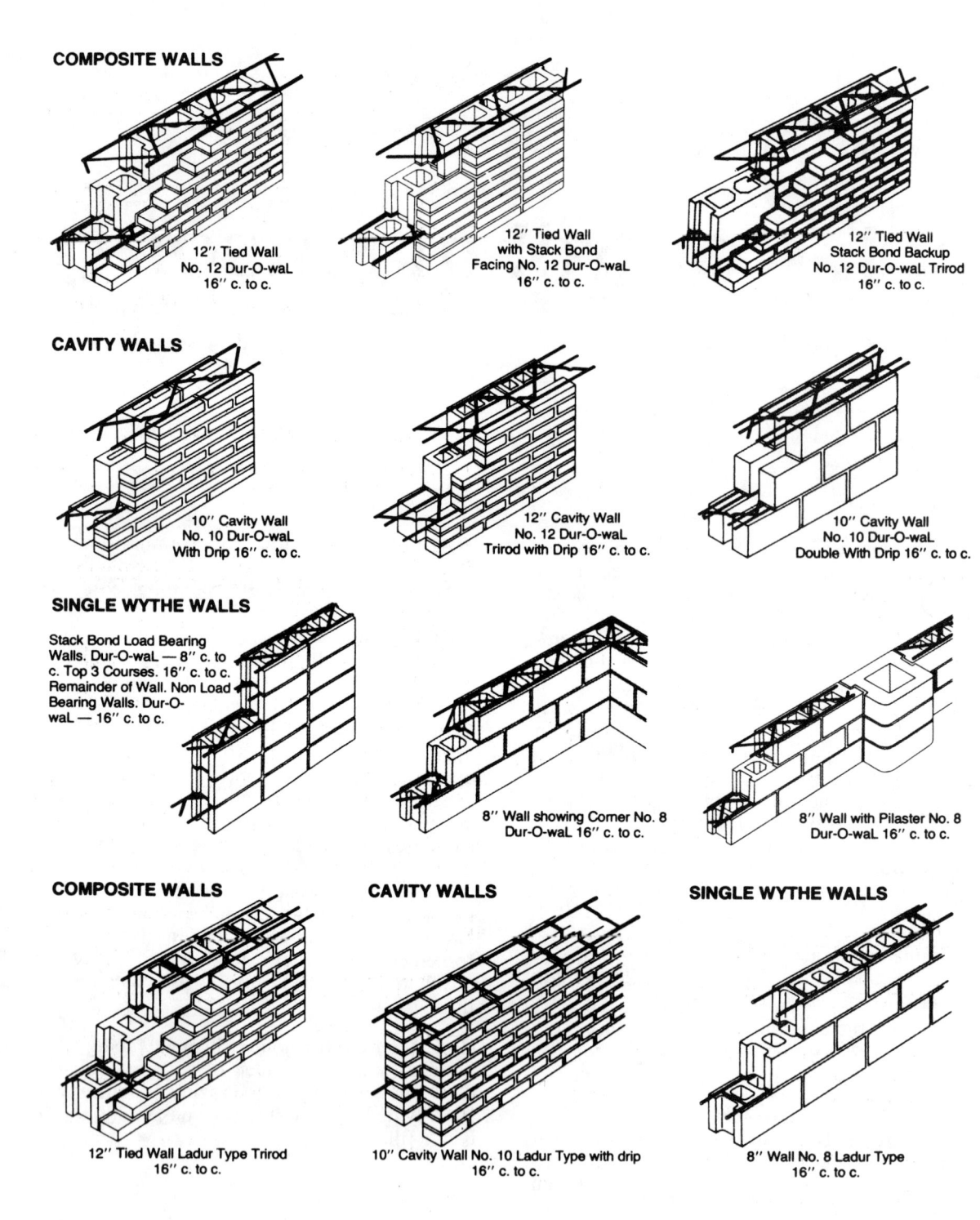

**Dur-O-wal truss and Ladur Type installations
Figure 10-1**

Placement

Out to out spacing side rods shall be approximately 2 inches less than the nominal thickness of the wall or wythe and shall be placed to insure a minimum of ⅝" mortar cover on exterior face of walls and ½" on the interior face.

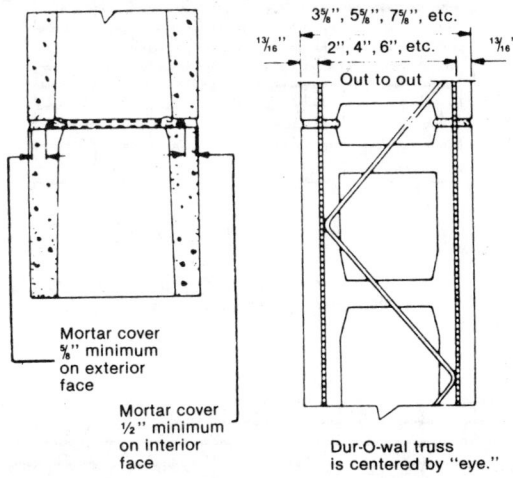

Corners and Tees

Prefabricated or job fabricated corner and tee sections shall be used to form continuous reinforcement around corners, and for anchoring abutting walls and partitions. Material in corner and tee sections shall correspond to type and design of reinforcement used.

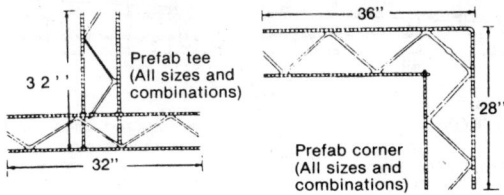

Splices

Side rods shall be lapped at least 6 inches at splices.

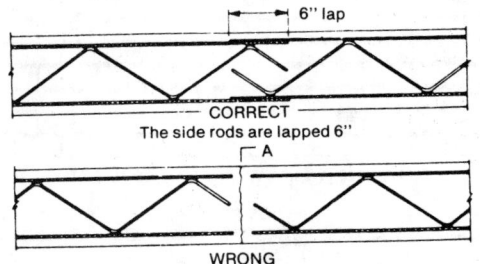

The side rods are not lapped. Reinforcement is ineffective in preventing a crack from starting and opening up at A.

Proper lapping of rods at splices is essential to the continuity of the reinforcement so that tensile stress will be transmitted from one rod to the other across the splice.

Faced and Composite Walls

Dur-O-wal shall be centered over both wythes and the galvanized diagonal cross rods shall serve as ties. Dur-O-wal shall be spaced 16" o.c. vertically and the collar joint shall be filled solidly with mortar.

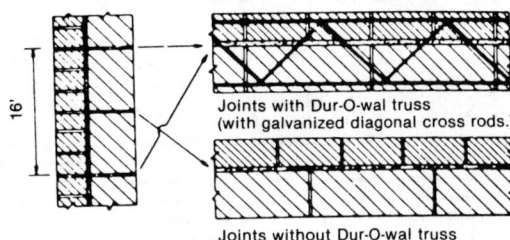

Walls Not Tied With Dur-O-wal Truss Cross Rods

Place Dur-O-wal in (each wythe) 16" o.c. of all faced, cavity or veneered wall not otherwise noted as being tied with the cross rods of Dur-O-wal.

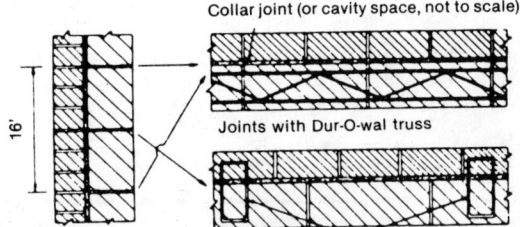

Extra Ties

Provide extra ties at all openings in masonry walls by bending and hooking side rod or cross rod of Dur-O-wal or by adding either regular or adjustable wall ties in alternate courses with Dur-O-wal.

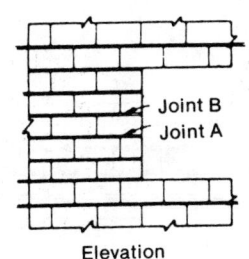

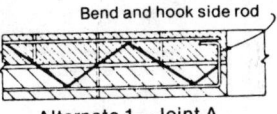

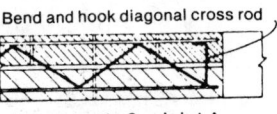

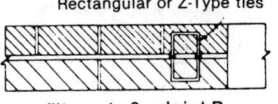

Installation and placement of Dur-O-wal truss and Ladur Type joint reinforcement
Figure 10-2

WALL OPENINGS — Unless otherwise noted, Dur-O-waL shall be installed in the first and second bed joints, 8 inches apart immediately above lintels and below sills at openings and in bed joints at 16-inch vertical intervals elsewhere. Reinforcement in the second bed joint above or below openings shall extend two feet beyond the jambs. All other reinforcement shall be continuous except it shall not pass through vertical masonry control joints.

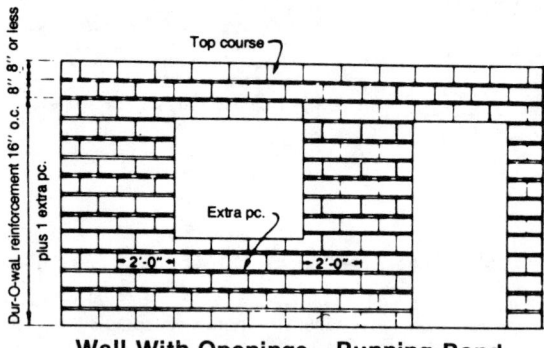

Wall With Openings—Running Bond

SINGLE WYTHE WALLS — Exterior and interior. Place Dur-O-waL 16" o.c. and in bed joint of the top course.

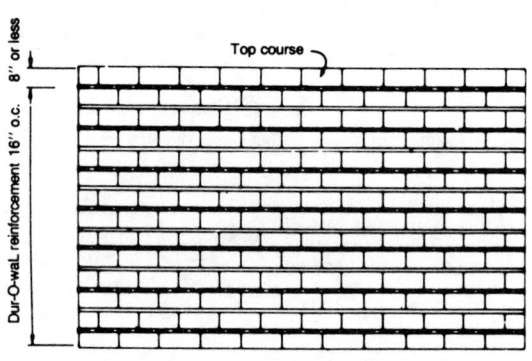

Wall With No Openings—Running Bond

FOUNDATION WALLS — Place Dur-O-waL 8" o.c. in upper half to two-thirds of wall.

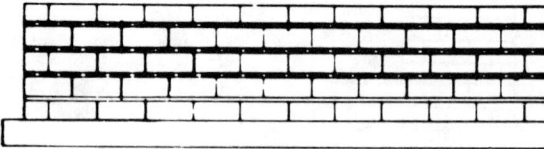

Foundation Wall

BASEMENT WALLS — Place Dur-O-waL in first joint below top of wall and 8" o.c. in the top 5 bed joints below openings.

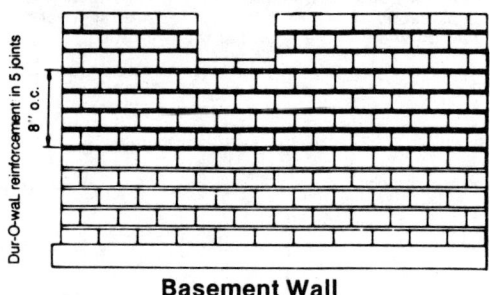

Basement Wall

STACK BOND — Dur-O-waL shall be placed 16" o.c. vertically in walls laid in stack bond except it shall be placed 8" o.c. for the top 3 courses in load bearing walls.

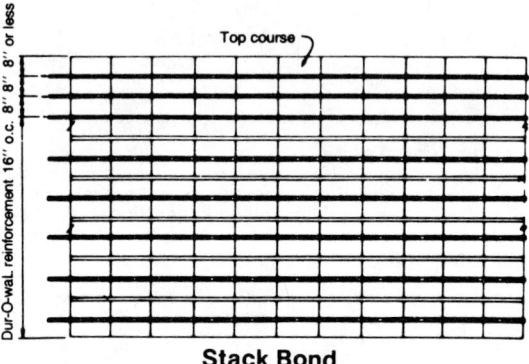

Stack Bond

CONTROL JOINTS — Unless as otherwise noted, all reinforcement shall be continuous except it shall not pass through vertical masonry control joints.

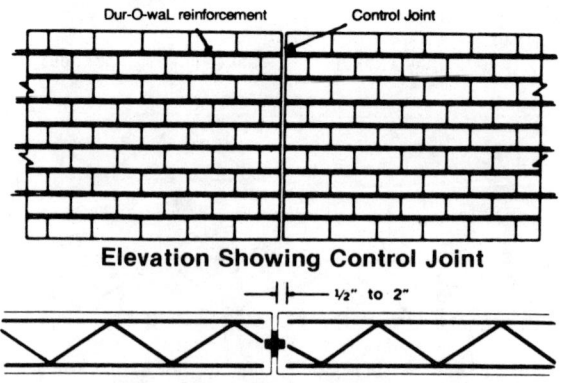

Elevation Showing Control Joint

Horizontal Section (Different Scale)
Dur-O-wal Reinforcement Should Not Cross Over Control Joint

Typical spacing of Dur-O-wal truss and Ladur Type joint reinforcement
Figure 10-3

Recommended Spacing of Control Joints	Nonreinforced Concrete Masonry (2) Vertical Spacing of Joint Reinforcement		Reinforced Concrete Masonry (3)
	16 in.	8 in.	
Expressed as Ratio of Panel Length to Height, L/H	3	4	4
With Panel Length (L) Not to Exceed: Regardless of Height (H)	50'	60'	100'

(1) American Concrete Institute Committee 531, "Concrete Masonry Structures — Design and Construction", Title No. 67-23, Vol. 67, 1970.

(2) Recommendations presume the use of Type 1 Moisture-Controlled Units. Limits should be reduced by one-half where nonmoisture-control units are employed. Where the wall is solidly grouted, limits should be reduced by one-third.

(3) Recommendations presume the use of Type 1 Moisture-Controlled Units. Limits should be reduced by one-half where nonmoisture-control units are employed. Minimum horizontal steel of 0.07 percent of the gross cross sectional area should be increased proportional to the volume of grout. A solid grouted wall should have a minimum horizontal reinforcement of 0.13 percent gross area. See page 24-25 for discussion of reinforced masonry construction.

Control Joint Spacing(1)
Table 10-1

Adjustable Wall Ties enable contractors to adopt labor-saving techniques and overcome problems encountered in laying up walls that are not conventionally designed. Some uses and advantages offered by Adjustable Wall Ties are:

(1) Alleviating coursing problems — adaptable to job conditions where the facing and backing do not course out at proper intervals or where it is desirable to build one wythe ahead of the other
(2) Increased productivity — allows the mason to concentrate on one wythe of the wall, to any given height, before changing to the other wythe
(3) May be used with rigid insulation — provides for the proper mechanical attachment of rigid insulation to the backing wythe
(4) Ease of inspection — architect or building inspector can easily see if the ties are installed as specified
(5) Eliminates bending of ties — the mason does not have to bend or reshape conventional rigid ties when misalignment occurs

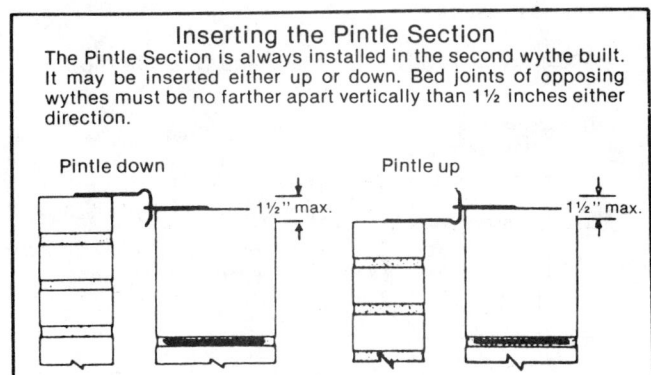

Inserting the pintle section
Figure 10-5

(6) Anchoring intersecting walls — provides a means of mechanically anchoring intersecting walls when masonry bond at intersections is not required
(7) Waterproofing — provides for better waterproofing of the backing wythe in cavity walls
(8) Parging — the wythe to be parged is erected first with the eye section set out just far enough to accommodate the pintle section after parging. This speeds up and improves the quality of the parging because of the absence of projecting rigid ties.
(9) Speeds up construction — the inside wythe can be built to fully or partially enclose the building so that work can be started inside. Exterior wythe can be built and tied in later.

Adjustable Wall Ties are a patented product of the Dur-O-Wal Company. Refer to Figures 10-4 through 10-6 for various types of wall ties and their applications.

Control Joint Materials
Rapid Control Joint is a patented product of the Dur-O-Wal Company. Its basic function is to provide a vertical stress-relieving joint in concrete masonry walls while

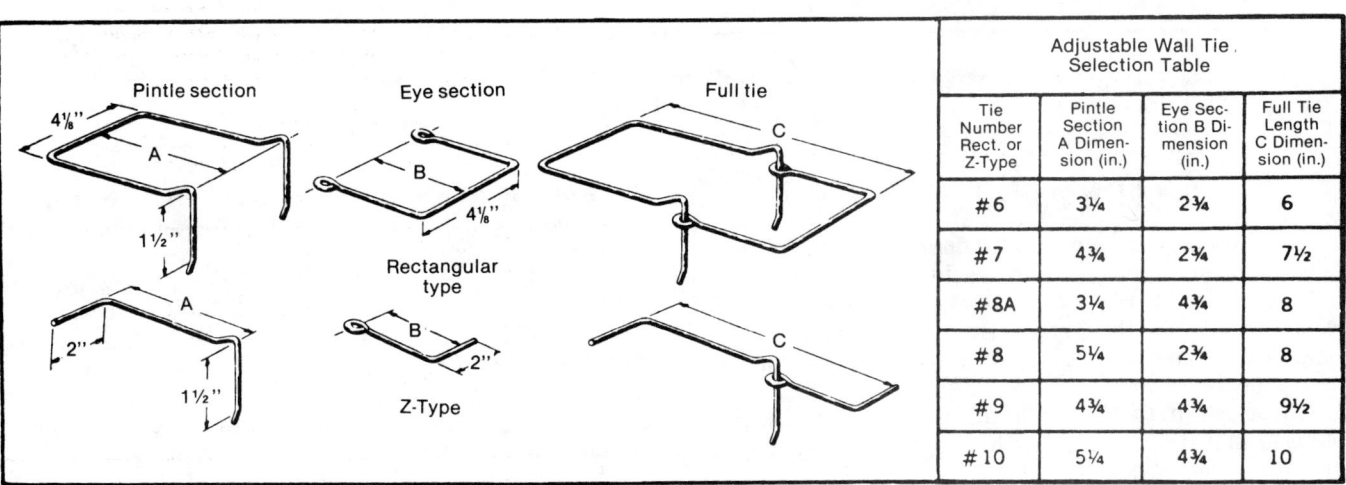

Adjustable Wall Tie Selection Table			
Tie Number Rect. or Z-Type	Pintle Section A Dimension (in.)	Eye Section B Dimension (in.)	Full Tie Length C Dimension (in.)
#6	3¼	2¾	6
#7	4¾	2¾	7½
#8A	3¼	4¾	8
#8	5¼	2¾	8
#9	4¾	4¾	9½
#10	5¼	4¾	10

Adjustable Wall Ties
Figure 10-4

Masonry & Concrete Construction

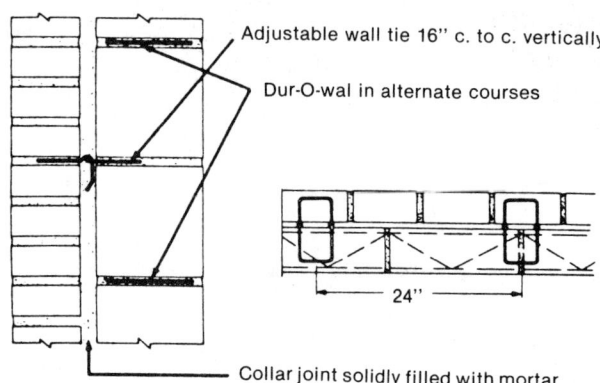

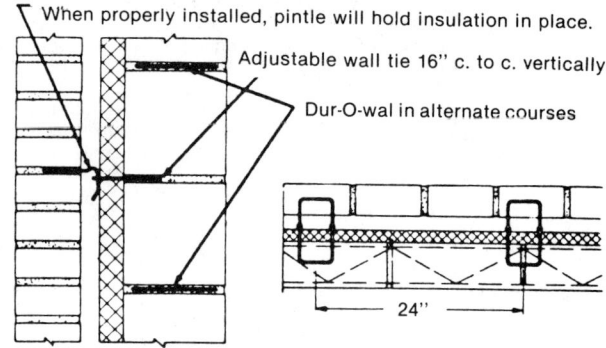

Overall Wall Width (In.)	Width of Cavity and Wythes			Tie Number Rect. or Z-Type
	Exterior A (In.)	Cavity B (In.)	Backup C (In.)	
10	4	2	4	No. 7
11	4	3	4	No. 8
12	4	2	6	No. 7
13	4	3	6	No. 8
14	4	2	8	No. 7
15	4	3	8	No. 8

**Recommended applications of Adjustable Wall Ties
Figure 10-6**

providing adequate shear strength for lateral stability of the wall. (See Figure 10-7.)

For control joints to perform adequately, the following three principles should be considered:

(1) Stress relief — the joint must cut the masonry wall completely from the top to bottom, so as to truly form a stress-relieving joint.
(2) Shear Strength — the joint must be structurally sound in that sufficient strength is developed to provide for lateral stability.
(3) Weathertight — the joint must be either self-sealing or one that can be easily caulked to prevent moisture penetration.

Reinforcing Masonry, The Ivany Block System

The Ivany system eliminates forming and stripping, skeletal framework, and tying of horizontal to vertical steel. It also improves coordination of the trades. The bricklayers lay the block, place the steel, and pour the concrete in one continuous process. No jurisdictional disputes arise. The Ivany system can be used to form bond beams, columns, and lintels. There are no snap-tie holes to fill and no rubdown as is the case with poured concrete walls. (See Figure 10-8.)

The blocks are a single-core modular unit, with special webs to accommodate horizontal reinforcing steel. There is an inner cell lip to assure proper vertical reinforcing steel placement. (See Figure 10-9.)

Crack Control in Masonry Walls

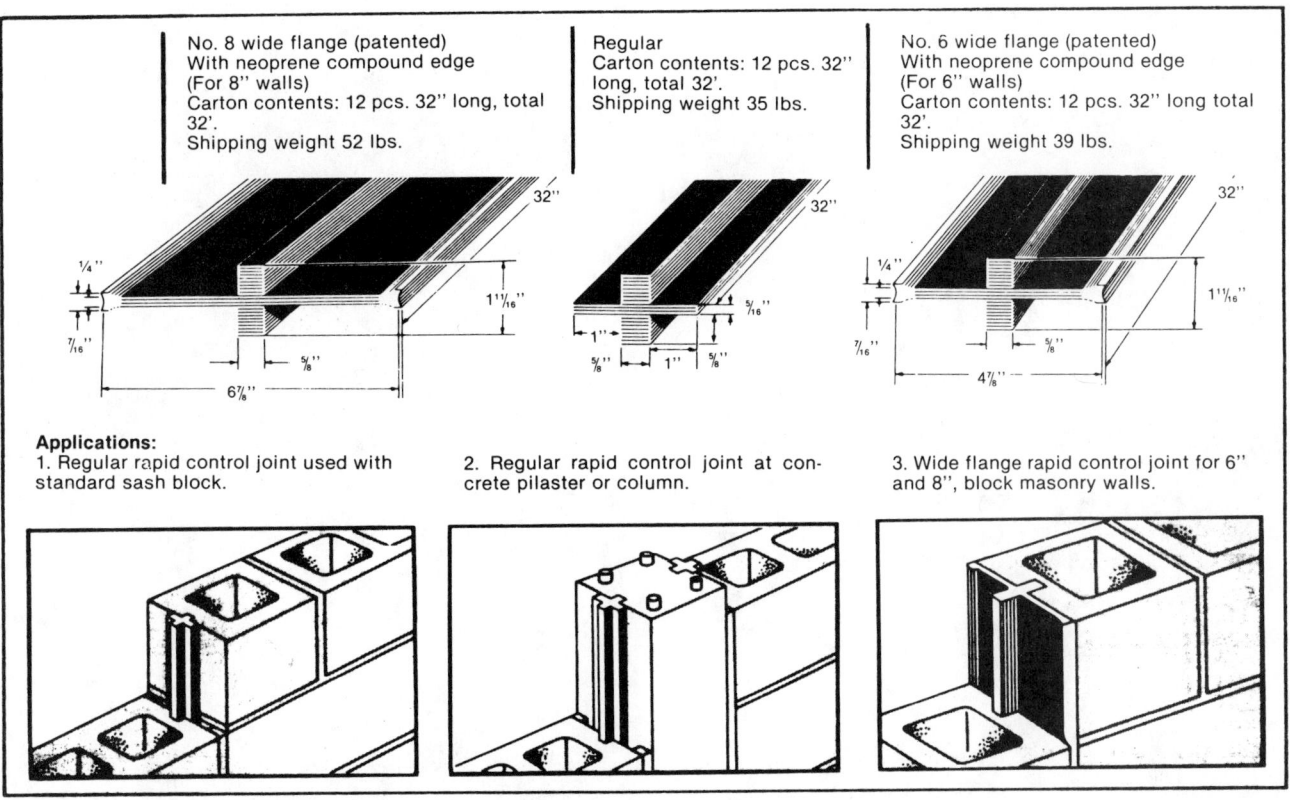

Rapid Control Joints and recommended applications
Figure 10-7

Wall being constructed by the Ivany Block System
Figure 10-8

Block and reinforcement used in the Ivany Block System
Figure 10-9

Chapter 11
Brick Wall Construction

For many years brick sizes were relatively standard, and only three were generally available: Standard, Roman, and Norman. The actual sizes of the last two varied only slightly from the standard. Except for the SCR brick which was introduced in 1952 and patented, all brick units had a 4-inch nominal bed depth or thickness dimension. The SCR brick had a 6-inch nominal bed depth. However, since most brick were a nominal 4 inches in thickness and 2¼ inches in height, the order in which brick dimensions were listed was of relatively little importance.

This situation no longer exists. Brick are now available in many varied sizes, ranging in thickness (bed depth) from a nominal 3 inches to 8 inches and even 10 and 12 inches. In height they range from a nominal 2 inches to 8 inches and are in lengths up to 16 inches. Consequently, it is important to list brick dimensions in an orderly manner to avoid misunderstandings in both shipments and construction procedures.

The sizes of brick shown in Tables 11-1 and 11-2 and in Figures 11-1 and 11-2 are most typical of those currently being produced. However, few manufacturers produce all of the sizes shown. Also, other sizes are produced by some manufacturers, the dimensions of which vary from those shown here. For these reasons, consult with manufacturers or distributors in your area before proceeding with a design incorporating a specific size of brick that may not readily be available in your locality.

Although the sizes of units indicated are typical, the names are not completely standard throughout the industry. Except for the Standard, Roman and Norman sizes, individual manufacturers may have their own names for certain sizes listed here, or may have adopted the name for a size listed for a unit with different dimensions. To avoid confusion, first identify the brick by size.

Except for the nonmodular Standard, oversize and 3-inch units, most brick are produced in modular sizes. The nominal dimensions of modular brick are equal to the manufactured dimensions plus the thickness of the mortar joint for which the unit is designed. In general, the joint thicknesses for use with brick are either 3/8 inch or 1/2 inch.

The actual manufactured dimensions of the units may vary, of course, from the specified dimensions, but by not more than the permissible tolerances for variation in dimension as prescribed in the applicable ASTM Specifications*.

The designated manufactured heights for the standard brick, the standard modular brick and all other modular brick designed to be laid three courses to 8 inches are the same (2¼ inches). Tables 11-3 and 11-4 give the vertical coursing dimensions for modular and nonmodular brick.

A major portion of the cost of construction is expended on labor. Many builders have found, however, that one method of minimizing these increased costs is through the use of larger-faced units. If by the act of laying a single unit a mason can place a square foot of wall area in the building, the resulting masonry unit cost is reduced. For veneer or other nonstructural brickwork, the use of the thinner brick units can result in material cost savings. The face size of the units selected should be of a scale to agree with the size and architectural tone of the building.

BONDS AND PATTERNS IN BRICKWORK
The word *bond*, when used in reference to masonry, may have three meanings:

---

* Standard Specifications for Facing Brick, ASTM Designation C 216; Standard Specifications for Building Brick, ASTM Designation C 62; and Standard Specifications for Hollow Brick, ASTM Designation C 652.

Masonry & Concrete Construction

Structural Bond: The method by which individual masonry units are interlocked or tied together to cause the entire assembly to act as a single structural unit.

Pattern Bond: The pattern formed by the masonry units and the mortar joints on the face of the wall. The pattern may result from the type of structural bond used or may be a purely decorative pattern unrelated to the structural bonding.

Mortar Bond: The adhesion of mortar to the masonry units or to reinforcing steel.

Structural Bonds

Structural bonding of masonry walls may be accomplished in three ways:
(1) By the overlapping (interlocking) of the masonry units
(2) By the use of metal ties embedded in the connecting joints
(3) By the adhesion of grout to adjacent wythes of masonry

The overlapping bond is based on variations of two traditional methods of bonding. The first is English bond and consists of alternating headers and stretchers. (See Figure 11-3A.) The second is Flemish bond and consists of alternating headers and stretchers in every course, so arranged that the headers and stretchers in every other course appear to be in vertical lines. (See Figure 11-3B.)

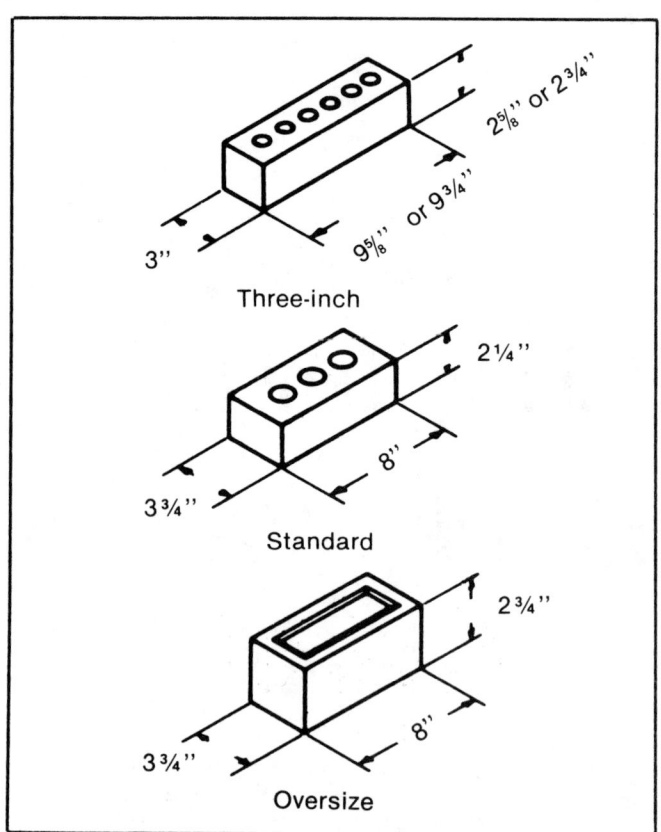

Actual dimensions of non-modular brick
Figure 11-1

Unit Designation	Manufactured Size, Inch		
	t	h	ℓ
Three-inch*	3	2-5/8	9-5/8
	3	2-3/4	9-3/4
Standard	3¾ **	2-1/4	8
Oversize	3¾ **	2-3/4	8

Sizes of non-modular brick
Table 11-1

* In recent years, the so-called "three-inch" brick has gained popularity in certain areas. The term "three-inch" designates its thickness or bed depth. The sizes shown in the table are the ones most commonly produced under the designation "King-size". Other sizes of 3-in. brick are also produced under such designations as "Big John", "Jumbo", "Scotsman" and "Spartan". Originally developed primarily for use as a veneer unit, it is also used to construct 8-in. cavity walls and 8-in. grouted walls.

** The manufactured thickness of standard or over-size non-modular brick will vary from 3½ in. to 3¾ in. Therefore, if other than a running bond is desired, the designer should check with the manufacturer of the brick selected.

Modern building codes require that masonry-bonded brick walls be bonded so that not less than 4 percent of the wall surface is composed of headers, with the distance between adjacent headers not exceeding 24 inches, vertically or horizontally.

Structural bonding of masonry walls with metal ties is used in both solid-wall and cavity-wall construction. (See Figure 11-4.)

Most building codes permit the use of rigid steel bonding ties in solid walls. Use at least one metal tie for each 4½ square feet of wall surface and stagger the ties in alternate courses. The distance between adjacent ties should not exceed 24 inches vertically nor 36 inches horizontally. Provide additional bonding ties at all openings, spaced not more than 3 feet apart around the perimeter and within 12 inches of the opening.

If ties less than 3/16 inch in diameter are used, reduce the tie spacing so that the tie area per square foot of wall is not less than specified above.

Structural bonding of solid and reinforced brick masonry walls is sometimes accomplished by pouring grout into the cavity or collar joint between wythes of masonry.

The method of bonding will depend on the use requirements, wall type and other factors. However, the metal-tie method is generally recommended for exterior walls. Some of the advantages of this method are greater resistance to rain penetration and ease of construction. Metal ties also allow slight differential

Brick Wall Construction

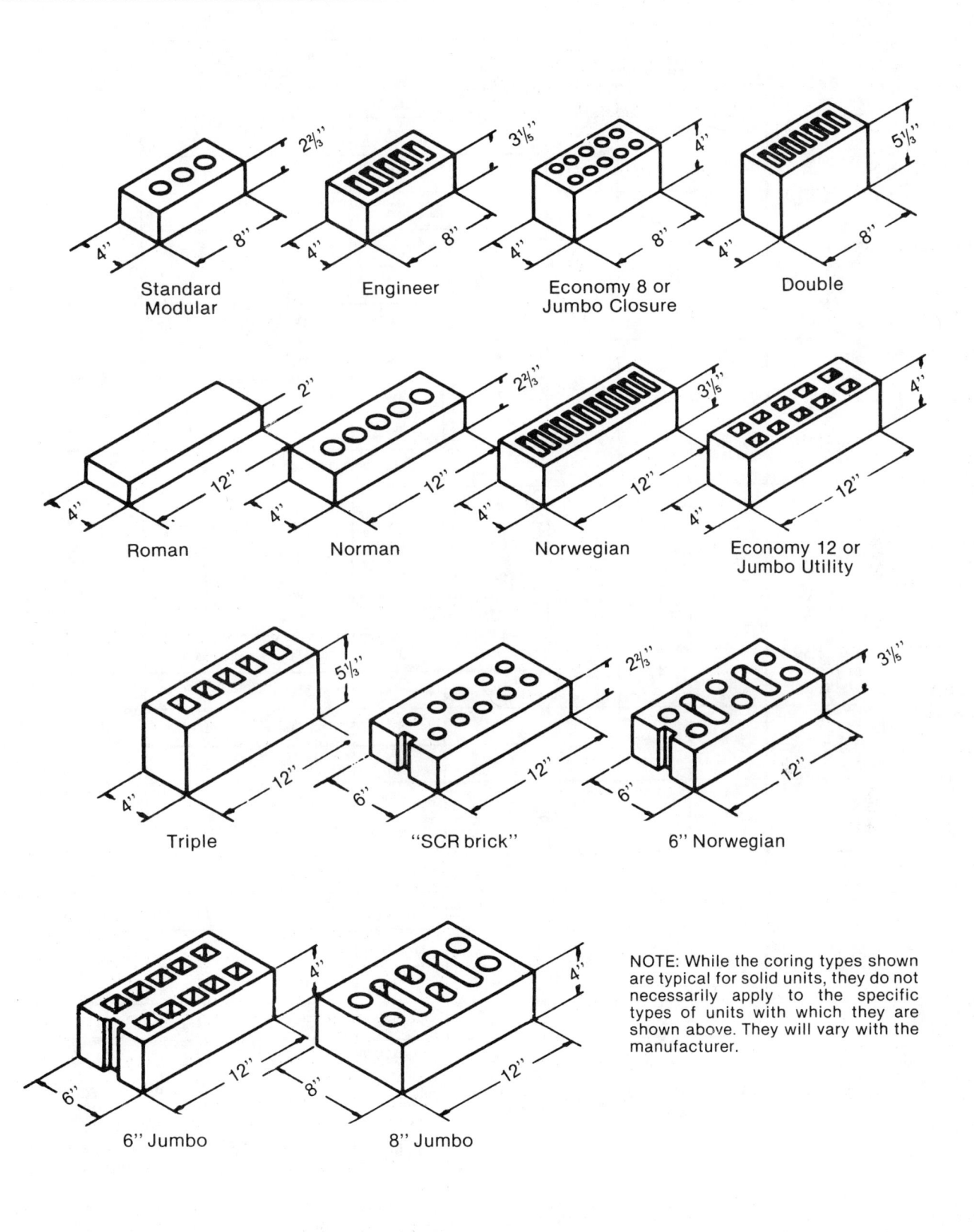

Nominal dimensions of modular brick
Figure 11-2

Unit Designation	Nominal Dimensions, in.			Joint Thickness in.	Manufactured Dimensions in.			Modular Coursing in.
	t	h	l		t	h	l	
Standard Modular	4	2⅔	8	⅜	3⅝	2¼	7⅝	3C = 8
				½	3½	2¼	7½	
Engineer	4	3⅕	8	⅜	3⅝	2 13/16	7⅝	5C = 16
				½	3½	2 11/16	7½	
Economy 8 or Jumbo Closure	4	4	8	⅜	3⅝	3⅝	7⅝	1C = 4
				½	3½	3½	7½	
Double	4	5⅓	8	⅜	3⅝	4 15/16	7⅝	3C = 16
				½	3½	4 13/16	7½	
Roman	4	2	12	⅜	3⅝	1⅝	11⅝	2C = 4
				½	3½	1½	11½	
Norman	4	2⅔	12	⅜	3⅝	2¼	11⅝	3C = 8
				½	3½	2¼	11½	
Norwegian	4	3⅕	12	⅜	3⅝	2 13/16	11⅝	5C = 16
				½	3½	2 11/16	11½	
Economy 12 or Jumbo Utility	4	4	12	⅜	3⅝	3⅝	11⅝	1C = 4
				½	3½	3½	11½	
Triple	4	5⅓	12	⅜	3⅝	4 15/16	11⅝	3C = 16
				½	3½	4 13/16	11½	
SCR brick [2]	6	2⅔	12	⅜	5⅝	2¼	11⅝	3C = 8
				½	5½	2¼	11½	
6-in. Norwegian	6	3⅕	12	⅜	5⅝	2 13/16	11⅝	5C = 16
				½	5½	2 11/16	11½	
6-in. Jumbo	6	4	12	⅜	5⅝	3⅝	11⅝	1C = 4
				½	5½	3½	11½	
8-in. Jumbo	8	4	12	⅜	7⅝	3⅝	11⅝	1C = 4
				½	7½	3½	11½	

[1] Available as solid units to ASTM C 216- or ASTM C 62-, or, in a number of cases, as hollow brick conforming to ASTM C 652-.

[2] Reg. U.S. Pat. Off., SCPI.

Sizes of modular brick[1]
Table 11-2

No. of Courses	*2¼-in. High Units*		*2⅝-in. High Units*		*2¾-in. High Units*	
	⅜" Joint	*½" Joint*	*⅜" Joint*	*½" Joint*	*⅜" Joint*	*½" Joint*
1	0'- 2⅝"	0'- 2¾"	0'-3"	0'- 3⅛"	0'- 3⅛"	0'- 3¼"
2	0'- 5¼"	0'- 5½"	0'-6"	0'- 6¼"	0'- 6¼"	0'- 6½"
3	0'- 7⅞"	0'- 8¼"	0'-9"	0'- 9⅜"	0'- 9⅜"	0'- 9¾"
4	0'-10½"	0'-11"	1'-0"	1'- 0½"	1'- 0½"	1'- 1"
5	1'- 1⅛"	1'- 1¾"	1'-3"	1'- 3⅝"	1'- 3⅝"	1'- 4¼"
6	1'- 3¾"	1'- 4½"	1'-6"	1'- 6¾"	1'- 6¾"	1'- 7½"
7	1'- 6⅜"	1'- 7¼"	1'-9"	1'- 9⅞"	1'- 9⅞"	1'-10¾"
8	1'- 9"	1'-10"	2'-0"	2'- 1"	2'- 1"	2'- 2"
9	1'-11⅝"	2'- 0¾"	2'-3"	2'- 4⅛"	2'- 4⅛"	2'- 5¼"
10	2'- 2¼"	2'- 3½"	2'-6"	2'- 7¼"	2'- 7¼"	2'- 8½"
11	2'- 4⅞"	2'- 6¼"	2'-9"	2'-10⅜"	2'-10⅜"	2'-11¾"
12	2'- 7½"	2'- 9"	3'-0"	3'- 1½"	3'- 1½"	3'- 3"
13	2'-10⅛"	2'-11¾"	3'-3"	3'- 4⅝"	3'- 4⅝"	3'- 6¼"
14	3'- 0¾"	3'- 2½"	3'-6"	3'- 7¾"	3'- 7¾"	3'- 9½"
15	3'- 3⅜"	3'- 5¼"	3'-9"	3'-10⅞"	3'-10⅞"	4'- 0¾"
16	3'- 6"	3'- 8"	4'-0"	4'- 2"	4'- 2"	4'- 4"
17	3'- 8⅝"	3'-10¾"	4'-3"	4'- 5⅛"	4'- 5⅛"	4'- 7¼"
18	3'-11¼"	4'- 1½"	4'-6"	4'- 8¼"	4'- 8¼"	4'-10½"
19	4'- 1⅞"	4'- 4¼"	4'-9"	4'-11⅜"	4'-11⅜"	5'- 1¾"
20	4'- 4½"	4'- 7"	5'-0"	5'- 2½"	5'- 2½"	5- 5"
21	4'- 7⅛"	4'- 9¾"	5'-3"	5'- 5⅝"	5'- 5⅝"	5'- 8¼"
22	4'- 9¾"	5'- 0½"	5'-6"	5'- 8¾"	5'- 8¾"	5'-11½"
23	5'- 0⅜"	5'- 3¼"	5'-9"	5'-11⅞"	5'-11⅞"	6'- 2¾"
24	5'- 3"	5'- 6"	6'-0"	6'- 3"	6'- 3"	6'- 6"
25	5'- 5⅝"	5'- 8¾"	6'-3"	6'- 6⅛"	6'- 6⅛"	6'- 9¼"
26	5'- 8¼"	5'-11½"	6'-6"	6'- 9¼"	6'- 9¼"	7'- 0½"
27	5'-10⅞"	6'- 2¼"	6'-9"	7'- 0⅜"	7'- 0⅜"	7'- 3¾"
28	6'- 1½"	6'- 5"	7'-0"	7'- 3½"	7'- 3½"	7'- 7"
29	6'- 4⅛"	6'- 7¾"	7'-3"	7'- 6⅝"	7'- 6⅝"	7'-10¼"
30	6'- 6¾"	6'-10½"	7'-6"	7'- 9¾"	7'- 9¾"	8'- 1½"
31	6'- 9⅜"	7'- 1¼"	7'-9"	8'- 0⅞"	8'- 0⅞"	8'- 4¾"
32	7'- 0"	7'- 4"	8'-0"	8'- 4"	8'- 4"	8'- 8"
33	7'- 2⅝"	7'- 6¾"	8'-3"	8'- 7⅛"	8'- 7⅛"	8'-11¼"
34	7'- 5¼"	7'- 9½"	8'-6"	8'-10¼"	8'-10¼"	9'- 2½"
35	7'- 7⅞"	8'- 0¼"	8'-9"	9'- 1⅜"	9'- 1⅜"	9'- 5¾"
36	7'-10½"	8'- 3"	9'-0"	9'- 4½"	9'- 4½"	9'- 9"
37	8'- 1⅛"	8'- 5¾"	9'-3"	9'- 7⅝"	9'- 7⅝"	10- 0¼"
38	8'- 3¾"	8'- 8½"	9'-6"	9'-10¾"	9'-10¾"	10'- 3½"
39	8'- 6⅜"	8'-11¼"	9'-9"	10'- 1⅞"	10'- 1⅞"	10'- 6¾"
40	8'- 9"	9'- 2"	10'-0"	10'- 5"	10'- 5"	10'-10"
41	8'-11⅝"	9'- 4¾"	10'-3"	10'- 8⅛"	10'- 8⅛"	11'- 1¼"
42	9'- 2¼"	9'- 7½"	10'-6"	10'-11¼"	10'-11¼"	11'- 4½"
43	9'- 4⅞"	9'-10¼"	10'-9"	11'- 2⅜"	11'- 2⅜"	11'- 7¾"
44	9'- 7½"	10'- 1"	11'-0"	11'- 5½"	11'- 5½"	11'-11"
45	9'-10⅛"	10'- 3¾"	11'-3"	11'- 8⅝"	11'- 8⅝"	12'- 2¼"
46	10'- 0¾"	10'- 6½"	11'-6"	11'-11¾"	11'-11¾"	12'- 5½"
47	10'- 3⅜"	10'- 9¼"	11'-9"	12'- 2⅞"	12'- 2⅞"	12'- 8¾"
48	10'- 6"	11'- 0"	12'-0"	12'- 6"	12'- 6"	13'- 0"
49	10'- 8⅝"	11'- 2¾"	12'-3"	12'- 9⅛"	12'- 9⅛"	13'- 3¼"
50	10'-11¼"	11'- 5½"	12'-6"	13'- 0¼"	13'- 0¼"	13'- 6½"
100	21'-10½"	22'-11"	25'-0"	26'- 0½"	26'- 0½"	27'- 1"

[1] Brick positioned in wall as stretchers. Vertical dimensions are from bottom of mortar joint to bottom of mortar joint.

Vertical coursing dimensions for non-modular brick[1]
Table 11-3

No. of Courses	*Nominal Height (h) of Unit*[2]				
	2"	2⅔"	3⅕"	4"	5⅓"
1	0'- 2"	0'- 2 11/16"	0'- 3 3/16"	0'-4"	0'- 5 5/16"
2	0'- 4"	0'- 5 5/16"	0'- 6 3/8"	0'-8"	0'-10 11/16"
3	0'- 6"	0'- 8"	0'- 9 5/8"	1'-0"	1'- 4"
4	0'- 8"	0'-10 11/16"	1'- 0 13/16"	1'-4"	1'- 9 5/16"
5	0'-10"	1'- 1 5/16"	1'- 4"	1'-8"	2'- 2 11/16"
6	1'- 0"	1'- 4"	1'- 7 3/16"	2'-0"	2- 8"
7	1'- 2"	1'- 6 11/16"	1'-10 3/8"	2'-4"	3'- 1 5/16'
8	1'- 4"	1'- 9 5/16"	2'- 1 5/8"	2'-8"	3'- 6 11/16'
9	1'- 6"	2'- 0"	2'- 4 13/16"	3'-0"	4'- 0"
10	1'- 8"	2'- 2 11/16"	2'- 8"	3'-4"	4'- 5 5/16"
11	1'-10"	2'- 5 5/16"	2'-11 3/16"	3'-8"	4'-10 11/16"
12	2'- 0"	2'- 8"	3'- 2 3/8"	4'-0"	5'- 4"
13	2'- 2"	2'-10 11/16"	3'- 5 5/8"	4'-4"	5'- 9 5/16"
14	2'- 4"	3'- 1 5/16"	3'- 8 13/16"	4'-8"	6'- 2 11/16"
15	2'- 6"	3'- 4"	4'- 0"	5'-0"	6'- 8"
16	2'- 8"	3'- 6 11/16"	4'- 3 3/16"	5'-4"	7'- 1 5/16"
17	2'-10"	3'- 9 5/16"	4'- 6 3/8"	5'-8"	7'- 6 11/16"
18	3'- 0"	4'- 0"	4'- 9 5/8"	6'-0"	8'- 0"
19	3'- 2"	4'- 2 11/16"	5'- 0 13/16"	6'-4"	8'- 5 5/16"
20	3'- 4"	4'- 5 5/16"	5'- 4"	6'-8"	8'-10 11/16"
21	3'- 6"	4'- 8"	5'- 7 3/16"	7'-0"	9'- 4"
22	3'- 8"	4'-10 11/16"	5'-10 3/8"	7'-4"	9'- 9 5/16"
23	3'-10"	5'- 1 5/16"	6'- 1 5/8"	7'-8"	10'- 2 11/16"
24	4'- 0"	5'- 4"	6' 4 13/16"	8'-0"	10'- 8"
25	4'- 2"	5'- 6 11/16"	6'- 8"	8'-4"	11'- 1 5/16"
26	4'- 4"	5'- 9 5/16"	6'-11 3/16"	8'-8"	11'- 6 11/16"
27	4'- 6"	6'- 0"	7'- 2 3/8"	9'-0"	12'- 0"
28	4'- 8"	6'- 2 11/16"	7'- 5 5/8"	9'-4"	12'- 5 5/16"
29	4'-10"	6'- 5 5/16"	7'- 8 13/16"	9'-8"	12'-10 11/16"
30	5'- 0"	6'- 8"	8'- 0"	10'-0"	13'- 4"
31	5'- 2"	6'-10 11/16"	8'- 3 3/16"	10'-4"	13'- 9 5/16"
32	5'- 4"	7'- 1 5/16"	8'- 6 3/8"	10'-8"	14' 2 11/16"
33	5'- 6"	7'- 4"	8'- 9 5/8"	11'-0"	14'- 8"
34	5'- 8"	7'- 6 11/16"	9'- 0 13/16"	11'-4"	15'- 1 5/16"
35	5'-10"	7'- 9 5/16"	9'- 4"	11'-8"	15'- 6 11/16"
36	6'- 0"	8'- 0"	9'- 7 3/16"	12'-0"	16'- 0"
37	6'- 2"	8'- 2 11/16"	9'-10 3/8"	12'-4"	16'- 5 5/16"
38	6'- 4"	8'- 5 5/16"	10'- 1 5/8"	12'-8"	16'-10 11/16"
39	6'- 6"	8'- 8"	10'- 4 13/16"	13'-0"	17'- 4"
40	6'- 8"	8'- 10 11/16"	10'- 8"	13'-4"	17'- 9 5/16"
41	6'-10"	9'- 1 5/16"	10'-11 3/16"	13'-8"	18'- 2 11/16"
42	7'- 0"	9'- 4"	11'- 2 3/8"	14'-0"	18'- 8"
43	7'- 2"	9'- 6 11/16"	11'- 5 5/8"	14'-4"	19'- 1 5/16"
44	7'- 4"	9'- 9 5/16"	11'- 8 13/16"	14'-8"	19'- 6 11/16"
45	7'- 6"	10'- 0"	12'- 0"	15'-0"	20'- 0"
46	7'- 8"	10'- 2 11/16"	12'- 3 3/16"	15'-4"	20'- 5 5/16"
47	7'-10"	10'- 5 5/16"	12'- 6 3/8"	15'-8"	20'-10 11/16"
48	8'- 0"	10'- 8"	12'- 9 5/8"	16'-0"	21'- 4"
49	8'- 2"	10'-10 11/16"	13'- 0 13/16"	16'-4"	21'- 9 5/16"
50	8'- 4"	11'- 1 5/16"	13'- 4"	16'-8"	22'- 2 11/16"
100	16'- 8"	22'- 2 11/16"	26'- 8"	33'-4"	44'- 5 5/16"

Brick positioned in wall as stretchers.

[2] For convenience in using table, nominal ⅓'', ⅔'' and ⅕'' heights of units have been changed to nearest 1/16''. Vertical dimensions are from bottom of mortar joint to bottom of mortar joint.

Vertical coursing dimensions for modular brick[1]
Table 11-4

Brick Wall Construction

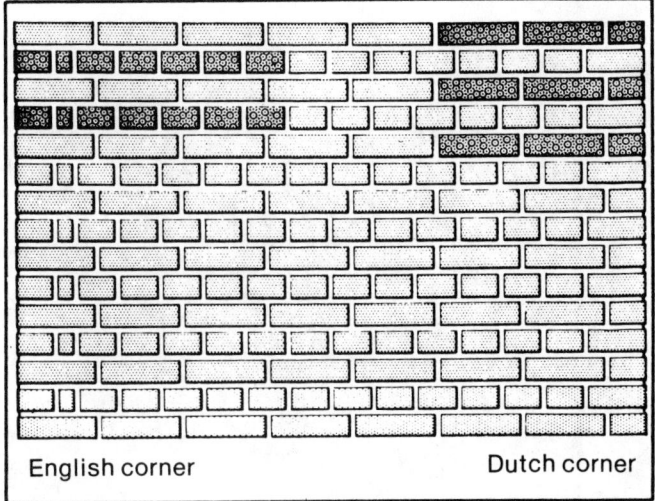

English bond
Figure 11-3A

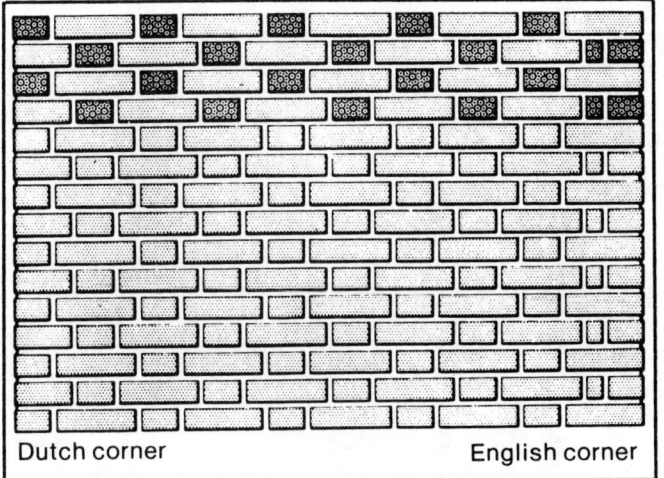

Flemish bond
Figure 11-3B

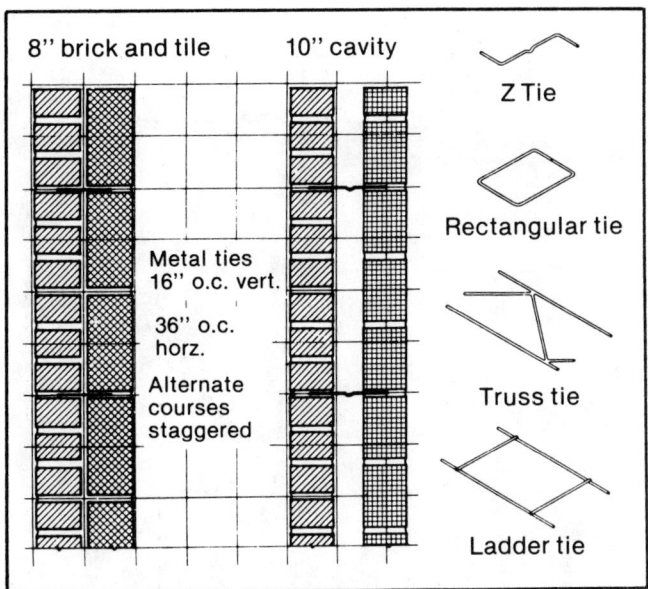

Metal-tied masonry walls
Figure 11-4

movements of the facing and backing which may relieve stresses and prevent cracking.

Pattern Bonds

Frequently, structural bonds such as English or Flemish or variations of these may be used to create patterns in the face of the wall. However, in the strict sense of the term, pattern refers to the change or varied arrangement of the brick texture or color used in the face. Therefore, it may be possible to secure many patterns using the same structural bond. Patterns also may be produced by the method of handling the mortar joint or by projecting or recessing certain brick from the plane of the wall, thus creating a distinctive wall texture that is not solely dependent upon the texture of the individual brick. (See Figure 11-5 and 11-6.)

There are five basic structural bonds commonly used today which create typical patterns. These are running bond, common or American bond, Flemish bond, English bond and block or stack bond. These are shown in Figures 11-5 through 11-11.

Through the use of these bonds and variations of color and texture of the brick and of the joint types, an almost unlimited number of patterns can be developed.

Patterns may be produced by projecting or recessing certain brick from the plane of the wall. Figure 11-5 is an example of a Flemish bond at the header course. The brick project in stack bond.

Figure 11-6 shows a pattern produced by projecting the headers of a Flemish-bond wall. Make sure the headers are not twisted and that they vary little in color. Use a slicker or a caulking trowel to finish off the joints above and below the projected brick.

Running Bond The simplest of the basic pattern bonds, the running bond consists of all stretchers. Since there are no headers in this bond, metal ties are usually used. Running bond is used largely in cavity wall construction and veneered walls of brick, and often in facing tile walls where the bonding may be accomplished by extra-wide stretcher tile. The wall in Figure 11-7 is finished with raked joints.

Common or American Bond is a variation of running bond with a course of full-length headers at regular intervals. (See Figure 11-8.) These headers provide structural bonding as well as pattern. Header courses usually appear at every fifth, sixth, or seventh course. In laying out any bond pattern, it is important to start the corners correctly. For common bond, start a three-quarter brick each way from the corner at the header course. Common bond may be varied by using a Flemish header course.

127

Masonry & Concrete Construction

**Flemish bond at the header course with brick projecting in stack bond
Figure 11-5**

**Flemish bond with projecting headers
Figure 11-6**

Brick Wall Construction

Running bond
Figure 11-7

Common or American bond
Figure 11-8

Masonry & Concrete Construction

**Flemish bond
Figure 11-9**

Flemish Bond consists of alternating headers and stretchers in every course, so arranged that the headers and stretchers in every other course appear in vertical lines. (See Figure 11-9.) The stretchers, laid with the length of the wall, develop longitudinal bonding strength while the headers, laid across the width of the wall, bond the wall transversely.

English Bond is composed of alternate courses of headers and stretchers. The headers are centered on the stretchers, and the joints between stretchers in all courses are aligned vertically. (See Figure 11-10.) Use snap headers in courses which are not structural-bonding courses.

Block or Stack Bond is purely a pattern bond. There is no overlapping of units since all vertical joints are aligned. (See Figure 11-11.) Usually this pattern is bonded to a backing with rigid ties, but you may use 8-inch bonder units when available. In large wall areas and in load-bearing construction, reinforce the wall with steel reinforcement placed in the horizontal mortar joints. In stack bond, you must use prematched or dimensionally accurate masonry units if the vertical alignment of the head joints is to be maintained.

English Cross or Dutch Bond is a variation of English bond which differs only in that vertical joints between the stretchers in alternate courses do not align vertically. These joints center on the stretchers themselves in the course above and below.

There are two methods used in starting the corners in Flemish and English bonds. Figure 11-12 shows the so-called "Dutch corner" in which a three-quarter brick closure is used, and the English corner in which a 2-inch or quarter-brick closure, called a "queen closure," is used. The 2-inch closure should always be placed 4 inches in from the corner and never at the corner.

Figures 11-13A and B show patterns that may be obtained by varying brick color. Figure 11-13A shows a double-stretcher garden-wall bond with the pattern units in diagonal lines. Figure 11-13B shows the garden-wall bond with the pattern units set in dovetail fashion.

Wall Texture Recently many contemporary modifications of the traditional bonds have been used by omitting units to form perforated walls or screens. A screen wall bond is shown in Figure 11-14.

English bond
Figure 11-10

Block or stack bond
Figure 11-11

Masonry & Concrete Construction

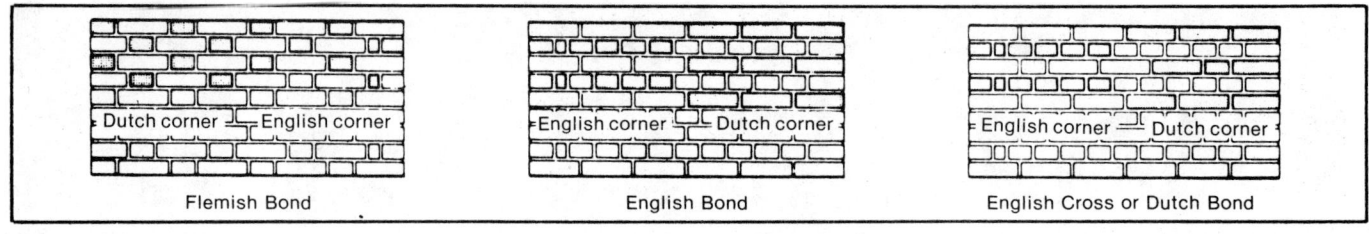

**Dutch and English corners
Figure 11-12**

**Double stretcher garden wall bond with units in diagonal lines
Figure 11-13A**

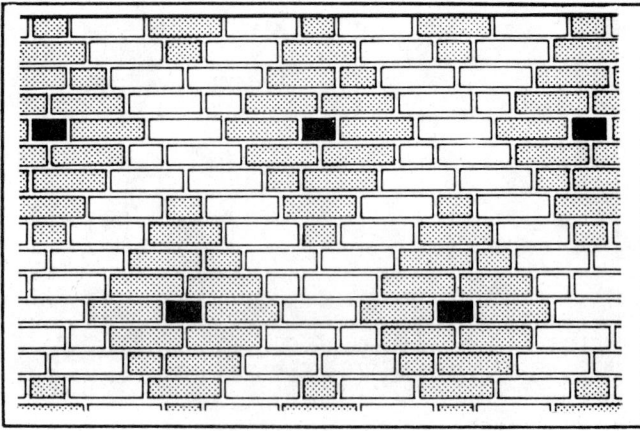

**Garden wall bond with units set in dovetail fashion
Figure 11-13B**

**Screen wall bond
Figure 11-14**

Making a tooled joint with a concave jointing tool
Figure 11-15

MORTAR JOINTS

The mortar serves four functions:

1. It bonds the units together and seals the spaces between.
2. It compensates for dimensional variations in the units.
3. It bonds to reinforcing steel, causing the steel to act as an integral part of the wall.
4. It provides a decorative effect on the wall surface by creating shadow or color lines.

As previously indicated, the treatment of mortar joints in the face of the wall affects the pattern and wall texture.

Mortar Joint Finishes

Mortar joint finishes fall into two classes: troweled (struck) and tooled joints. In the troweled joint, the excess mortar is simply cut off (struck) with the trowel and finished with the trowel. For the tooled joint, a special tool other than the trowel is used to compress and shape the mortar in the joint. Figure 11-15 shows a concave jointing tool being used to make a strong water-repellent joint.

Struck or Troweled Joint This is a common joint in ordinary brickwork, and it is an easy joint to strike with a trowel. (See Figure 11-16.) The mason works from the inside of the wall. Some compaction occurs, but the small ledge does not shed water readily, resulting in a less watertight joint.

Weathered Joint The weathered joint is similar except that it has to be worked from below the brick. It is better than the struck joint as the mortar is compacted by pressing the trowel in at the top of the joint. This type of joint sheds water more readily than the struck joint.

Rough-Cut or Flush Joint This is the simplest joint for a mason to do, since it is made by holding the edge of the trowel flat against the brick and cutting in any direction. This produces an uncompacted joint with a small hairline crack where the mortar is pulled away from the brick by the cutting action. (See Figure 11-17.)

Raked Joint This joint is made by removing the surface of the mortar while it is still soft. While the joint may be compacted, it is difficult to make weathertight and is not recommended where heavy rain, high wind, or freezing is likely to occur. This joint produces marked shadows and tends to darken the overall appearance of the wall. Figure 11-18 shows the first operation in

Masonry & Concrete Construction

Struck or troweled joint
Figure 11-16

Rough-cut or flush joint
Figure 11-17

Brick Wall Construction

Raked joint being made with a special tool
Figure 11-18

Smoothing out the joint with a slicker
Figure 11-19

Masonry & Concrete Construction

**Weeping joints
Figure 11-20**

producing raked joints. A tool called a joint raker or skate wheel is used to rake the mortar to an even depth.

The second operation in raking joints is usually to smooth out the joint after it is raked. This helps to compress the joint and also makes a neater appearance. Slickers or caulking trowels are used for this operation. (See Figure 11-19.)

Colored mortar may be used to enhance the patterns in masonry. There are two ways this usually is done: (1) The entire mortar joint may be colored, or (2) the tuck point method may be used. This technique is best for tooled joints. Complete the entire wall with a 1-inch deep raked joint and carefully fill in the colored mortar later.

Weeping Joints This type of joint is not recommended for cold climates because of the problem of water seeping into the joints. It is also a difficult joint to construct. The mortar must be left hanging from the wall to form the weeping effect. (See Figure 11-20.) Many masons have trouble with it because the tendency is to cut the joint as the brick is laid in the wall.

Chapter 12
Brick Veneer Construction

Brick-veneer construction consists of a nominal 3-inch or 4-inch (100mm) thick exterior brick wythe tied to a backup system with metal ties in such a way that a 1-inch clear space is provided between the veneer and the backup system. The backup system may be wood frame, metal stud, or masonry. The brick veneer is supported on the foundation and is designed to carry no vertical loads other than its own weight.

NEW CONSTRUCTION

Until recently, brick-veneer construction was limited to houses. It is now being used on low-rise construction and is often considered for high-rise buildings. The application of brick veneer in high-rise construction will be discussed later. This chapter covers design of brick veneer in buildings limited to three stories in height. The height limitations shown in Table 12-1 are based on successful past performances of brick veneer and various backup systems.

There are six requirements for satisfactory performance of brick veneer: (1) an adequate foundation, (2) a sufficiently strong, rigid, well-braced backup system, (3) proper attachment of the veneer to the backup system, (4) proper detailing, (5) the use of proper materials, and (6) good workmanship in construction.

Properties of Brick Veneer

Strength Factors that affect the strength of brick veneer are the height of the brick veneer, the stiffness of the backup, and the tie system. In addition to its own weight, the only weight that the brick veneer should carry is a proportionate share of the lateral load. Due to the relatively low stiffness normally achieved in wood-frame and metal-stud backup systems, the brick veneer usually carries the majority of any lateral load. This occurs even though the brick veneer is not usually designed to carry lateral loads.

Fire Resistance The typical brick-veneer wall assemblies have fire ratings of up to 2 hours. Figure 12-1 shows a typical brick-veneer wall assembly with a 2-hour fire rating.

Moisture Resistance Brick-veneer wall assemblies are drainage-type walls. Walls of this type, which include cavity walls, are recommended where maximum resistance to rain penetration is desired. It is essential to maintain the 1-inch clear space between the brick veneer and the backup to ensure proper drainage.

Resistance to Heat Transmission Brick-veneer wall assemblies provide capacity insulation and resistance to the transmission of heat. The mass of the brick veneer provides the capacity insulation. It effectively lowers and delays the peak heating and cooling loads.

Acoustical Properties Brick-veneer wall assemblies reduce sound transmission by two means. The mass reduces sound transmission by absorbing the energy of the sound vibrations. The discontinuity of construction prevents vibrations of the exterior brick wythe from directly vibrating the rest of the wall assembly, thereby retarding sound transmission to the interior.

Although there are no specific data available on the sound-transmission characteristics of brick-veneer wall assemblies, the brick-veneer wall system shown in Figure 12-1 has an estimated STC of 40 to 44.

Selection of Materials

Brick The brick for veneer should conform to ASTM C 62 or C 216. Because the brick wythe is isolated from the remainder of the wall, use Grade SW when the brick masonry is to be used in freezing climates.

Do not use salvaged brick. In general, masonry constructed with salvaged brick is weaker and less durable than masonry constructed with new brick.

Masonry & Concrete Construction

Nominal Thickness of the Brick Veneer, in (mm)	Empirical Height Limitations		
	Stories	Height at Plate, ft (m)	Height at Gable, ft (m)
3 (75)	2	20 (6.10)	28 (8.53)
4 (100)	3	30 (9.14)	38 (11.58)

**Empirical height limitations for brick veneer
Table 12-1**

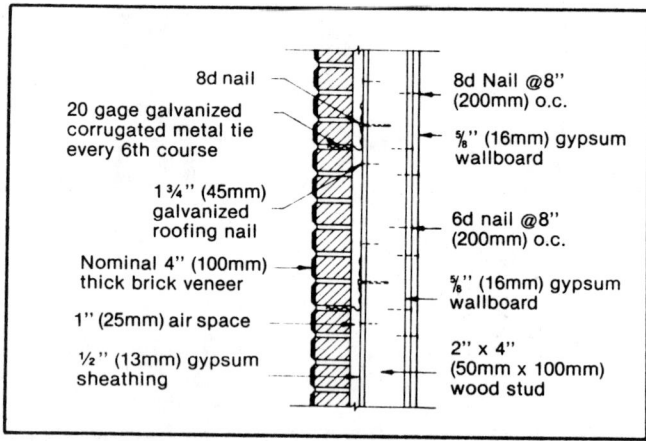

**Typical brick veneer wall assembly
Figure 12-1**

Mortars Portland cement-lime mortar is recommended for brick veneer because its performance is usually better and more predictable than that of masonry-cement mortars. Many masonry cements contain air-entraining agents or other ingredients which sacrifice good bond for workability, color, etc.

Type N mortar is suitable for most brick veneer; however, Type S or Type M may be required. Use Type S mortar where a high degree of flexural resistance is required and Type M where the brick is in contact with earth.

Anchors and Ties Empirically designed brick veneer is usually supported on the foundation with lateral support provided by the ties and backup system. The ties providing the lateral support for the veneer must be capable of resisting tension and compression resulting from forces perpendicular to the plane of the wall, but still permit slight vertical and horizontal movement parallel to the plane of the wall.

The selection of ties depends on the backup system. Typical ties are shown in Figure 12-2.

Ties for Wood-Frame Backup Corrosion-resistant corrugated metal ties, at least 22 gauge, 7/8 inch wide, 6 inches long, as shown in Figure 12-2 (a), are used to attach brick veneer to wood-frame backup. Fasten the ties to the wood frame with corrosion-resistant nails that penetrate the sheathing and drive them a minimum of 1½ inches into the studs.

Ties for Metal-Stud Backup Use corrosion-resistant wire ties at least 9 gauge to attach brick veneer to metal studs. Typical wire ties, as shown in Figure 12-2 (b) and (c) may be used to attach brick veneer to metal studs. Use wire ties with a minimum diameter of 3/16 inch, as shown in Figure 12-2 (d), to attach brick veneer to structural steel. A method of securely attaching the ties to the steel frame is shown in Figure 12-3.

Ties for Masonry Backup Rectangular ties, U ties or Z ties used to attach brick-veneer masonry to masonry backup systems are shown in Figure 12-2 (g) and (h). Do not use Z ties in conjunction with hollow units. Use continuous horizontal joint reinforcement with tab ties for use with concrete masonry backup. All ties should be of at least 9-gauge corrosion-resistant wire. In most instances, design brick veneer with masonry backup systems as cavity walls.

Ties for Concrete Backup Where brick veneer is attached to concrete, use corrosion-resistant wire or flat-bar dovetail anchors. Wire anchors should be at least 6 gauge and 4 inches wide with the wire looped and closed. Flat-bar dovetail anchors should be at least 16 gauge with a minimum width of 7/8 inch. Fabricate flat-bar dovetail anchors so that the end embedded in masonry is turned up 1/4 inch. Embed dovetail anchors at least 2 inches into the bed joint of the veneer, and anchor them to dovetail slots placed in the concrete. Dovetail anchors are shown in Figure 12-2 (e) and (f). One method for the attachment of brick veneer to a concrete structural member is shown in Figure 12-3.

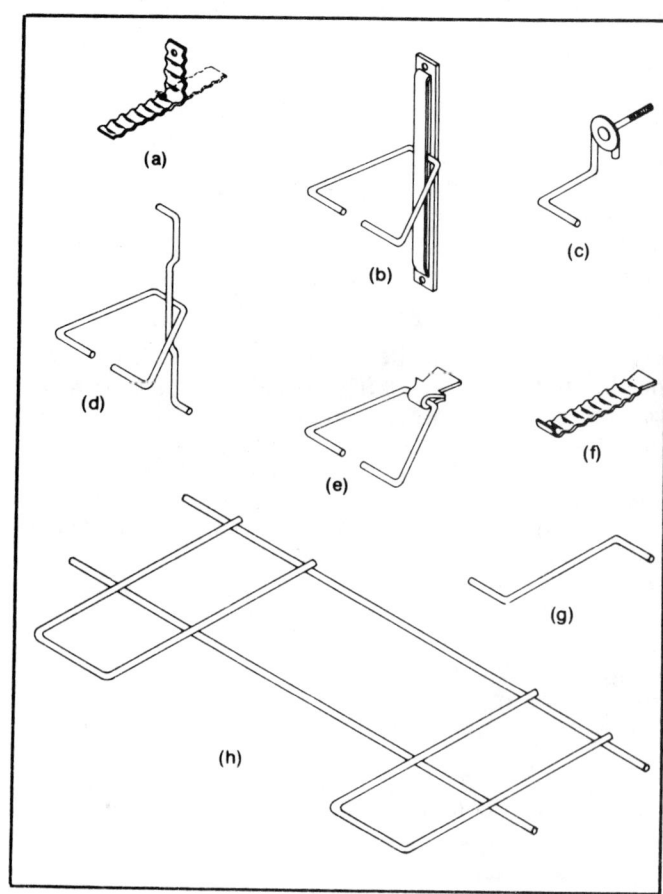

**Typical ties for brick veneer
Figure 12-2**

Brick Veneer Construction

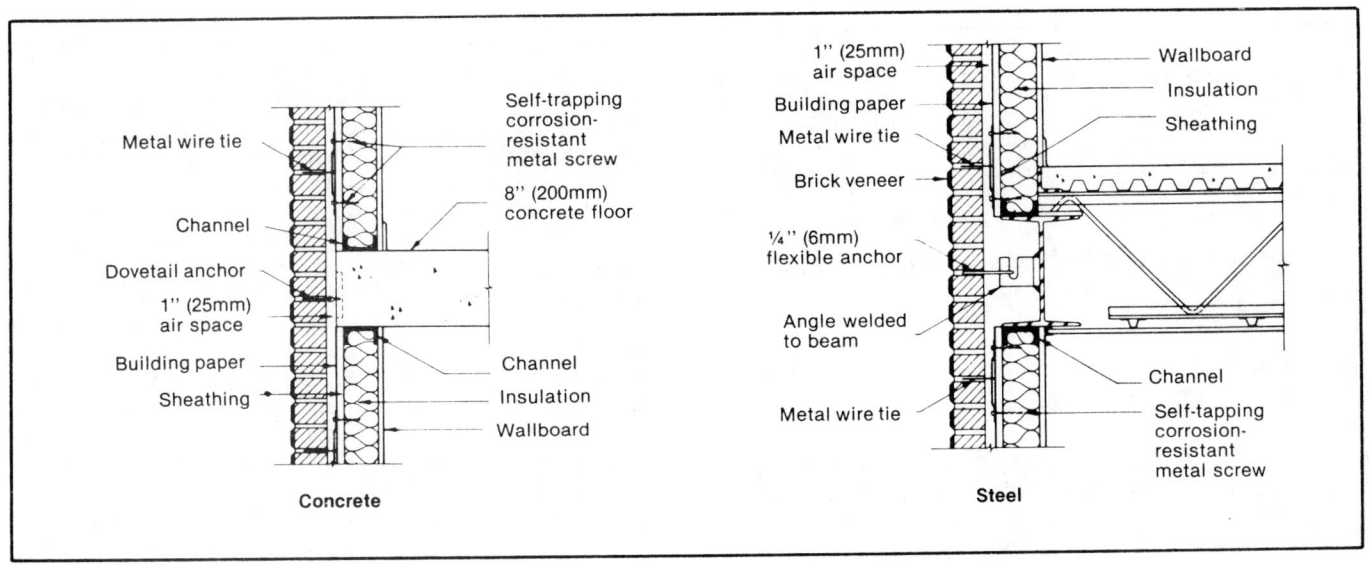

Attachment of ties to steel and concrete frames
Figure 12-3

Flashing and Weep Holes There are many types of flashing available which are suitable for use in brick veneer walls. Sheet metals, bituminous membranes, plastics or combinations of these have proven to be successful. Acceptable bituminous membranes do not include asphalt-impregnated felt. Selection of flashing is often determined by cost; however, use only superior materials because replacement in the event of failure is exceedingly expensive, if not impossible.

Form weep holes by omitting mortar from all or part of the head joint when constructing the veneer. They may also be formed by using a removable forming material such as a well-oiled rod, which will leave an unobstructed opening, or by using plastic tubing, rope wicks, or other materials which are left in place. See Chapter 14 for more information.

Movement Provisions

Design provisions for movement which include bond breaks, expansion joints, and joint reinforcement are usually required in residential and low-rise brick-veneer construction. However, they are required in other specific situations.

Bond Breaks Through-the-wall flashing at the base of the wall between the veneer and the foundation provides sufficient bond break.

Expansion Joints Expansion-joint materials for horizontal and vertical movement are required in brick veneer when there are long walls, wall returns, or shelf angles.

Horizontal Joint Reinforcement Fabricate horizontal joint reinforcement from wire meeting ASTM A82 or ASTM A185. It should have a corrosion-resistant coating which complies with ASTM A 116, Class 3, or ASTM A 153 B-2.

Lintel Materials

Reinforcement for reinforced brick-masonry lintels should be steel bars manufactured from steel in accordance with ASTM A 615, Grades 40 or 60, and should conform to ASTM A 36. Steel angle lintels should be at least 1/4 inch thick with a horizontal leg of at least 3½ inches for use with nominal 4-inch-thick brick veneer, and 2½ inches for use with nominal 3-inch-thick brick veneer. The maximum clear span for 1/4-inch-thick steel angles is 8 feet. The minimum required bearing length is 4 inches. The maximum clear span may be restricted by the fire-protection requirements of some building codes.

Design and Construction

Foundations for Brick Veneer Figure 12-4 shows three typical foundation details for brick veneer. Construct the thickness of the foundation or foundation wall supporting the brick veneer at least equal to the total thickness of the brick-veneer wall assembly. Many building codes permit a nominal 8-inch foundation wall under a single-family dwelling constructed of brick veneer, provided the top of the foundation wall is corbeled as shown in Figure 12-4(c). The total projection of the corbel should not exceed 2 inches with individual corbels projecting not more than one third the thickness of the unit nor one half the height of the unit. The top corbel course should not be higher than the bottom of the floor joist and should be a full header course.

Anchors and Ties Construction details for the various tie systems used in brick-veneer construction are shown in Figures 12-3 through 12-6. There should be one brick-veneer tie for every 2⅔ square feet of wall area, a maximum spacing of 24 inches on center. For one- and two-family wood-frame construction, the tie spacing may be modified to one tie for each 3¼ square feet of wall area with a maximum spacing of 24 inches on center.

Embed all ties to a minimum depth of 2 inches into the bed joints of the brick veneer. Place ties so they are completely surrounded by the mortar.

Masonry & Concrete Construction

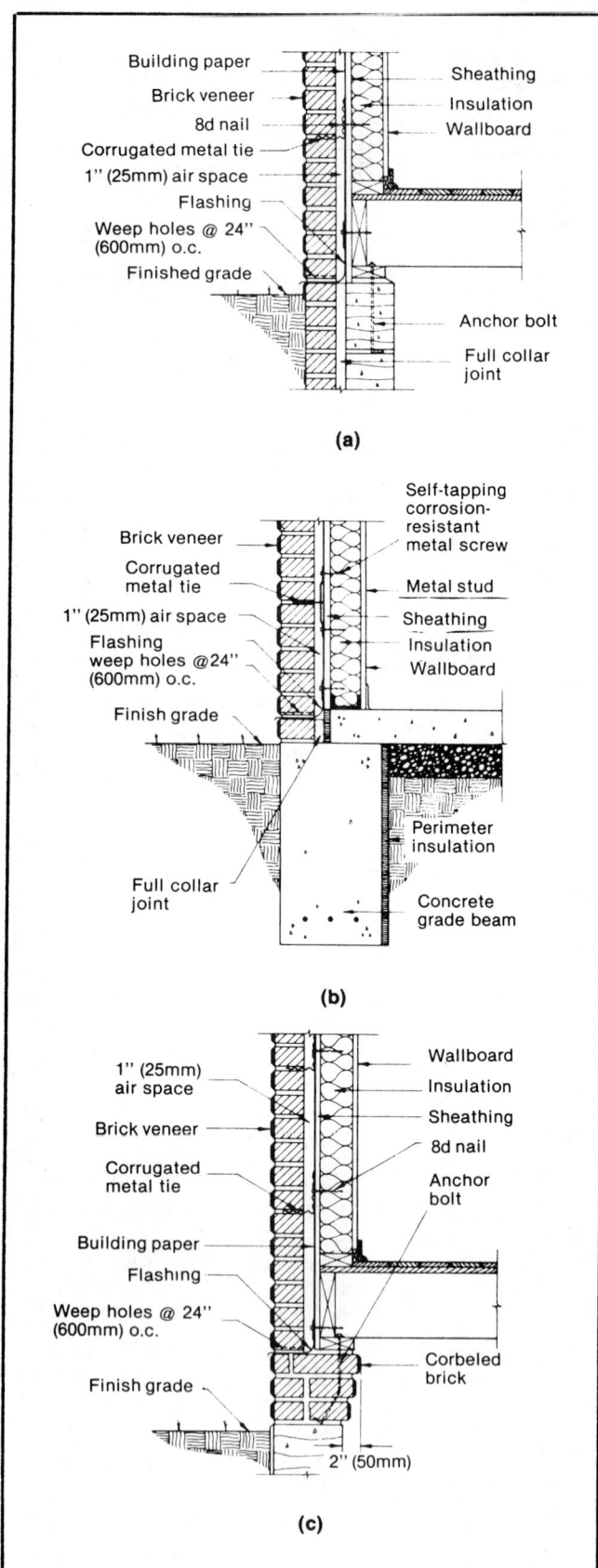

Typical foundation details for brick veneer
Figure 12-4

Flashing and Weep Holes Locate flashing and weep holes above and as near to grade as possible at the bottom of the wall and at all openings so that the wall will drain properly. Space weep holes no more than 24 inches on center and locate them in the head joints immediately above all flashings. Weep holes formed with wick material should be a maximum of 16 inches on center. If veneer continues below the flashing at the bases of the wall, grout the space between the veneer and the backup to the height of the flashing. Securely fasten the flashing to the backup system and extend through the face of the brick veneer. Typical flashing details are shown in Figures 12-4 and 12-5. Carefully install flashing to prevent punctures or tears. Where several pieces of flashing are required to flash a section of the veneer, lap the ends of the flashing and properly seal the joints.

Lintels, Sills and Jambs Typical construction details for lintels, sills, and jambs are shown in Figure 12-5. The advantages of using reinforced brick lintels are more efficient use of materials, built-in fireproofing, elimination of differential movement, and no required painting or other maintenance.

Eave Details Two typical eave details are shown in Figure 12-6. These are suggested design details for the area at the top of the veneer.

Movement Provisions

Bond Breaks In areas where significant differential foundation movement may cause severe cracking in walls rigidly attached to the foundation, bond breaks help relieve the stresses caused by horizontal movements between the wall and the supporting foundation. Differential movement in excess of 1/4 inch per 15 feet is considered significant movement and must be considered in detailing.

Expansion Joints Expansion joints are rarely required in residential construction; however, in some designs of commercial buildings, expansion joints are required. Check the building code.

When calculating the placement of expansion joints, use the maximum and minimum mean temperatures of the isolated brick wythe rather than the mean wall temperature.

Horizontal Joint Reinforcement Masonry materials, such as concrete masonry, that are subject to initial shrinkage stresses require horizontal joint reinforcement for stability against such movement. Brick is not subject to initial shrinkage; therefore, horizontal joint reinforcement is never required in brick masonry for this purpose. It may be beneficial to use limited amounts of horizontal reinforcement in brick veneer for added transverse strength or to increase the spacing between expansion joints.

Single-wire horizontal joint reinforcement may be used to add integrity to the veneer in such places as corners. When using horizontal joint reinforcement, it must be discontinuous at all movement joints.

Caulking and Sealants Exterior caulking joints at the perimeter of the exterior door and window frames should not be less than 1/4 inch nor more than 3/8 inch. Fill these joints solid with an elastic caulking compound or sealant forced into place with a pressure gun. Proper-

Brick Veneer Construction

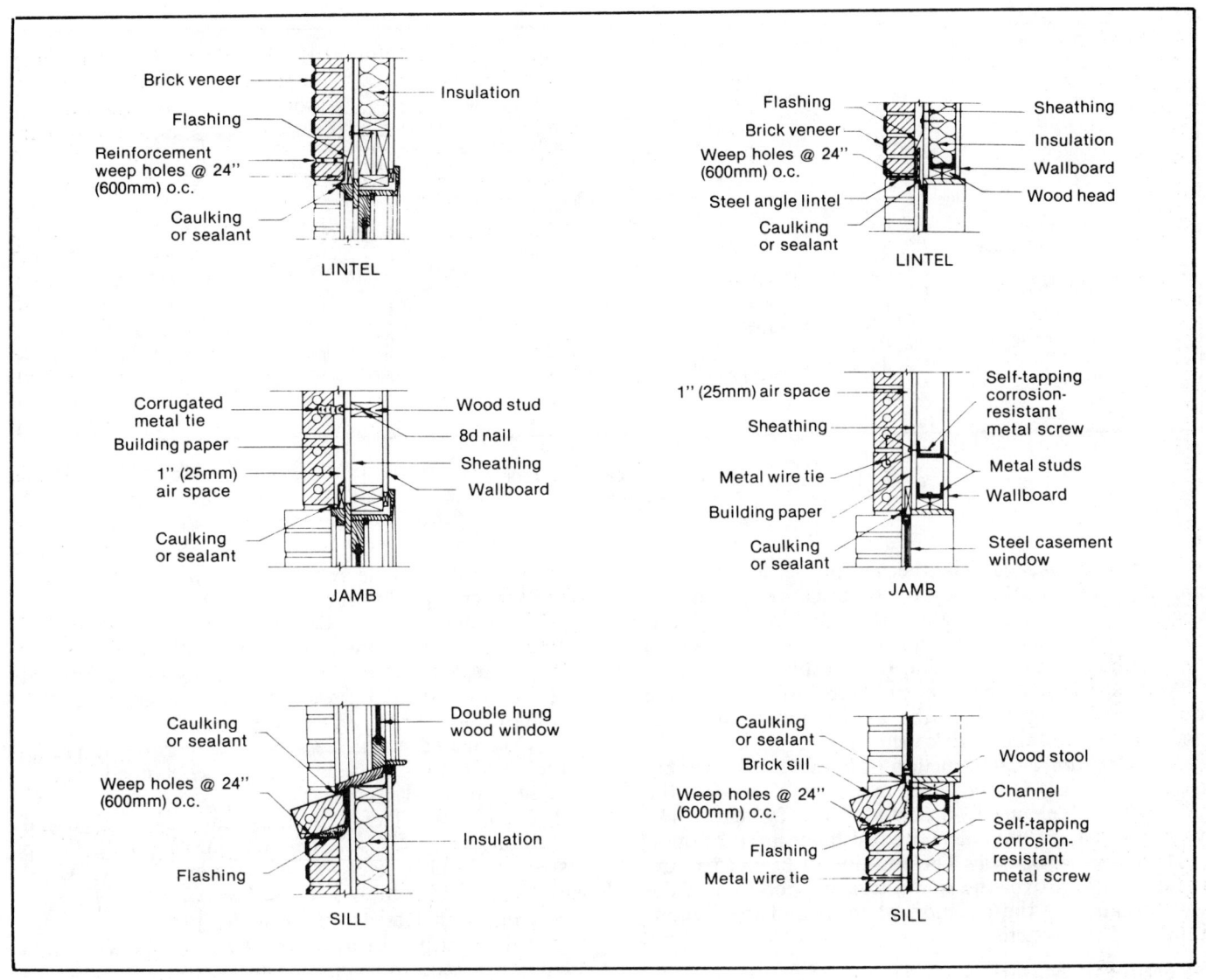

Typical lintel, jamb, and sill details
Figure 12-5

ly prime all joints before placing caulking compounds or sealants. Use compressible backer rope materials for joints deeper than 3/4 inch or wider than 3/8 inch.

Workmanship Good workmanship is as essential in constructing brick veneer as it is in all other types of brick-masonry construction. Completely fill all joints intended to receive mortar. Keep joints or spaces not intended to receive mortar clean and free of droppings so that the wall assembly performs as a drainage wall. If mortar blocks the air space, it may provide a bridge for water to travel to the interior.

Tool the joints with a jointer as soon as the mortar has become thumbprint hard; this firmly compacts the mortar against the edges of the adjoining brick. The types of joints recommended for use with brick veneer are Concave, V, and Grapevine. Other joints are not recommended because they do not provide the necessary resistance to moisture penetration.

EXISTING CONSTRUCTION

The application of brick veneer to existing structures is popular because it enhances the appearance and improves the performance of existing walls. The most common application is in refinishing the exterior of one- and two-family dwellings and also in refacing the fronts of commercial buildings.

Brick veneer over existing construction consists of a nominal 3-inch or 4-inch thick brick wythe attached to an existing wall with metal ties in such a way that a 1-inch air space is maintained between the veneer and the existing wall. New brick veneer can be applied to wood-frame, metal, concrete, or masonry structures.

Properties

In addition to improving the appearance of the existing structure, the application of brick veneer may also enhance many of the performance properties of the wall and the structure to which it is applied.

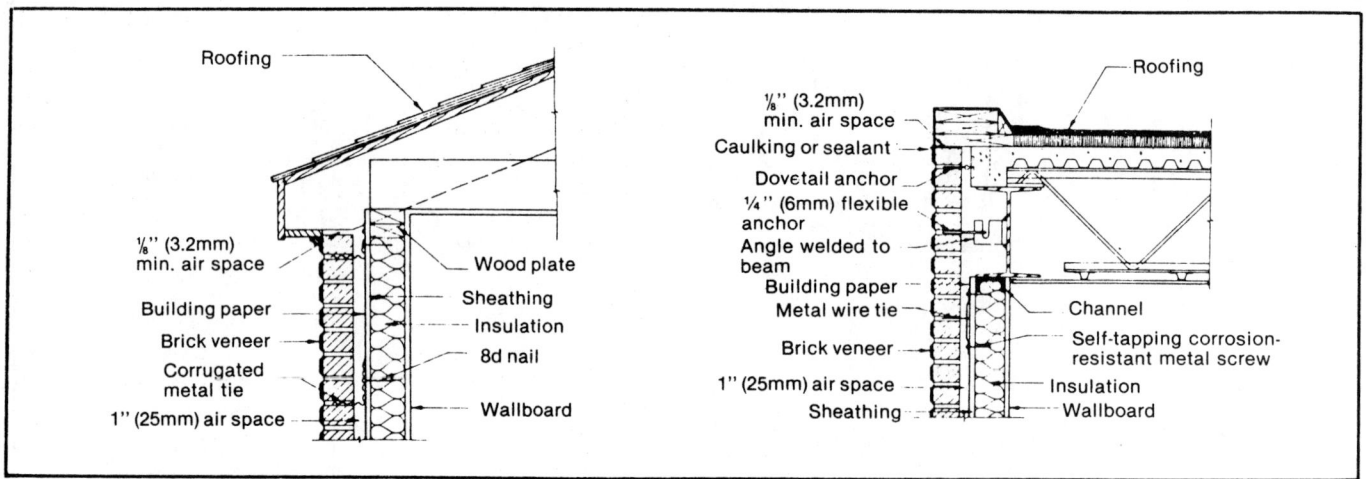

**Typical eave details
Figure 12-6**

Thermal Properties The thermal properties of a wall are improved by the addition of brick veneer in two ways: the addition of mass and the reduction of infiltration. The application of brick veneer also provides an opportunity to add insulation if desirable.

Moisture Resistance The moisture resistance of the wall is also improved by the application of properly detailed and installed brick veneer.

Fire Resistance Construction of brick veneer over existing walls of combustible materials decreases the possibility of externally initiated fires. Typical brick-veneer wall assemblies have fire resistance up to 2 hours.

Acoustical Properties The addition of brick to an existing wall improves the sound transmission loss of the wall because of the addition of mass and the discontinuity of the system.

Selection of Materials

The proper selection of quality materials is essential to the satisfactory performance of a brick-veneer wall assembly. No amount of design detailing or construction can compensate for the improper selection of materials.

Brick Use nominal 3-inch- or 4-inch-thick brick, conforming to ASTM C 62 or ASTM C 216 for brick veneer. Grade SW brick is recommended because the brick wythe is isolated from the remainder of the wall by the air space, thus exposing the brick to maximum temperature extremes.

Do not use salvaged brick because they may not provide the strength and durability necessary for satisfactory performance.

Mortar The use of the correct mortar is very important to the successful performance of the brick veneer. Portland cement-lime mortars are recommended because they have a long history of proven performance.

Type N portland cement-lime mortar is recommended for brick veneer, except that Type M portland cement-lime mortar should be used for brick veneer below grade, where the brickwork is in contact with the earth.

Ties The type of tie system that should be used with brick veneer depends on the construction of the existing wall. Corrugated-metal ties may be used with wood-frame backup. Use metal wire ties elsewhere. Several types of ties which may be used in brick veneer applied to existing construction are shown in the section of this chapter on new brick-veneer construction.

1. Corrugated-metal Ties—Should be corrosion resistant. They should be at least 22 gage, 7/8 inch wide and 6 inches long.

2. Metal Wire Ties—Should be at least 9 gage and corrosion resistant. Use 3/16-inch-diameter metal ties to fasten the brick veneer to the structural frame. Metal wire ties should comply with ASTM A 28 or A 185.

3. Corrosion Resistance—Usually provided by copper or zinc coating, or by using stainless steel. To ensure adequate resistance to corrosion, coatings or materials should conform: Zinc-Coating of Flat Metal—ASTM A153, Class B-1, B-2, or B-3; Zinc-Coating of Wire—ASTM A116, Class 3; Copper-Coated Wire—ASTM B 227, Grade 30 HS; Stainless Steel—ASTM A 167, Type 304.

Tie Fasteners The type of fastener used to attach the ties to the existing wall also depends on the construction of the existing wall.

1. Wood Frame—Use corrosion-resistant nails to attach the corrugated-metal ties to wood-frame construction. Hammer the nails at least 1¼ inches into the wood studs.

2. Metal—Use corrosion-resistant, self-tapping metal screws to attach metal wire ties to metal construction. The screws should penetrate at least 1/2 inch into the metal.

3. Concrete or Masonry—There are several methods of attaching the metal wire ties to existing concrete or masonry walls. Attach them with lag bolts and expansion shields or masonry nails. The fasteners and anchors should be corrosion resistant.

Design and Detailing

Proper design and detailing of brick veneer applied to

Brick Veneer Construction

existing construction is very important to ensure that the wall assembly acts as it is intended. Areas of concern in design and detailing are structural performance, support of the veneer, attachment of the veneer to the existing structure, flashing and weep holes, movement provisions, framing around openings, and the top of the veneer.

Structural Design Brick veneer is a nonload-bearing component of the wall assembly. In addition to its own weight, the only load the brick veneer should support is a proportionate share of any lateral loads. Wide differences between the stiffness characteristics of the brick veneer and those of the existing wall result in the brick veneer carrying a disproportionate share of the lateral loads not considered in the design.

Steel Angles When a continuous steel angle is used to support the new brick veneer at the foundation wall, it should be of steel conforming to ASTM A 36, and should be treated or coated to resist corrosion. Bolts or other fasteners should be corrosion resistant.

The sizing of the angle and the sizing and spacing of the bolts should be determined by structural analysis or according to the specifications.

Steel angles for lintels should be a minimum of 1/4 inch thick with at least 3-inch legs and the steel should conform to ASTM A 36.

Flashing Flashing materials for use with brick veneer may be bituminous membranes, plastics, sheet metals, or combinations of these. Select only superior materials because replacement in the event of failure will be costly, if not impossible. Do not use asphalt-impregnated felt paper as a flashing material.

Weep Holes Form weep holes by inserting a material into the mortar joint or by omitting all or part of the head joint. Forming materials, such as well-oiled rods, are removed to leave an unobstructed opening. Other forming materials, such as plastic tubes or rope wicks, may be left in place. Sometimes metal screening, fibrous glass, or other materials are placed in open weepholes, but this should not be done indiscriminately. Materials such as metal screening can corrode and cause staining of the masonry.

The height limitations for brick veneer are based on the past history of successful performance. Empirical height limitations are provided in Table 12-1.

Supporting Brick Veneer

Brick veneer may be supported directly on either existing or new concrete foundations. Alternatively, it may be supported on steel angles anchored to existing concrete or masonry walls.

Foundations The brickwork should extend down to the existing foundation where possible, as shown in Figure 12-7 (a). If the existing foundation is not wide enough to support the entire thickness of the brick wythe, a new foundation, as shown in Figure 12-7 (b), can be installed at the same depth as the existing foundation. Install bond break between the existing and new foundations to allow for any differential movement.

Steel Angles An alternate method of supporting the brick veneer is shown in Figure 12-7 (c). This requires attaching a continuous corrosion-resistant steel angle to the existing foundation wall. Install the angle at or

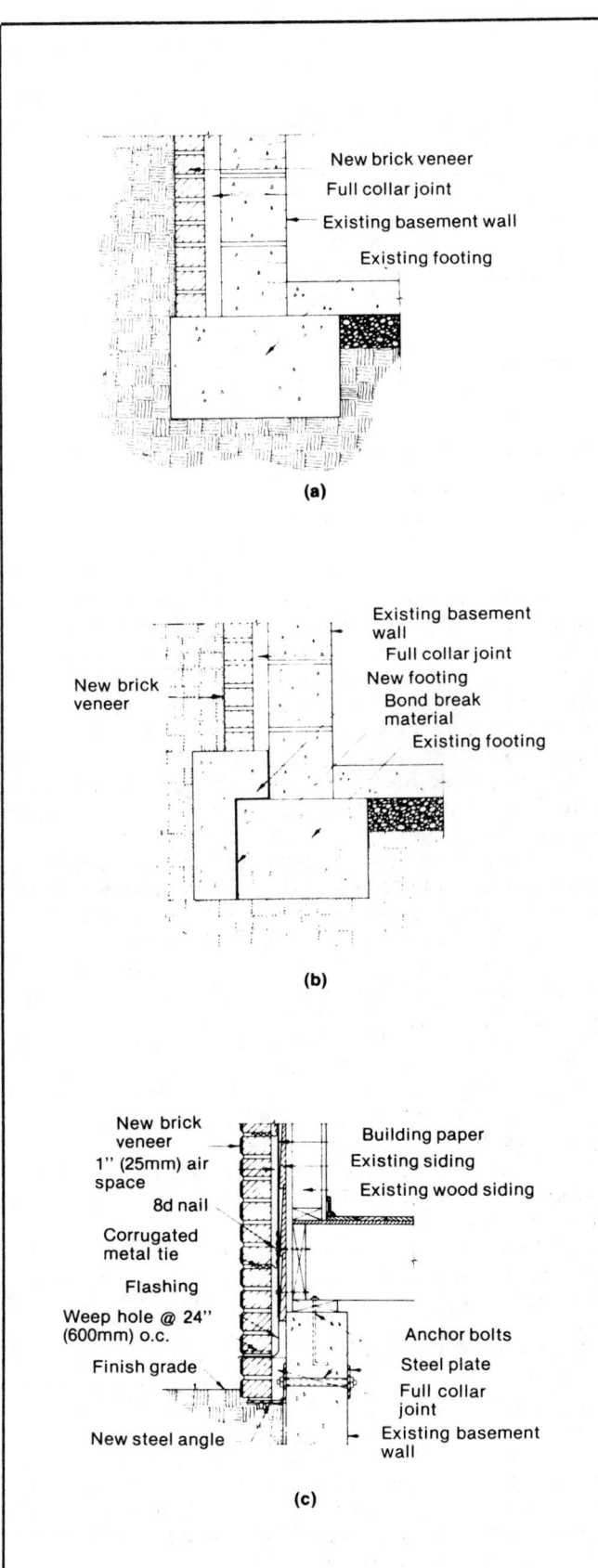

Typical foundation details
Figure 12-7

143

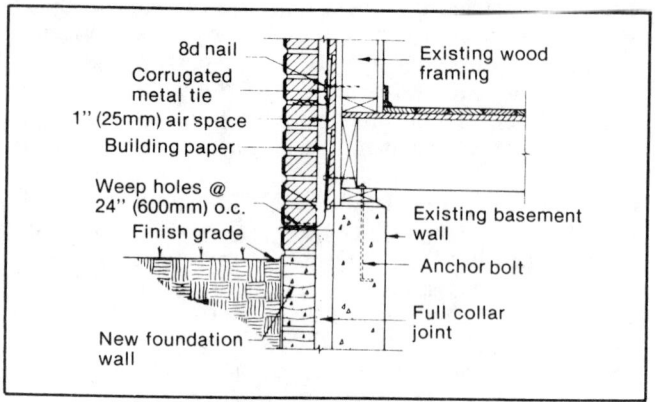

**Base detail
Figure 12-8**

slightly below grade. Installing the angle below frost line decreases the possibility of deleterious effects resulting from freeze-thaw actions. Attach the angles to existing basement or foundation walls constructed of concrete or masonry. Never attach or anchor angles to wood plates or framing members.

Use this method of support with caution. Make a careful analysis of the loads being applied to the angles and give special consideration to the eccentricities of the applied loads. Carefully compute the sizing and spacing of bolts, taking into account not only the loads to be carried and their resulting eccentricities, but also the strength of the foundation wall itself. In general, confine this method of support to one-story structures where the total height of the plate does not exceed 14 feet.

Attachment

The brick veneer must be securely attached to the existing structure. Provide one tie for each 2⅔ square feet of wall area. The maximum spacing of ties, whether horizontally or vertically, should not exceed 24 inches on center. This tie spacing applies above and below grade. The above-grade spacing may be reduced to one tie for each 3¼ square feet of wall area for one- and two-family dwellings not exceeding one story in height.

Flashing and Weepholes Good flashing details, similar to those shown in Figures 12-2 and 12-3, are essential to brick veneer construction. In order to divert the moisture out of the air space through the weepholes, install continuous flashing at the bottom of the air space. The flashing must be at or above grade. Where the veneer continues below grade, completely fill the space between the veneer and the existing structure with mortar or grout. Install flashing at the heads and sills of all openings and wherever the air space is interrupted. Extend the flashing through the face of the brick veneer to form a drip. Where the flashing is not continuous, such as at heads and sills, turn up the ends approximately 1 inch.

Locate weepholes in the head joints immediately above all flashings as shown in Figure 12-8. The maximum spacing of the weepholes should be 24 inches on center. When wick materials are used in the weepholes or when the flashing does not extend through the face of the brick veneer, the spacing of the weepholes should not exceed 16 inches on center.

Movement Provisions

Provisions to accommodate differential movement due to temperature, moisture, shrinkage, and creep are not ordinarily required in small brick-veneer buildings. For structures larger than single-family houses, include in the design considerations of potential differential movements and proper details to accommodate them.

Design and details for differential movement may include expansion joints, flexible anchorage, joint reinforcement, bond breaks, and sealants.

Framing Around Openings

Typical lintel, jamb, and sill details are shown in Figure 12-9. New brick sills can usually be constructed so that the existing sill overlaps the new brick sill.

Install new moulding at the existing jambs and heads of openings so that the framing is extended enough to permit the air space between the brick veneer and the existing construction to be sealed properly.

Lintels may be of reinforced brick masonry, steel angles, or precast concrete. Reinforced brick masonry and steel-angle lintels are the most commonly used in brick-veneer construction.

The minimum required bearing length for steel-angle lintels is 4 inches. The spans and sizes of steel-angle lintels may be modified by fireproofing requirements in local building codes.

Top of the Veneer

A typical detail for the top of the brick veneer at an existing eave is shown in Figure 12-10. There should be at least a 1/8-inch clear space between the top of the last course and the bottom of the soffit. This space should be covered with a new moulding strip and sealant or caulking. If there is insufficient eave to properly cover the top of the veneer, make provisions to extend the eave.

Construction

Foundation Supports Supporting brick veneer on new or existing foundations requires excavating down to the existing foundation. Make the excavation sufficiently wide for the brick mason to work. Prior to placing the masonry on an existing foundation, brush the foundation clean of loose soil and debris.

Angles When constructing brick veneer on continuous corrosion-resistant steel angles, lay the first course of brick in a mortar-setting bed. This provides a means to compensate for any variations and misalignment of the steel angles.

Installing Additional Insulation

Applying brick veneer over existing construction offers

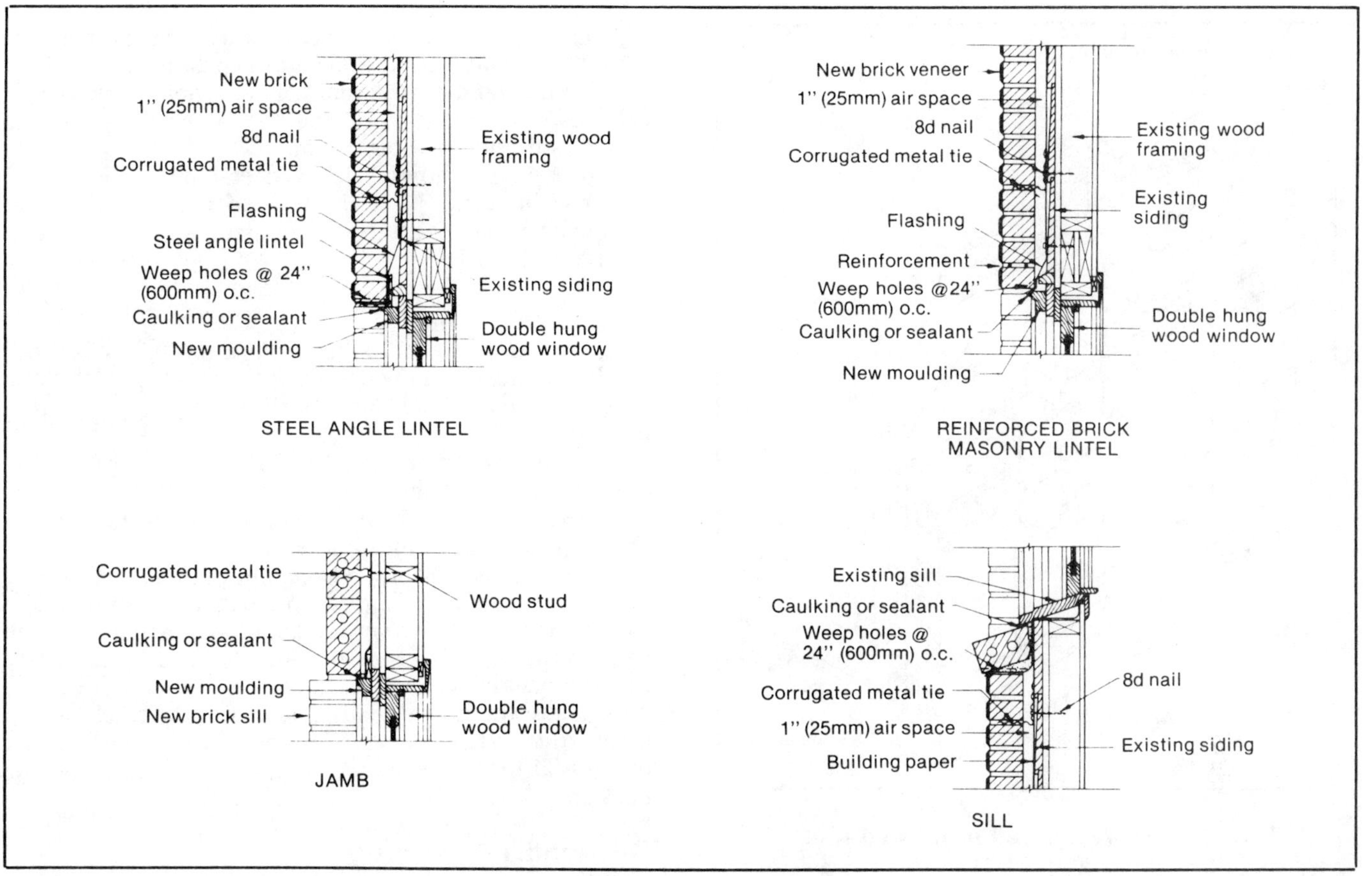

Brick Veneer Construction

Typical lintel, jamb, and sill details
Figure 12-9

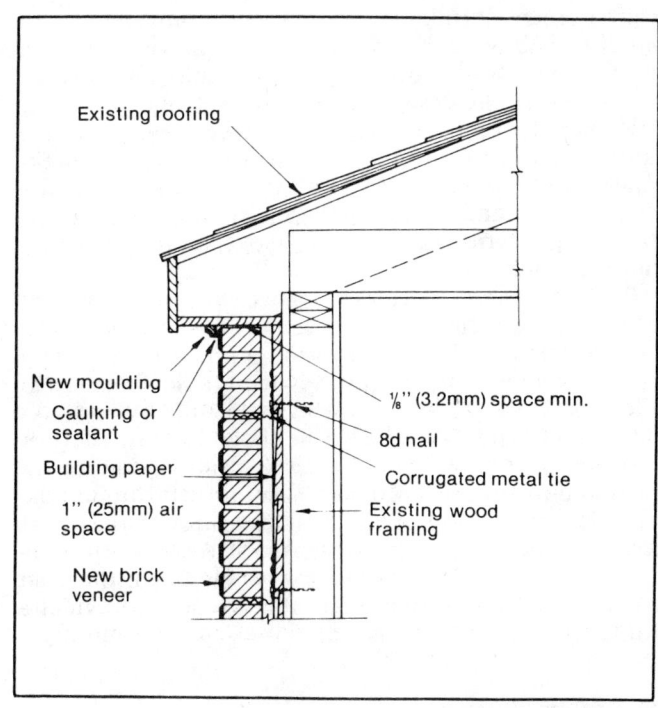

Typical eave detail at the top of brick veneer
Figure 12-10

an opportunity to better insulate the existing exterior walls. Install the insulation materials directly over the existing finish prior to erecting the new brick veneer. Maintain a 1-inch air space between the brick veneer and the rigid insulation. If the existing wood frame or metal-stud walls contain little or no insulation, the existing siding of the wall may be removed so that insulation can be installed within the wall. The materials removed from the existing wall may be reapplied.

PANEL AND CURTAIN WALLS

Over the past seven or eight years a new exterior brick-masonry wall system has come on the market and has been widely used. The wall system is brick veneer over metal-stud backup with gypsum sheathing. (See Figure 12-11.)

The new exterior brick wall system is an innovation in application of the conventional brick veneer over wood studs that has been used successfully for one- and two-family houses for decades. This innovation has been used without benefit of much study and understanding of the nature and behavior of the materials. Problems can occur with the application of this system to buildings above three stories which are of steel and concrete frame, unless properly designed, detailed and constructed.

Masonry & Concrete Construction

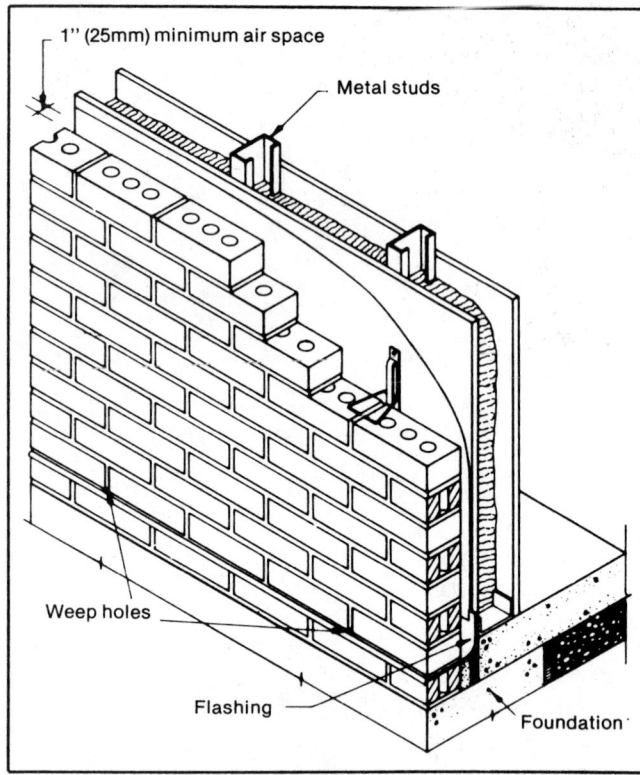

Typical brick veneer and metal stud wall
Figure 12-11

However, when this wall system is properly designed, detailed, and constructed, it performs quite satisfactorily. It offers several advantages: (1) It provides a space for insulation. (2) It permits the building to get into the "dry" and the brickwork to continue at a convenient time when weather permits. (3) It provides the appearance and attributes of a brick building.

Brick Veneer Over Metal Studs

A veneered wall is by definition "a wall having a facing of masonry units or other weather-resistant noncombustible materials securely attached to the backing, but not so bonded as to intentionally exert common action under load." A brick-veneer wall consists of an exterior wythe of brick, isolated from the backup by a minimum of 1 inch of air space, and attached to the backup by corrosion-resistant ties. (See Figure 12-11.)

Over the years, as this wall system has gained in popularity, observations and experience have indicated that satisfactory performance can be achieved by attention to the following.

1. Deflection characteristics of the wall system and components under lateral wind loads.
2. Understanding and providing for the movements that can be expected with the materials used.
3. Careful and proper detailing of flashing and weep holes.
4. Proper selection of materials for strength and durability.
5. Use of proper ties and their adequate spacing.
6. The climatic conditions and exposure.
7. Understanding the manner in which brick veneer behaves.
8. Good construction techniques.

Figure 12-12 shows brick veneer construction on a high-rise building. Note the type of masonry tie used. This brick veneer is attached to a metal-stud system with insulating boards.

Behavior Of Veneer

Brick veneer is subject to numerous imposed loads, both axial and lateral. In the design of the veneer, all of these must be taken into account for the system to perform as it should.

Axial Loads Brick veneer should not be subjected to any axial loading other than its own weight. It should be designed as nonload-bearing, and it must be detailed so that this is actually the situation.

Lateral Loads Many codes imply that the veneer backup should be designed to carry all lateral loads. This works when the veneer material is applied to the backup, as in the case of mosaic tile. The brick veneer and the backup system are separate and, consequently, must share the lateral load. Therefore, the brick-veneer metal-stud system must be designed and detailed so that the lateral loads are shared by both the veneer and the backup.

Differential Movement

Building-Frame Movements Building frames are subject to many movements. Steel frames and members are subject to deflections, thermal movements, and elastic shortening due to imposed loads. Concrete frames and members, in addition to the above movements, are subject to creep, shrinkage, and moisture movements. In buildings above three stories, these movements can result in a vertical shortening of the frame that must be considered in the design of the wall system.

Veneer Movements The brick-veneer wythe is also subject to various normal movements. Since the brick-veneer wythe is separated from the backup, is usually unrestrained, and is usually thermally isolated from the stabilizing interior, it may have movements far greater than expected.

Backup-System Movements Since the backup system is built between floors and is securely attached both top and bottom, any shortening of the frame will be directly transferred into the backup. As the frame shortens, it will produce axial loads on the metal-stud system. As these loads are increased, the backup may deflect laterally. This bowing of the stud system, whether inward or outward, will produce a similar deflection in the brick veneer as the load is transferred through the metal ties. The additional flexural tensile stress in the veneer caused by this deflection, along with the flexural tensile stress caused by the external lateral loading from the wind, may become excessive.

Recommendations

Movement Joints In order to prevent distress in the masonry, provisions must be made for the movements

**Brick veneer construction on a high-rise building
Figure 12-12**

described previously. The best method of providing for these movements is through the use of vertical and horizontal expansion joints.

Provisions must be made for the total differential movements, i.e., the shortening and/or deflection of the structural frame plus the growth from thermal changes and possibly moisture growth in the brick masonry.

No single recommendation for the positioning and spacing of the expansion joints can be applicable to all structures. Analyze each building to determine the potential movements, and make provisions to relieve excessive stress which might be expected to result from such movement. The movement of the brick veneer due to thermal and moisture expansion may be greater than the movement in solid or composite walls exposed to the same environment. This is due to the greater differences between the mean maximum and the mean minimum temperatures of the brick veneer and the absence of restraint usually provided by dead and live loads, masonry bonders, or filled collar joints in solid walls. The isolated brick veneer will undergo maximum changes, and as a result, the number of expansion joints required will be increased. Exercise special care in design, detailing, and construction.

Vertical Expansion Joints Expansion joints in brick masonry oriented vertically are designed to accommodate potential horizontal movements in the masonry. The expansion joint must be fully compressible to permit the movement to occur without distress from excessive or concentrated stress. (See Figure 12-13.)

Design and locate expansion joints so as not to impair the integrity of the wall.

For parapet walls, all vertical expansion joints should be carried through the wall. Space additional expansion joints through the parapet approximately halfway between those running full height, unless the parapet is reinforced.

Horizontal Expansion Joints When the height of the veneer or the number and location of openings necessitates supporting the veneer on shelf angles attached to the structural frame, place horizontal pressure-relieving joints immediately beneath each angle, as shown in Figure 12-14. This is particularly important in reinforced concrete-frame buildings.

Construct pressure-relieving joints by either leaving an airspace or by placing a fully compressible material under the shelf angle. In either case, seal the joint with a permanent elastic sealant.

Horizontal Joint Reinforcement Horizontal joint reinforcement is not usually required in brick veneer since brick masonry is not subject to initial drying shrinkage stresses as is concrete masonry. However, use limited amounts of horizontal joint reinforcement in the brick veneer to improve continuity at corners, offsets, or at intersecting walls.

Single-wire horizontal joint reinforcement may be used to add integrity to the veneer in such places as corners. Horizontal joint reinforcement must be broken at all movement joints. All joint reinforcement must be adequately protected from corrosion.

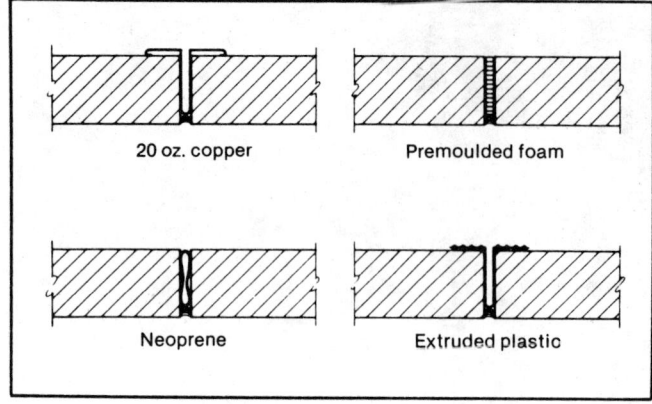

Vertical expansion joint fillers
Figure 12-13

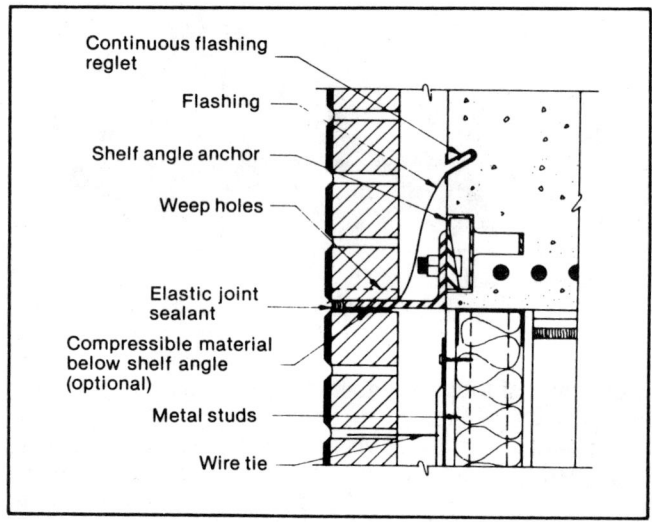

Steel shelf angles
Figure 12-14

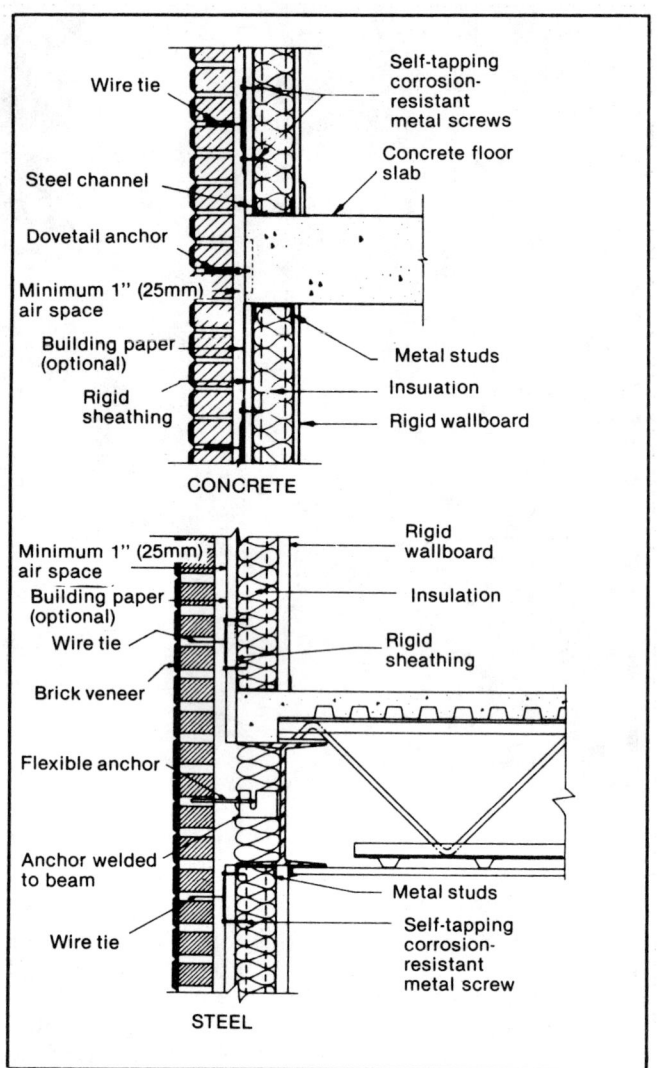

Attachment of brick veneer to structural frame
Figure 12-15

Anchorage When brick veneer is used to enclose skeleton-frame structures, be careful to anchor the masonry veneer to the skeleton frame in a manner that will permit each to move freely in-plane and relative to the other. (See Figure 12-15.) Skeleton frames are more flexible than brick veneer and undergo greater deflection under load. The frame and enclosing veneer differ in their reaction to moisture and in the magnitude of the thermal movement.

Where anchors tie the veneer to the structural frame to provide lateral support, they should be flexible, resisting tension and compression, but not shear. This flexibility permits in-plane differential movements between the frame and the veneer without cracking or distress.

Shelf Angles Where building codes or other factors do not permit the brick to be self-supporting for its full height, support the veneer at each floor or at least at every other floor by shelf angles, as shown in Figure 12-14. Ensure proper anchorage and shimming of these angles to prevent deflections and/or rotations which may induce high concentrated stresses in the masonry. The shelf angles should be made of structural steel and properly sized and anchored to carry the imposed loads so that total deflections and rotations are less than 1/16 inch. For severe climates and exposures, consider using galvanized or stainless-steel shelf angles. Even where galvanized or stainless-steel shelf angles are used, install continuous flashing to cover the angle. Regardless of the type, do not install shelf angles as one continuous member. Provide a space at intervals to allow thermal expansion and contraction to occur without causing distress to the wall.

Windows The proper framing and attachment of windows can prove to be an extremely difficult problem if the differences in the movement between the brick veneer and the frame or backup are not taken into account. Attach the window to either the brick veneer or to the backup system, but do not attach it rigidly to both. (See Figure 12-16.)

Brick Veneer Construction

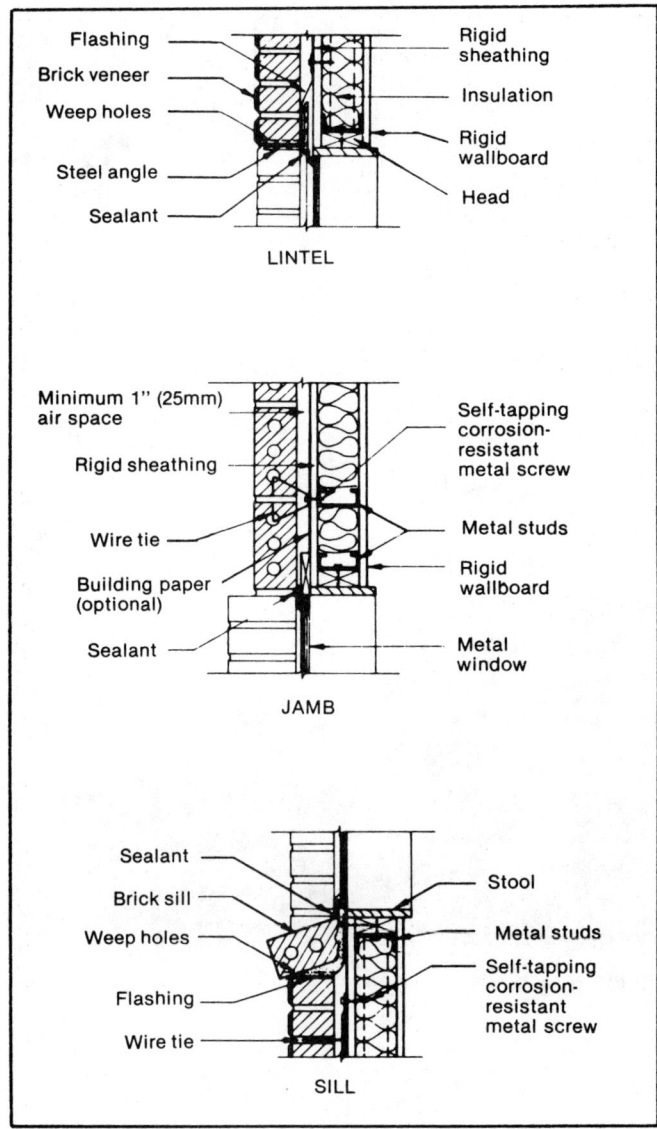

**Typical lintel, jamb, and sill details
Figure 12-16**

Structural Adequacy

Lateral Loading All lateral loading, such as wind loads, must be initially resisted by the exterior wall system and then transferred to the building frame and, eventually, to the foundation. To satisfactorily accomplish this, several things must be considered.

1. Load Distribution—Connect the brick veneer to the backup with metal ties in sufficient number and of sufficient stiffness so that under lateral loading, the system deflects equally. Since the deflections of both the veneer and the backup must be equal, the load will be distributed in accordance with their relative stiffness. This becomes crucial when the material is relatively rigid and brittle compared to the more flexible and elastic metal-stud backup.

2. Deflection Considerations—To achieve a satisfactory distribution of loads, take into account the differences in the shear modulus of the brick veneer and metal-stud backup. This must be done by limiting the allowable lateral deflection of the system.

3. Adequate Moment Resistance—In order to provide for full lateral support of the metal studs, and thus be able to use the full allowable stresses in the studs, properly attach sheathing on both sides of the studs. If this is not clearly detailed and specified and properly constructed, problems may arise. As an example, the practice of loading the scaffold from the inside of the building has often resulted in exterior sheathing being left out of the system in various locations. In addition, proper attachment of the backup system at the top and bottom is mandatory.

The interior of a metal-stud backup system is shown in Figure 12-17. The proper framing and attachment of the windows can prove to be an extremely difficult problem if the differences in the movement between the veneer and the frame or backup are not taken into account. Attach the window to either the brick veneer or to the backup system, but do not attach it rigidly to both.

Recommendations

Support of Veneer In order to alleviate the many problems associated with the use of shelf or relief angles at each floor, design the brick veneer to support its own dead weight on the foundation, unless heights (in excess of 100 feet) or number and location of openings in the veneer make it mandatory that the walls be vertically supported by the structural frame. This may be in conflict with local building codes, so check before designing in this manner.

By following this suggestion, the veneer can be designed as a continuous vertical member spanning several supports, thus reducing the deflection and the tensile stresses induced by the flexural moments, as shown in Figure 12-18.

Allowable Deflection This wall system has not been tested as a complete system. Because of this lack of established data on the question of relative rigidities of brick-veneer facing and metal-studs backup, the recommended limitations on deflection criteria are based on analysis and engineering judgement. The Brick Institute of America and the Metal Lath/Steel Framing Association do not agree on the maximum allowable deflection of the backup system. Further research is planned to study this question.

Metal Studs To provide lateral support and to permit the use of the full allowable stress in the design of the metal studs, securely attach sheathing to both sides of the studs. This sheathing must be rigid, properly detailed, and properly attached for it to be effective.

Horizontally brace the studs at mid-height for added strength, stiffness and fire resistance. Design studs at all jambs, headers and sills of windows, doors and other openings with loads based on the tributary area of the opening with adequate transfer of loads to the structure within deflection criteria.

Ties Provide one tie for each $2\frac{2}{3}$ square feet of wall area, spaced a maximum of 24 inches on center vertically and horizontally, as shown in Figure 12-19. These ties

Masonry & Concrete Construction

Interior of a metal-stud backup system
Figure 12-17

should be minimum 9-gauge corrosion-resistant wire ties of the type shown in Figure 12-20. Do not use corrugated metal veneer ties. Embed all ties at least 2 inches into the bed joints of the brick veneer. Securely attach them to the metal studs through the sheathing and not to the sheathing alone.

Mortar The mortar has an important effect on the strength of the brick-veneer wall. Transverse strength tests of full-scale walls have shown that the bond between the mortar and the brick units is the most important single factor affecting wall strength when the load is applied so that failure occurs through a horizontal joint.

These tests indicate that portland cement-lime mortars, under ASTM C 270 or BIA M1-72, Type S, provide maximum bond between masonry units and mortar. Use Type S mortar in brick-veneer walls at locations where wind loads are expected to exceed 25 psf. For other locations, Type N mortar may be used. In any case, select the lowest-strength mortar that is compatible with the structural requirements. Use only portland cement-lime-based mortars for brick masonry with veneer panel and curtain walls above three stories.

Brick Veneer If it is not possible to increase the stiffness of the metal-stud backup system sufficiently to take its share of the lateral load, several methods are available to adequately size the walls.

1. Thickness—Increase the thickness of the brick veneer itself sufficiently so that it is capable of resisting all of the lateral loads.
2. Reinforced Brick Masonry—The use of hollow brick units with reinforced and grouted cells is a possibility. This type of wall is most beneficial in severe earthquake zones.
3. Backup System—Consider changing the backup system from metal studs to masonry. It may be more advantageous to use a cavity wall or insulated cavity wall design instead of a brick-veneer design.

Parapet Walls Avoid parapet walls unless absolutely necessary. (See Figure 12-21.) If they must be used, they should be properly designed, detailed and constructed. (See Figure 12-22.)

Water Penetration Another area of concern involves the problems of water penetration. The majority of these problems occur because there are no standard accepted details available for this type of construction. Improperly detailed flashing, weep holes, movement joints, ties and anchors, and projections of floor slabs to the outside face of the veneer, may lead to water penetration problems.

Brick Veneer Construction

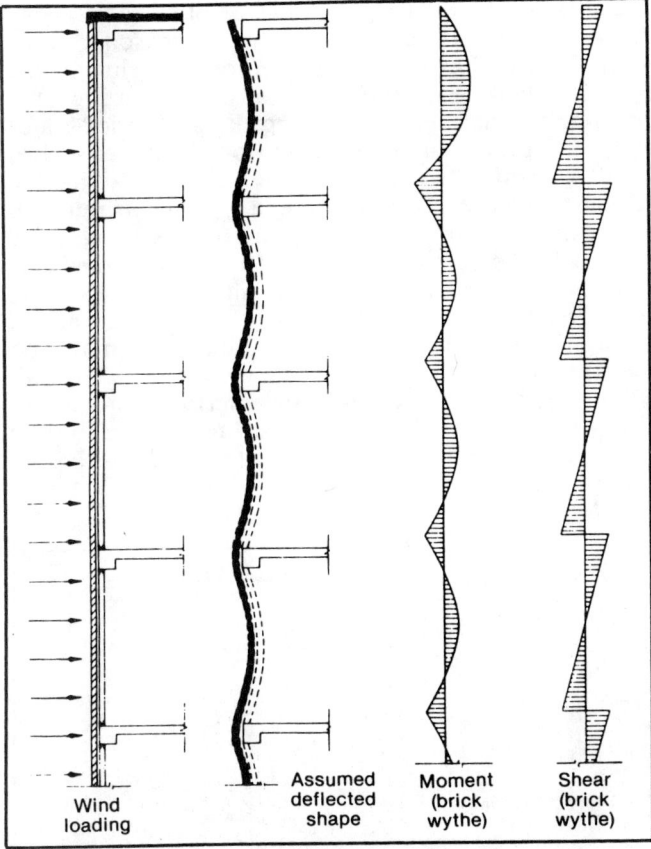

Lateral wind loading effects on brick veneer
Figure 12-18

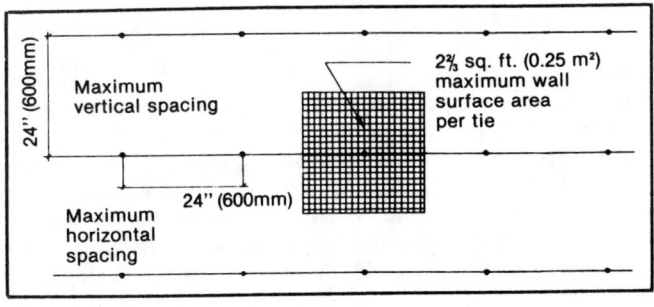

Spacing of metal ties
Figure 12-19

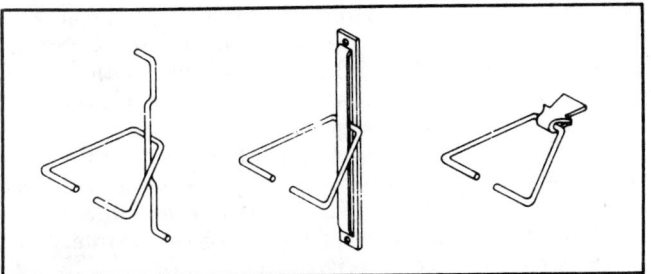

Typical ties
Figure 12-20

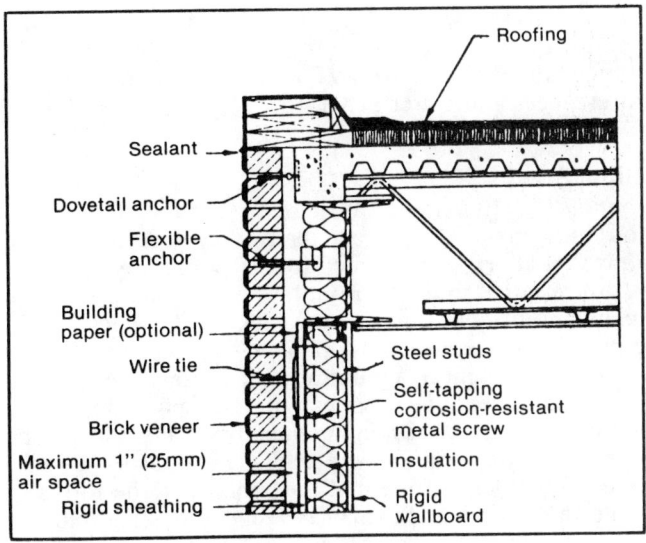

Non-parapet wall
Figure 12-21

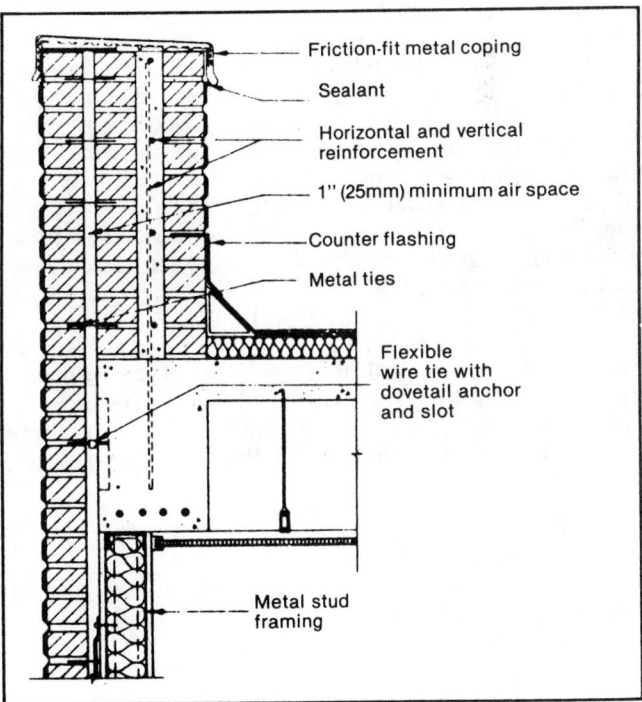

Reinforced parapet wall
Figure 12-22

Clear Coatings Other problems have occurred because of the misconception that clear coatings, such as silicones, for brick masonry walls are cureall's. This is not true. In many instances, clear coatings trap moisture or salts in the walls and lead to spalling of the brick faces.

BRICK-MASONRY CAVITY WALLS

Every structure must be detailed to meet particular requirements. Details that are satisfactory on one struc-

151

ture may not be workable on another. However, certain details can usually be found that minimize the possibility of damage to masonry walls from cracking, efflorescence, and water penetration.

Bond Breaks

Foundations In many areas there are significant foundation movements which can cause severe cracking in walls that are rigidly attached to the foundation. If these walls are left free of the foundation, they tend to span the low points and thus reduce cracking. In general, differential movements in foundations supporting cavity walls must be kept to a minimum, or serious distress will follow. Differential movement of 1/4 inch in 15 feet is sufficient to cause cracking in masonry walls. However, observations on cavity-type and other masonry walls have shown that differential movements in the foundation of more than 1/2 inch in 15 feet could occur and yet the walls remain in good shape and have no cracks.

Figure 12-23 illustrates a typical foundation detail. In this case, the bond is broken between the base of the cavity wall and the top of the concrete beam by building paper. The transfer of movements in the foundation to the wall is thus minimized. Bond breaks also permit differential thermal and moisture movements without distress to either the brick wall or the concrete foundation. In addition, a bond beam or tie beam can be formed at the bottom of the wall by placing reinforcing bars and filling the cavity with grout. This ties the inner and outer wythes of masonry together and distributes any strain over a longer length of wall. This can also be accomplished by a closer spacing of the horizontal joint reinforcement at the bottom of the wall. The above procedures tend to contain any vertical cracks that may originate at the bottom of the wall.

When it is necessary to anchor the masonry wall to the foundation, it is still possible to detail the wall in a manner which allows some differential movement. Such anchorage may be required for load-bearing structures of high slenderness ratio or in earthquake-design areas.

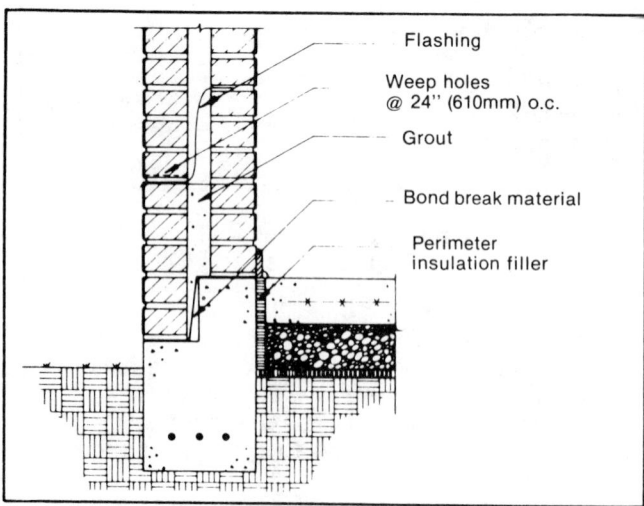

Typical foundation detail
Figure 12-23

Concrete Slabs Thermal strains or other movements are often blamed for the cracking in masonry walls when the actual cause is the expansion or curling of the concrete slabs bearing on the walls. The curling of a concrete slab has been known to pick up the brick bonded to it. Unfortunately, this behavior of concrete is frequently overlooked by the designer in detailing the structure. Figure 12-24 illustrates a typical concrete slab detail that relieves this condition. In this design, the bond is broken between the concrete slab and the brick wall by building paper. This permits the slab to have some freedom in movement with respect to the wall. In addition, it permits the longitudinal thermal and moisture movements to occur without distress. The slab is thickened onto a beam over the interior wythe to help stiffen the slab and minimize curling. Under certain climatic conditions, provisions must be made for insulation (not shown in illustration).

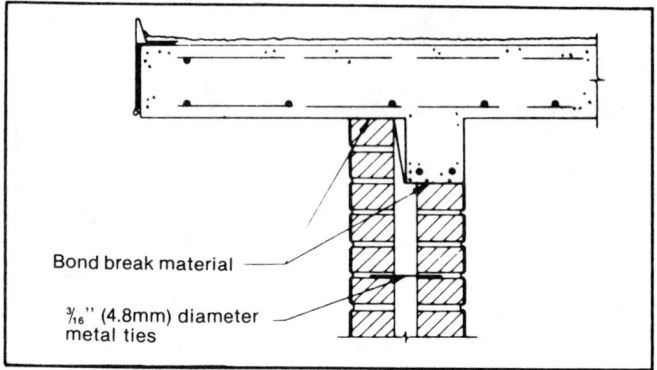

Concrete roof slab detail
Figure 12-24

Bearing

Structural Steel Steel has a coefficient of expansion approximately twice that of brick masonry. If the temperature difference in the materials is large, and the steel is firmly anchored to, or confined within, the masonry, then cracking of the masonry will probably occur. Normal practice has been to anchor the joists or steel positively in the masonry. To improve this design, lubricate the bearing surfaces and provide slotted holes in the seats of the steel members. Tighten the anchor bolts by hand only, or friction will prevent the necessary movement. Figure 12-25 illustrates a structural system using steel joists bearing on a masonry wall.

Wood Floor Joists Wood floor joists normally have a 3-inch fire-cut end and will bear only on the interior wythe of a cavity wall. If the ends project into the cavity, they can form a ledge which may create a moisture bridge across the cavity. All building codes require joists to be anchored to the masonry walls at specified intervals in a prescribed manner. Codes generally require an anchor at the end of every fourth joist. Where the joists are parallel to a wall, anchors engage 3 joists at intervals not exceeding 8 feet. Cavity-wall ties are usually required within 8 inches of joist bearing level. With such construction, the floor provides lateral support for the walls. (See Figure 12-26.)

Brick Veneer Construction

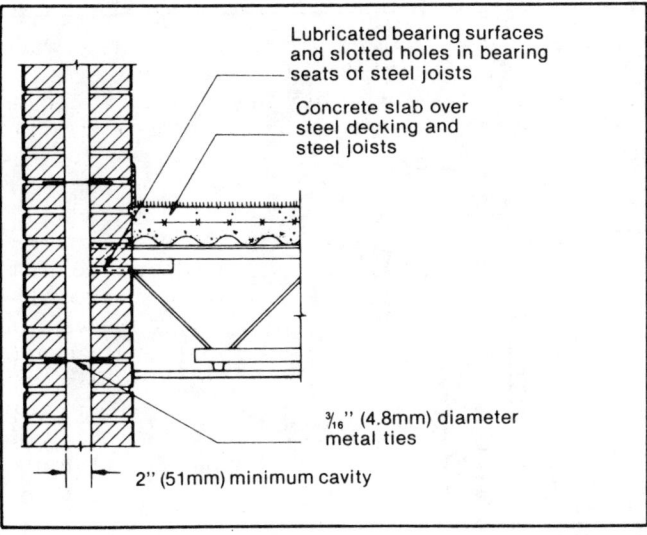

Steel joist structural floor assembly
Figure 12-25

Wood Rafter Plates Wood roofs can be anchored to cavity walls by many methods, two of which are shown in Figure 12-27. The detail on the left illustrates a method using solid units in both wythes. The detail on the right may be used with vertical-cell backup units. Grout anchor bolts into the hollow cells to provide positive anchorage. Regardless of the method, extend anchor bolts holding roof plates into the masonry a minimum of 16 inches, which is normally about six standard-size brick courses. After the wood plate is installed, tighten the nut by hand.

Anchorage and Ties When masonry walls are used to enclose a skeleton frame structure, anchor the masonry walls to the frame in a manner which will permit each to move relative to the other. Skeleton frames are more flexible than brick walls and will undergo greater deflections under load. The frame and enclosing wall differ in their reaction to moisture and the magnitude of their thermal movement.

Where anchors tie the walls to the structure frame to provide lateral support, they should be flexible, resisting tension and compression, but not shear. This flexibility permits differential movements between the frame and the wall without cracking or distress. Figures 12-28 through 12-33 show typical methods for anchoring masonry walls to columns and beams with corrosion-resistant metal ties. These anchorage methods will permit both horizontal and vertical differential movements.

Movement Joints

Vertical Expansion Joints No single recommendation for the positioning and spacing of vertical expansion joints can be applicable to all structures. Analyze each building to determine the potential horizontal movements, and make provisions to relieve excessive stress which might be expected to result from such movements. The extent to which precautions should be taken to prevent brick masonry from cracking depends upon the exposure, character, and intended use of the structure. In some instances, it may be economically desirable to provide less than maximum protection as a calculated risk.

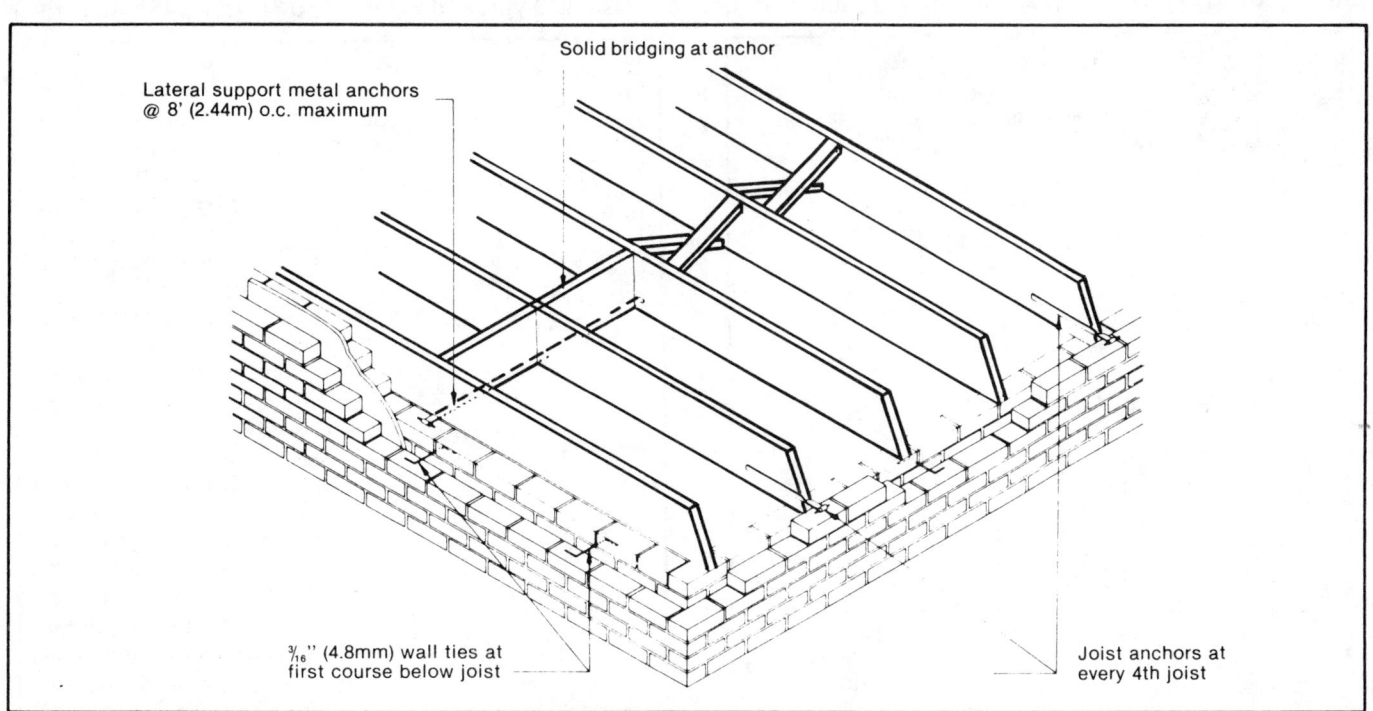

Anchorage of wood floor to cavity wall
Figure 12-26

Masonry & Concrete Construction

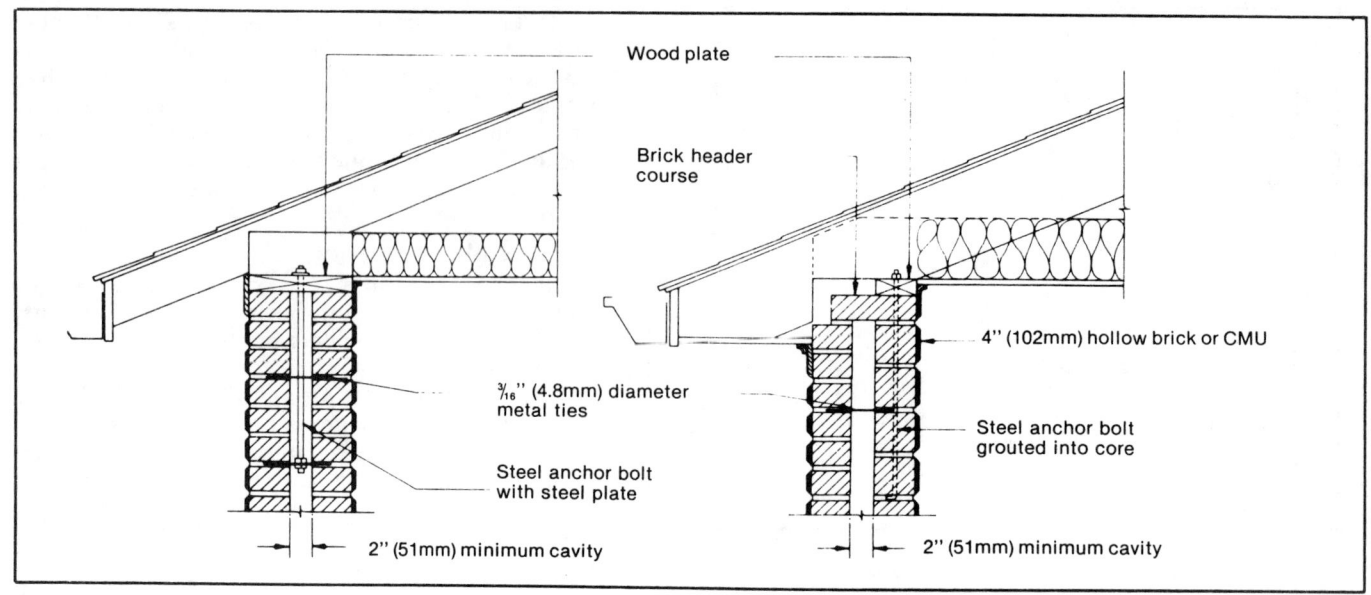

Anchorage of wood roof framing to cavity walls
Figure 12-27

One additional consideration of extreme importance is the distinction between control joints and expansion joints. Control joints are placed in concrete or concrete masonry walls, along with suitable joint reinforcement, to control cracking by reducing restraint and accommodating wall movement from shrinkage due to initial drying. Shrinkage due to drying is not found in clay masonry construction. This becomes obvious when one considers that the clay units, which comprise 70% or more of the total volume of a solid brick masonry wall, are manufactured by a firing process which drives off all moisture. As a result, control joints are not necessary in brick masonry walls. Expansion joints are placed to accommodate the movement of brick masonry walls due to change in temperature and moisture. Concrete masonry walls also experience expansion owing to changes in temperature and moisture, but they experience their shrinkage because of initial drying first. The control joints then act in both contraction and expansion. Typical details of expansion joints and their

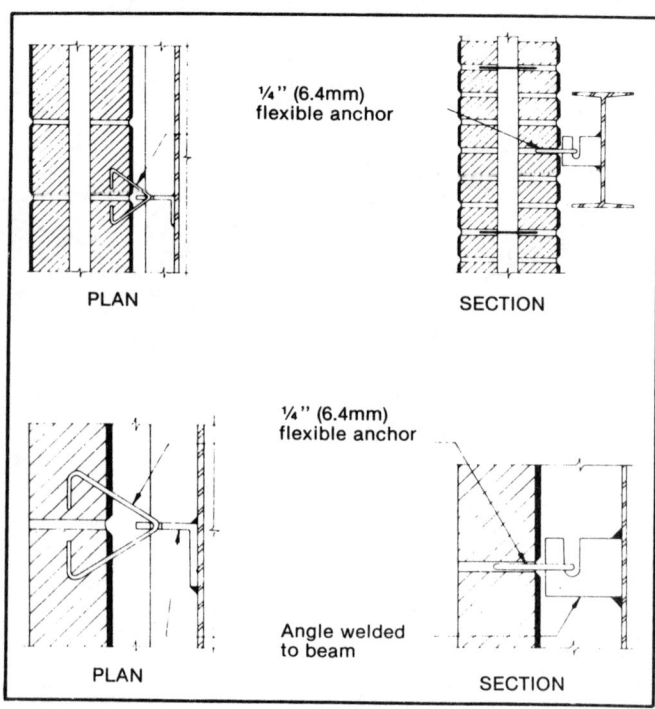

Wall anchorage to concrete beams
Figure 12-28

Wall anchorage to steel beams
Figure 12-29

Brick Veneer Construction

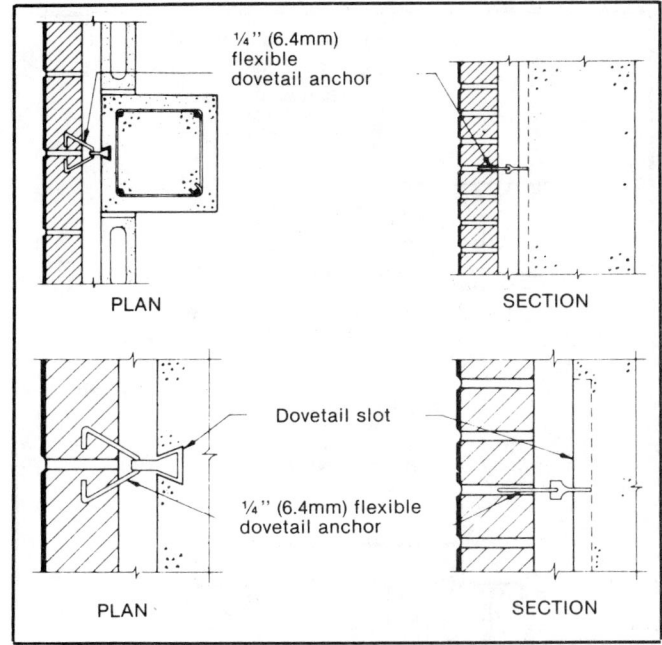

Wall anchorage to concrete columns
Figure 12-30

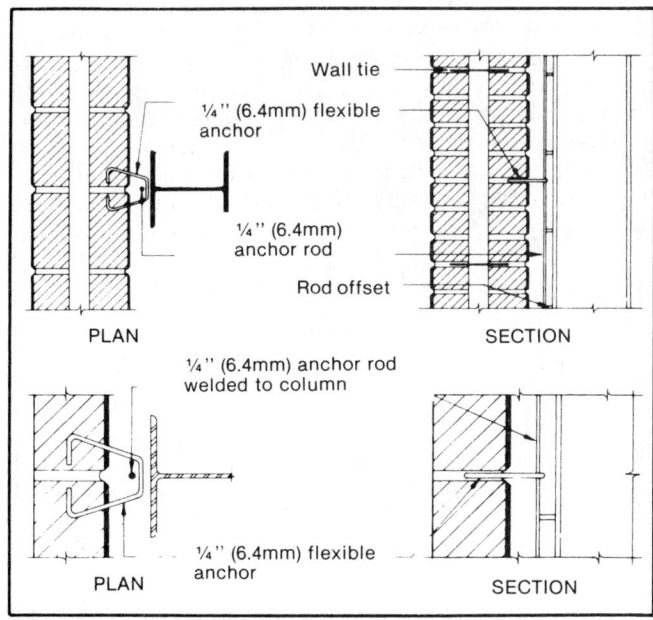

Wall anchorage to steel columns
Figure 12-31

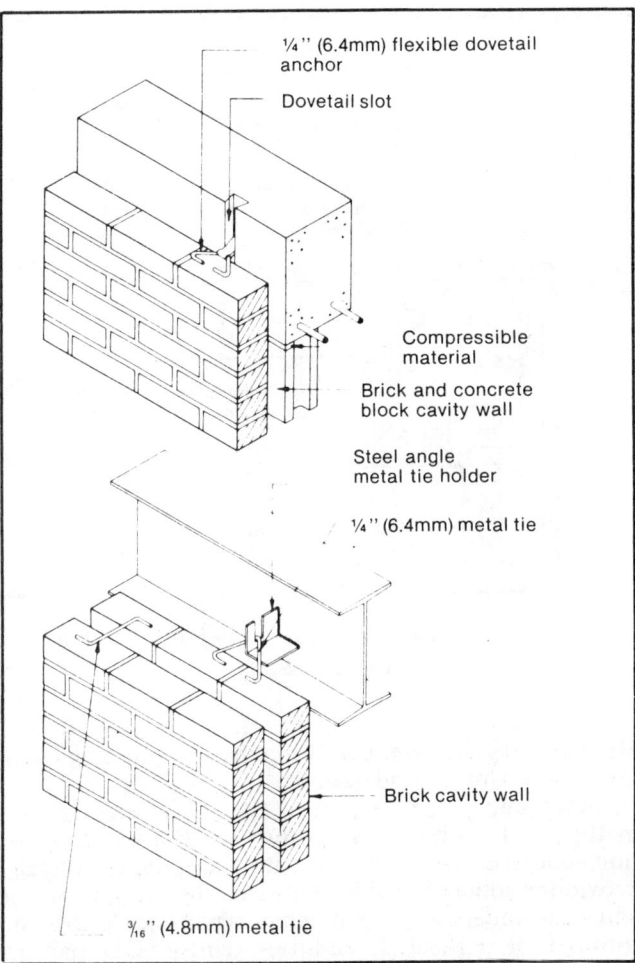

Brick cavity wall anchorage detail
Figure 12-32

the shelf angle, which supports the outer wythe of masonry at each floor, to the spandrel by one of several ways. Be careful to ensure proper anchorage and shimming of the angle to prevent deflections which might induce high concentrated stresses in the masonry. Design angles so that total deflections are less than 1/16 inch. Even if galvanized shelf angles are used, install continuous flashing as one continuous piece. Provide a space at intervals to permit thermal expansion and contraction to occur without damage to the wall.

Where shelf angles are used in this manner, place horizontal expansion joints at shelf angles. This is very important in concrete frame buildings. Seal the joints with a permanently elastic sealant of a color which will closely match the mortar joints.

Parapets

Of all the masonry elements used in buildings, probably the most difficult to detail adequately is the parapet wall. Designers have tried many different ways to design parapets to minimize cracking, leaking, and displacement. Experts generally agree that the only sure way to avoid parapet problems is to eliminate the parapet.

locations are shown in Figures 12-34 and 12-35.

Horizontal Expansion Joints Cavity walls are successfully used as curtain walls in concrete- and steel-frame buildings. When cavity walls are so used, support the inner masonry wythe by the frame at each floor level and lay it to the column faces. Tie the outer wythe, supported by shelf angles, to the structure by metal ties to the inner masonry wythe and the building frame. Secure

155

Masonry & Concrete Construction

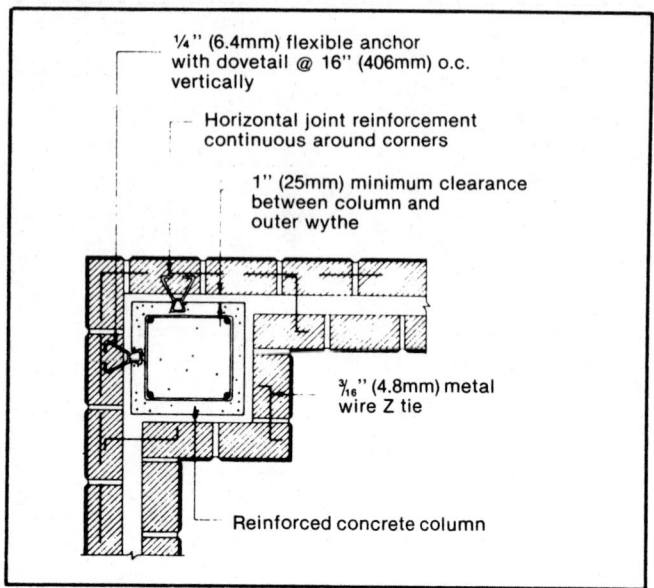

Reinforced concrete column anchorage detail
Figure 12-33

However, they are frequently required by building codes or architectural considerations.

The detail shown in Figure 12-36 is suggested as a method of building parapets. For cavity-wall construction, continue the cavity up into the parapet, thereby providing some flexibility between the outside wythe and the inner wythe. Extend expansion joints up through the parapet. In addition, reinforce the parapet wall and dowel it to the structural frame or space an additional expansion joint between those in the wall below. Also place expansion joints near corners to avoid displacement of the parapet. Parapet copings should provide a drip on both sides of the wall. Metal, stone, and fired clay copings of various designs usually provide this feature. Construct the back side of the parapet of durable materials, preferably of the same material that you used in the front side of the parapet. Do not paint or coat them; they must be left free to "breathe." Unless copings are impervious to watertight joints, place through flashings in the mortar bed immediately beneath them and firmly attach the coping to the wall below with anchor bolts.

Flashing and Weepholes

Moisture may enter masonry at vulnerable spots. Install flashing to divert this moisture to the outside. In areas of severe or moderate exposures, provide flashing under horizontal masonry surfaces, such as roof and parapet or roof and chimney; at overheads of openings, such as doors and windows; and frequently at floor lines, depending upon the type of construction.

To be most effective, extend the flashing through the outer face of the wall and turn it down to form a drip. Provide weep holes at intervals of 16 inches to 24 inches maximum to permit water accumulated on the flashing to drain outside.

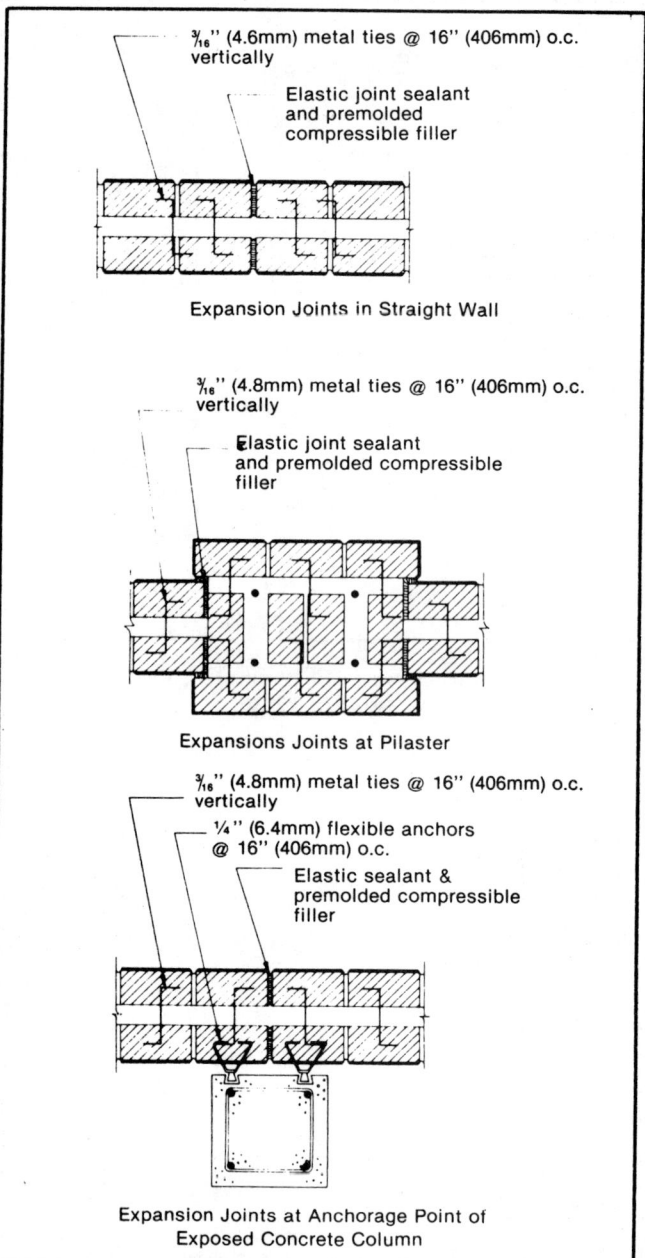

Expansion joint locations
Figure 12-34

If, for esthetic reasons, it is necessary to conceal the flashing, the number and spacing of weep holes are even more important. In this case, the spacing should not exceed 16 inches on center. Concealed flashing with tooled mortar joints can retain water in the wall for longer periods of time, thus concentrating the moisture at one spot.

To prevent any possible moisture infiltration and to promote cavity drainage, place the bottom of the cavity wall above the finished grade, and avoid placing earth over the weep holes during landscaping. With basement construction, use through-wall flashing at the bottom of the cavity to prevent moisture from penetrating the in-

Brick Veneer Construction

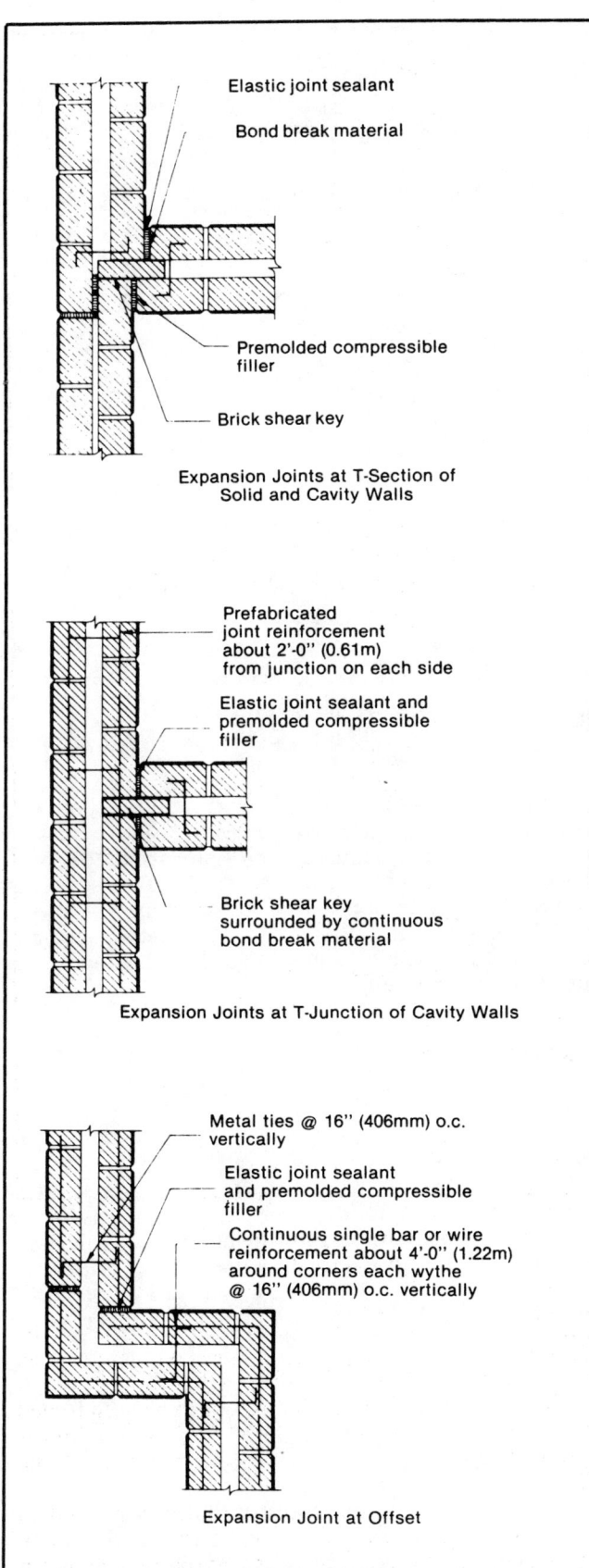

**Expansion joint locations
Figure 12-35**

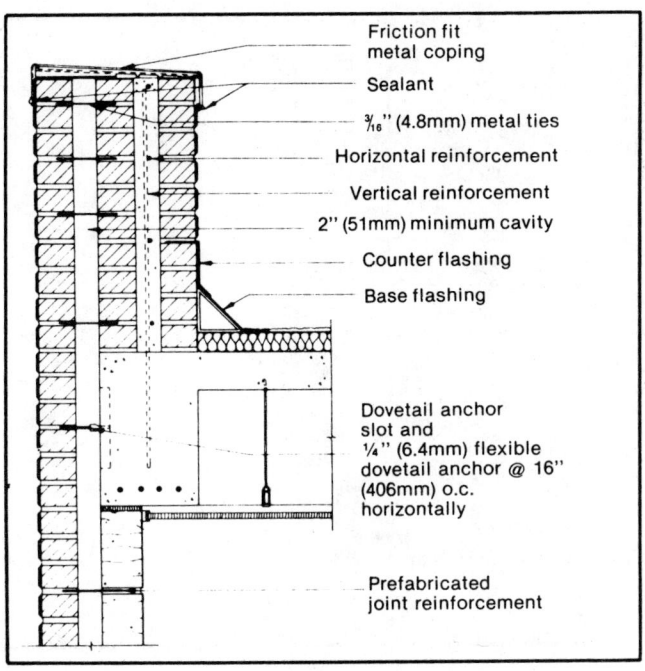

**Reinforced parapet wall
Figure 12-36**

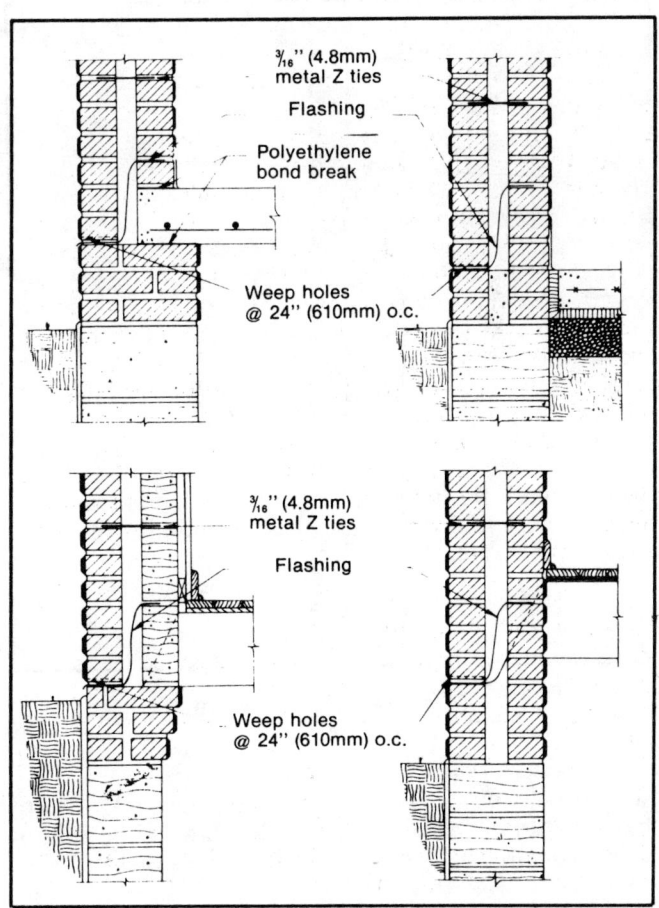

**Use of flashing and weep holes in basement walls
Figure 12-37**

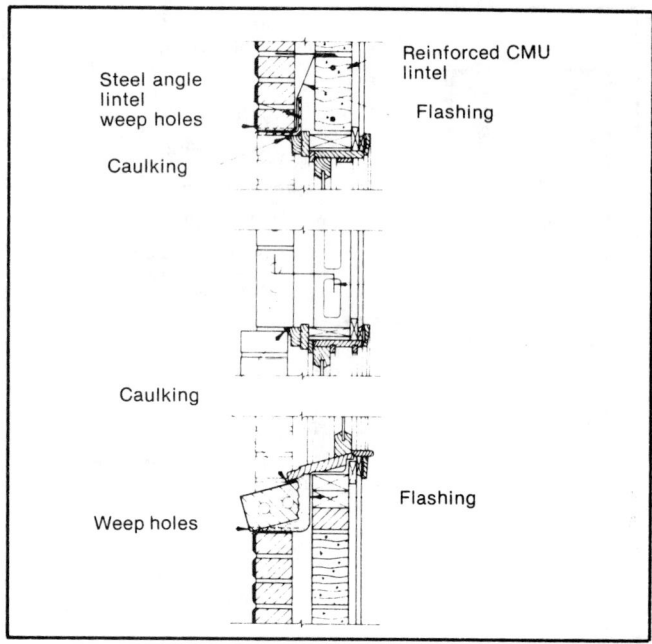

Double hung wood window
Figure 12-38

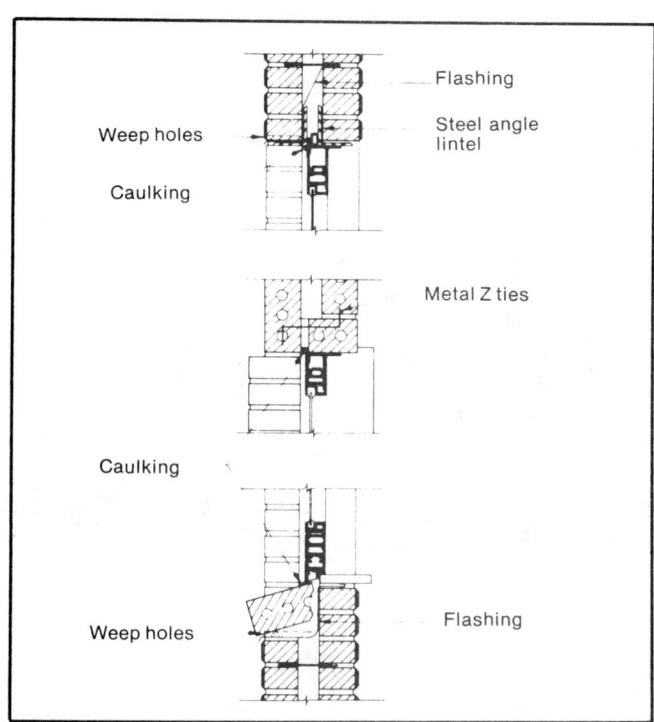

Commercial metal window
Figure 12-40

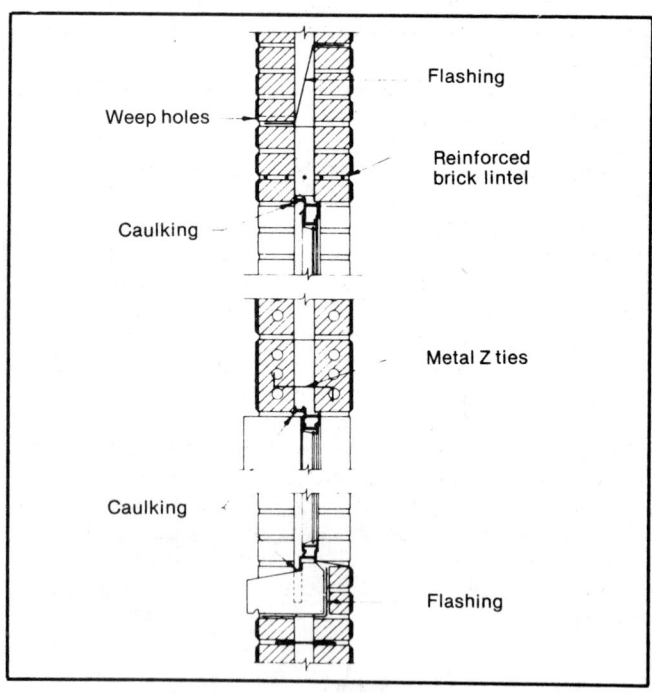

Metal casement window
Figure 12-39

side surface of the basement wall. (See Figure 12-37.) In basementless construction, the flashing at the dampproof course may also serve as a termite shield.

Doors and Windows
Stock sizes of windows and door frames are used in cavity walls, although sometimes additional blocking is needed for anchorage. Avoid solid masonry jambs at windows and doors in cavity walls. However, for steel windows, the jamb must be partially solid to accept most standard jamb anchors. Use wood or steel surrounds to adapt nonmodular steel casement windows to modular cavity walls. Place cavity-wall ties spaced at 3 feet or less around all openings and not more than 12 inches from each opening.

Caulking and Sealants
Too frequently, caulking is considered a means of correcting or hiding poor workmanship rather than as an integral part of construction. Detail and install it with the same care as the other elements of the structure.

Joints at masonry openings for door and window frames, expansion joints, and other locations where caulking may be required are the most susceptible areas for rain penetration. Give these areas proper attention during detailing and construction. (See Figures 12-38 through 12-40.) Also, provide maintenance programs to inspect and replace sealants or caulking which may have dried out or otherwise have become ineffective. In all cases, use a good-grade polysulfide, butyl or silicone rubber sealant. Do not use oil-based caulks. Regardless of the type used, proper priming and backing rope are very important.

Conduit
It is possible to get double duty out of the cavity by using it to carry short runs of conduit. Use this feature with caution so that a moisture bridge across the cavity is not formed.

Chapter 13
Salvaged Brick

Consider three factors in selecting a building material: esthetics, engineering, and economics. Salvaged brick are occasionally selected for their "rugged appearance," or for their low initial cost. Rare is the case when salvaged brick are chosen for their engineering properties, however, for in general they produce weaker, less durable masonry than new brick.

Most salvaged brick are obtained from demolished buildings which have stood 40 to 50 years or more. In fact, it may be next to impossible to salvage brick from a modern structure. Brick today are laid in portland cement-based mortar and the cement sticks almost permanently to the brick.

Fifty years ago manufacturing methods were markedly different from those of today. De-aired brick were unknown; coal- and wood-fired periodic and stove kilns were commonplace. The modern gas-fired tunnel kilns with accurate temperature controls throughout were also unknown. Manufacturing conditions then generally were such that large volumes of ware were fired under greater kiln-temperature variations than could be tolerated today. These conditions resulted in a wide variance in finished products. Brick from the high-temperature zones were hard-burned, high-strength and durable; those from low-temperature zones were under-burned, low-strength and of low durability. These temperature variations resulted in a wide range of absorption properties and color. The under-burned brick were more porous, slightly larger and lighter colored than the harder-burned brick.* Their usual pinkish-orange color resulted in the name "salmon brick."

During these bygone years, prevalent methods of construction made production of both hard-burned and salmon brick economically feasible. Most buildings had load-bearing brick walls which were 12 inches or more in thickness. The hardest, most durable units were used in the exterior wythes; the salmons (and others) were used as backup. Much sorting and grading of brick was performed by the mason on the construction site, although the brick manufacturers eventually assumed this responsibility.

The advent of skeleton frames marked the beginning of high-rise construction and the gradual demise of load-bearing masonry.† Architects and engineers began to design nonload-bearing walls, and gradually decreased wall thickness. This evolution in construction procedures necessitated a change in brick manufacturing procedures. Slowly the demand for salmon brick dwindled. And after the use of hollow backup units became prevalent, the need for salmon brick became practically nil. At the same time, having invented the thinner, lighter-weight panel wall, designers focused their attention on wall strength which they equated to compressive strength in individual brick.

Because the principal demand was for high compressive strength and durability, manufacturers had to produce a high proportion of well-burned brick. To do this necessitated a change in manufacturing methods. Thus, an evolution in construction procedures wrought a significant and beneficial evolution in the production of brick.

MATERIAL SELECTION
Physical Properties
Several arguments are often advanced in favor of salvaged brick:

---

* It is the nature of ceramic products to shrink during firing. Generally, for a given raw clay, the greater the firing temperature, the greater the shrinkage and the darker the color.

† Despite the reduction in its use, load-bearing masonry remains a very economical method of constructing low-rise buildings.

It is virtually impossible to sort and grade salvaged brick
Figure 13-1

1. Because brick are extremely durable, they can be salvaged and used again.
2. If the brick were satisfactory at the time they were first used, they are satisfactory now.

Both arguments are fallacious.

When brick are initially placed in contact with mortar, they absorb some water and some particles of cementitious materials. The initial rate of absorption (suction) is an important factor which greatly affects the bond between mortar and brick. Brick with extremely high or extremely low suction do not develop good bond.

With salvaged brick, more factors influence bond. Pores in brick are filled with particles of lime, dirt, and other deleterious matter. Many bedding surfaces of salvaged brick will not be thoroughly clean, but will instead be covered with mortar. The bond between the old mortar and the new mortar is not very strong. If the original mortar bond was weak, the new bond will be adversely affected. It has been demonstrated many times by comparative tests that the bond to salvaged brick is considerably less than to similar new brick.

Most authorities agree that water penetration through masonry units results from incompletely filled joints and incompletely bonded joints. That is, water penetrates through flaws in joints rather than directly through materials. Thus, masonry of salvaged brick, with its inferior mortar bond, is likely to be more susceptible to water penetration and to be weaker under lateral loading than similar masonry of new units. Its ultimate compressive strength will also be lower if salmon brick are present.

The durability of masonry depends upon the quality of the materials and the mortar bond. Generally, salmon brick are not durable when exposed to weathering. Yet, with the thinner masonry walls of today, brick are used primarily as a facing material. Thus, many salmon brick are eventually placed in exposed faces of salvaged-brick masonry. Even where solid brick walls are used, many salmons are likely to be exposed to weathering because it is impossible to accurately sort and grade salvaged brick. (See Figure 13-1.) With soft, highly absorbent salmon brick exposed to weather, and with poor mortar bond permitting excessive water penetration, it is quite likely that masonry of salvaged brick will spall, flake, pit, and crack due to freezing in the presence of excessive moisture. Figure 13-2 shows a close-up of a wall indicating the excessive spalling that is likely to occur where salvaged brick are exposed to weathering.

One common characteristic of most manufactured building materials is a reasonable degree of uniformity within a particular grade or within a given manufactured lot. Salvaged brick lack this distinction. Hard-burned and soft-burned brick, hopelessly mixed during wrecking operations, effectively create a material stockpile of two widely differing grades. A sample of the material will contain specimens of each grade. If

Excessive spalling is likely to occur when salvaged brick are exposed to weathering.
Figure 13-2

tested for absorption or compressive strength, the sample will show widely diversified characteristics. The average absorption or strength will not approximate the true values of either grade, but will lie somewhere in between. In effect, it is difficult to determine whether salvaged brick meet contemporary material specifications or building code requirements.

Esthetics

Salvaged brick may satisfy the desire for a rugged, colorful masonry surface. Architects often desire the extreme range of colors from dark red to the whites and grays of units still partially covered with mortar. But most frequently the light-pink color of the salmon creates the desired effect. Unfortunately, the pink in salmons results from under-burning which produces units that must not be exposed to weathering. Excessive disintegration due to weathering can soon drastically alter the appearance originally sought.

All pink brick are not necessarily under-burned. During recent years the architectural demand for a variety in colors has led to the extensive use of raw clays which burn other than dark red when fired to maturity. Today, among other colors, many hard-burned, pink bricks are available. These units may conform to the requirements for highest quality under applicable ASTM Specifications, i.e. many pink brick, conform to grades SW or MW (severe or moderate weathering) under ASTM C 216 or C 62.

Many manufacturers blend different colored brick to provide a rustic appearance similar to salvaged brick. There are advantages to using new brick. The architect may specify any desired color proportions and may specify the desired grade under ASTM specifications. Thus, he can obtain the desired esthetic effect without sacrificing durability or strength, a feat which is nearly impossible to accomplish when using salvaged brick.

Chapter 14

Cleaning and Painting Brick Masonry

The final appearance of a brick masonry wall depends primarily on the attention given to masonry surfaces during construction and cleaning. Many of the problems of brick masonry walls result from improper cleaning methods. Some walls have been irreparably damaged from lack of attention to cleaning procedures.

CLEANING BRICK MASONRY

Cleaning failures generally fall into one of three categories:

1. Failure to thoroughly saturate the brick masonry wall surface with water before and after application of chemical or detergent cleaning solutions. A dry wall permits absorption of the cleaning solution and may result in "mortar smear," "white scum," or the development of efflorescence or "green stain." Saturation of the wall surface prior to cleaning reduces the absorption rate, permitting the cleaning solution to stay on the surface rather than to be absorbed into the wall.

2. Failure to properly use chemical solutions. Improperly mixed or overly concentrated acid solutions can etch or wash out cementitious materials from the mortar joints. They have a tendency to discolor masonry units, particularly lighter shades, producing an appearance frequently termed "acid burn" and can also promote the development of green and brown stains. Also, chemical cleaning solutions are generally more effective when the outdoor temperature is 50 degrees F (10 degrees C) or above.

3. Failure to protect windows, doors, and trim. Many cleaning agents, particularly acid solutions, have a corrosive effect on metal. If permitted to come into contact with metal frames, the solutions may cause pitting of the metal and staining of the masonry surface and trim materials such as limestone and cast stone.

Before the actual cleaning of the projects begins, test and evaluate all cleaning procedures and solutions on a test area approximately 20 square feet in size. The indiscriminate use of muriatic acid or the wrong proprietary compound can cause unsightly, difficult-to-remove stains. Minute quantities of some minerals found in some fired clay masonry units and materials added to color brick, such as manganese, may react with some solutions and cause staining. The effectiveness of the method on the sample should not be judged for at least one week.

Keep the wall as free as possible from mortar smears. However, in modern construction where speed is important, even the most skilled of bricklayers find this is difficult. The following are some general precautions that can be taken to promote a cleaner wall:

1. Protect the base of the wall from rain-splashed mud and mortar spatter. Use straw, sand, sawdust, or plastic sheeting spread on the ground and extended 3 to 4 feet from the wall surface and 2 to 3 feet up the wall.

2. Turn scaffold boards near the wall on edge at the end of the day to prevent possible rainfall from spattering mortar and dirt directly on the wall.

3. Cover walls at the end of the workday to prevent mortar-joint washout and entry of water into the wall.

4. Protect stored brick from mud. Cover and store brick off the ground.

5. Practice careful workmanship to prevent excessive mortar droppings. Tool the joints when they are "thumbprint" hard. After tooling, cut off mortar trailing with a trowel and brush excessive mortar and dust from the surface. Use a bricklayer's brush of medium-soft bristle. Avoid any motion that will result in rubbing or pressing mortar particles into the brick faces.

Cleaning New Masonry

Refer to Table 14-1 as a general cleaning guide for new

Masonry & Concrete Construction

Brick Category	Cleaning Method	Remarks
Red and red flashed	Bucket and brush hand cleaning High pressure water Sandblasting	Hydrochloric acid solutions, proprietary compounds, and emulsifying agents may be used. **Smooth Texture:** Mortar stains and smears are generally easier to remove; less surface area exposed; easier to presoak and rinse; unbroken surface, thus more likely to display poor rinsing, acid staining, poor removal of mortar smears. **Rough Texture:** Mortar and dirt tend to penetrate deep into textures; additional area for water and acid absorption; essential to use pressurized water during rinsing.
Red, heavy sand finish	Bucket and brush hand cleaning High pressure water	Clean with plain water and scrub brush, or *lightly* applied high pressure and plain water. Excessive mortar stains may require use of cleaning solutions. *Sandblasting is not recommended.*
Light colored units, white, tan, buff, gray, specks, pink, brown and black	Bucket and brush hand cleaning High pressure water	*Do not use muriatic acid!!* Clean with plain water, detergents, emulsifying agents, or suitable proprietary compounds. Manganese colored brick units tend to react to muriatic acid solutions and stain. Light colored brick are more susceptible to "acid burn" and stains, compared to darker units.
Same as light colored units, etc., plus sand finish	Bucket and brush hand cleaning High pressure water	Lightly apply either method. (See notes for light colored units, etc.). *Sandblasting is not recommended.*
Glazed brick	Bucket and brush hand cleaning	Wipe glazed surface with soft cloth within a few minutes of laying units. Use soft sponge or brush plus ample water supply for final washing. Use detergents where necessary and acid solutions only for *very difficult mortar* stain. For dilution rate, see Step 1d, *Select the Proper Solution*, under Bucket and Brush Hand Washing. Do not use acid on salt glazed or metallic glazed brick. Do not use abrasive powders.
Colored mortars	Method is generally controlled by the brick unit	Many manufacturers of colored mortars do not recommend chemical cleaning solutions. Most acids tend to bleach colored mortars. Mild detergent solutions are generally recommended.

Cleaning guide for new masonry
Table 14-1

masonry. Cleaning methods for new masonry may be classified into three catagories: (1) bucket and brush hand cleaning, (2) high-pressure water, and (3) sandblasting. As a general precaution provide workmen with protective clothing and accessories.

Bucket and Brush Hand Cleaning This is the most popular but most misunderstood of all the methods used for cleaning brick masonry. Its popularity is due to the simplicity of execution and the ready availability of muriatic acid and proprietary cleaning compounds. A recommended general procedure using detergent, acid, or proprietary compound solutions is as follows:

1. Select the proper solution.

a. Hydrochloric acid dissolves mortar particles. Mix a 10 percent solution of muriatic acid in a nonmetallic container. Pour the acid into the water. Do not allow tools to come in contact with acid solutions.

b. For proprietary compounds make sure that the one selected is suitable for the brick and follow the manufacturer's recommended dilution instructions. Many proprietary cleaning solutions perform in a satisfactory manner for their intended cleaning jobs. However, their formulas are not generally disclosed and may be subject to change. Therefore, test each product that you are considering on a panel or inconspicuous wall area.

c. Use detergent or soap solutions to remove mud, dirt and soil accumulated during construction. A suggested solution is 1/2 cup dry measure of trisodium phosphate (Calgon or equal) and 1/2 cup dry measure of laundry detergent (All or equal) dissolved in one gallon of clean water.

d. For very difficult mortar stains on glazed brick and tile use no more than 1 part high-grade acid (chemically pure) to 25 parts clean water.

2. Schedule cleaning last, if possible. Mortar must be thoroughly set and cured. Avoid prolonged time periods elapsing between the completion of the masonry work and the actual cleaning whenever possible. It has been observed that mortar smears and spatters left over a long period of time (6 months to a year) can cure on the wall surface and become very difficult to remove. If this condition occurs, do not mix acid solutions stronger than recommended in order to dissolve the tightly bonded mortar particles.

3. Dry clean. Remove the larger particles by hand with wooden paddles and nonmetallic scrape hoes or chisels. Acid solutions, no matter how effective they may be, cannot be expected to completely remove or loosen large particles.

4. Protect metal, glass, wood, limestone, and caststone surfaces. Mask or otherwise protect windows, doors, and ornamental trim from acid solutions.

5. Presoak or saturate the area to be cleaned. Flush the wall with water from the top down. A saturated brick masonry wall will not absorb dissolved mortar particles. Saturate the area immediately below in order to prevent the absorption of the cleaning solution.

6. Starting at the top of the wall, apply the cleaning solution. Use a long-handled stiff-fiber brush or other type recommended by the cleaning solution manufacturer. Allow the solution to remain on the wall five to ten minutes. For proprietary compounds follow the manufacturer's instructions for application and scrubbing. Use wooden paddles or the other nonmetallic tools to remove stubborn particles. Do not use metal scrapers or chisels. Metal marks will oxidize and cause staining. When cleaning glazed brick or tile, do not use metal cleaning tools or brushes.

7. Clean a small area, preferably not more than approximately 20 square feet. The size of the wash-down area should be determined in advance after a trial run. Heat, direct sunlight, warm masonry and drying winds affect the drying time and reaction of the acid solutions. By working on the shady side of the building the cleaning crew can keep just ahead of the sunshine to avoid rapid evaporation. This also permits the crew to examine their work for initial results.

8. Rinse thoroughly. Flush walls with large amounts of clean water from top to bottom before they are allowed to dry. Acid solutions generally lose their strength after 5 to 10 minutes of contact with mortar particles. Failure to completely flush the wall of cleaning solution and dissolved matter from top to bottom may result in the formation of "white scum."

High-Pressure Water Cleaning To cut labor costs, many cleaning contractors utilize high-pressure water. Some high-pressure systems feature a high-pressure gun and nozzle equipped with a control switch. This setup permits the operator to apply solutions to a wall from the base unit. Other systems have two separate hoses—one with plain water and one with a cleaning solution. Nozzle pressures generally range between 400 psi and 700 psi at a flow rate of 3 to 8 gallons per minute.

Cleaning compounds used with this method should be compatible with the equipment. Some cleaning manufacturers are careful to recommend that only specific cleaning compounds be pumped through their equipment. Others build pumps that will resist hydrochloric acid solutions for reasonable lengths of time.

Equipment should be as portable as possible. Units may be on wheels, skids, trailers, or pickup-truck beds. More elaborate systems include pumps, engines, acid containers, and water-storage tanks fixed on truck beds. To clean with high-pressure water:

1. Test clean the selected solution on a sample area. The scouring of acid-type cleaning solutions on mortar smears may erode mortar joints. Check the equipment for cleaning solution compatibility. For proprietary compounds, mix in accordance with the manufacturer's instructions.

2. Dry clean, using wooden paddles or nonmetallic scrape hoes or chisels.

3. Protect metal, glass, wood, limestone, and caststone surfaces. Mask windows, doors, and ornamental trim from acid solutions.

4. Presoak the area to be cleaned by flushing with water from the top down.

5. Apply the cleaning solutions with a low-pressure orchard sprayer 30 to 50 psi, or directly through the high-pressure cleaning unit.

6. Permit the cleaning solution to remain on the wall for approximately five minutes.

7. Starting at the top, flush the wall with water to rinse thoroughly. CAUTION: It is possible for solutions

to be driven into the masonry when applied under high pressure and to become a cause of future staining. However, if the walls are sufficiently saturated with water before the solution is applied, the risk of penetration is reduced.

Sandblasting Dry sandblasting has been around for many years for restoration work and is one method that eliminates the dangers of mortar smear, acid burn and efflorescence inherent in acid cleaning. However, there is the possibility that, through improper execution, the face of brick units and mortar joints may be scarred. This method is sometimes preferred over conventional wet-cleaning since it eliminates the problem of chemical reaction with vanadium salts and other foreign matter.

Sandblasting by a qualified operator, in conjunction with proper specifications and job inspection, can be satisfactory. Basically, it involves a portable air compressor, blasting tank, blasting hose, nozzle, and protective clothing and hood for the operator. The air compressor should be capable of producing 60 to 100 psi at a minimum air-flow capacity of 125 cubic feet per minute. The inside orifice or bore of the nozzle may vary from 3/16 to 5/16 inch in diameter. The sandblasting machine (tank) should be equipped with controls to regulate the flow of abrasive materials to the nozzle at a minimum rate of 300 pounds per hour. A suggested procedure for sandblasting is as follows:

1. Select sandblast materials that are clean, dustfree and abrasive. They may be of mined silica sand, crushed quartz, granite, white urn sand (round particles), crushed nut shells or other abrasives. Mined silica sand and crushed quartz should be of a type "A" or "B" graduation. (See Table 14-2.) Note: There are various degrees of cutting or cleaning desired, and consequently, many types of abrasive material are available.

Type Gradation	U. S. Sieve Size	Percent Passing
Type "A"	30 Mesh	98 - 100
Fine	40 Mesh	75 - 85
Texturing[b]	50 Mesh	44 - 55
	100 Mesh	0 - 15
	200 Mesh	0
Type "B"	16 Mesh	87 - 100
Medium	18 Mesh	75 - 95
Texturing[c]	30 Mesh	20 - 50
	40 Mesh	0 - 15
	50 Mesh	0 - 15

[a] The screen analysis, as listed above is suggested primarily for mined silica sands and crushed quartz. Reference source: "Good Practic for Cleaning New Brickwork," produced by the Brick Association of North Carolina, P.O. Box 6305, Greensboro, NC 27405.

[b] Type "A" gradation is suggested for very lightly soiled brick masonry or where very light, fine texturing of the masonry surface is permitted.

[c] Type "B" gradation is suggested for heavy mortar stains, or where a medium texture on the masonry surface is permitted.

Typical screen analysis for sandblasting sand abrasives[a]
Table 14-2

2. Brick masonry should be dry and well cured.
3. Remove all large mortar particles as in previous methods.
4. Protect nonmasonry surfaces adjacent to cleaning areas. Use plastic sheeting, duct tape or other covering materials.
5. Test-clean several areas at varying distances from the wall and at several angles of application. Use working distances and angles that afford the best cleaning job without damaging brick or mortar joints. Instruct workmen to direct abrasive at the units and not at the mortar joints. CAUTION: Do not clean light and heavily sanded, coated, and slurry-finished brick by sandblasting.

Cleaning Existing Masonry

High-Pressure Steam This method lends itself readily and satisfactorily to various types of masonry, as compared to other methods. Always use steam to clean buildings with smooth, hard brick or brick with glazed surfaces. The more impervious a brick unit, the easier it should clean.

In most cases, buildings may be cleaned satisfactorily with plain high-pressure steam. For stains it is sometimes necessary to use a chemical or a detergent solution.

Sandblasting Dry sandblasting is the most commonly used method of sandblasting. Sand and compressed air are combined in a tank and forced through a hose and nozzle at various pressures. Use this method only when brick will not be damaged and when certain types cannot successfully be cleaned with high-pressure steam. Certain softer abrasives other than sand may be used, such as crushed pecan shells. Regardless of the type of brick in the building to be sandblasted, clean a sample area before proceeding with the entire job.

Hand-Washing Many buildings of smaller size have been cleaned successfully by hand-washing. It is a slower method and does not give the added advantage of heat as does high-pressure steam. Usually this work is done by using soap or detergent with cold water. This method generally is more costly because it is slower, does not lend itself to a job of any size, and must be repeated more often.

High-Pressure Cold Water This method usually results in a satisfactory job. An ample water supply is very necessary. However, disposing of the large volumes of water used is sometimes a problem.

Chemical and Steam Chemicals and high-pressure steam are used primarily to remove applied coatings to masonry, such as paint. This is a highly specialized field, and frequently, the proper cleaning agent can be determined only after an analysis of the various factors involved in a particular project.

Wet Sand Cleaning This method is used on hard brick and depends on water-cushioned abrasive action for its effectiveness. It is useful for removal of paint or other surface coatings where abrasion of the surface is permissible. Wet sand cleaning employs water in the cleaning action to eliminate dust.

Wet Aggregate Cleaning This is a special process developed by Western Waterproofing Company for use on soft brick and soft stone materials. It is particularly effective on surfaces with flutings, carvings, and other ornamentation. This gentle but thorough process uses a mixture of water and a friable aggregate free of silica and is delivered at low pressure through a special nozzle with a scouring action, which cleans effectively without damage to the surface.

Removing Efflorescence
The removal of efflorescent salts is relatively easy compared to removing other stains. Efflorescent salts are water soluble and generally will disappear of their own accord with normal weathering. This is particularly true of "new-building bloom." White efflorescent salts can be removed by dry-brushing or with clear water and a stiff brush. Heavy accumulation or stubborn deposits of white efflorescent salts may be removed with a solution of muriatic acid (1 part acid to 12 parts water) and scrubbing. Saturate the wall with water before and after the solution is applied. Some proprietary compounds have been developed to remove efflorescence.

Removing "Green Stain" (Vanadium Salts)
Brick units can develop yellow and green efflorescence resulting from vanadium salts. These stains can be found on red, buff, or white brick. The vanadium salts responsible for these stains originate in the raw material used to manufacture certain brick units.

As water travels through brick, it dissolves the vanadium oxide and sulfates. Chloride salts of vanadium require highly acidic leaching solutions and are usually the result of washing brick with acid solutions. Thus, the problem incurred with "green staining" often does not exist until the wall is washed down with an acid solution.

To minimize the occurrence of green stain:
1. Store brick off the ground under protective covering.
2. Do not use acid solutions to clean light-colored brick.
3. Follow the recommendations of the brick manufacturer for the proper cleaning compounds and procedures.

Should the walls, for any reason, be cleaned with an acid solution and green staining appears, the following procedure may be used to neutralize the acid:
1. Immediately following the acid wash, flush the wall with water.
2. Wash or spray the wall with a solution of potassium or sodium hydroxide, consisting of 1/2 pound hydroxide to 1 quart water or 2 pounds per gallon. Allow this to remain on the wall for two or three days.
3. Hose off the white salt remaining on the wall from the hydroxide after two or three days.

For removal of green stain, apply sodium hydroxide with a paint brush and allow plenty of time for it to work. Scrubbing is probably a wasted effort. Do not pass judgment on the results for at least three days.

A convenient way to use sodium hydroxide is in the form of Drano. The mixture used successfully in field trials is one 12-ounce can per quart of water. Sodium hydroxide leaves white salts on the wall that can be washed off with water after a few days.

Various proprietary cleaning compounds have been developed to remove green stain. Their effectiveness on a particular wall can be determined only by a test.

Removing "Brown Stain" (Manganese Stain)
Under certain special conditions this stain may occur on mortar joints of brickwork containing manganese-colored units. It appears as tan, brown, nearly black, or sometimes gray-colored staining. The brown stain has an oily appearance and may streak down over the face of the brick. It appears to be running down from the brick-mortar interface and is a result of manganese dioxide used in some brick as a coloring agent. When the solution reaches the mortar joints, the salts are deposited upon neutralization by the cement or the lime.

During firing in the manufacturing process of some brick, manganese coloring agents experience several chemical changes. This results in compounds that are not soluble in water, but are soluble in weak acid solutions. Since brick can take up acid by absorption, such weak acid solutions can prevail in brick washed with hydrochloric acid. Rainwater may also be acidic in some industrial areas.

To minimize this problem, do not use hydrochloric acid solutions on tan, brown, black, or gray brick. There are special proprietary cleaning compounds available for cleaning manganese brick. Test these for effectiveness. Follow the advice of the brick manufacturer.

The permanent removal of manganese stain may be difficult. After initial removal it often returns. The following method has been effective in removing brown stain and preventing its return:
1. Carefully mix a solution of 1 part acetic acid (80% or stronger), 1 part hydrogen peroxide (30-35%) and 6 parts water. CAUTION: Although this solution is very effective, it is a dangerous solution to mix and use. Acetic acid-hydrogen peroxide may also be available in a premixed form known as peracetic acid. This acid, a textile chemical, is also dangerous and may be difficult to purchase.
2. After wetting the wall, brush or spray the solution on the wall. Do not scrub. The reaction is usually very rapid and the stain quickly disappears. After the reaction is complete, rinse the wall thoroughly with water.
3. A proprietary compound, Brick Klenz, (manufactured by Magnus Division Economics Laboratory, Inc., St. Paul, Minnesota) has been used and sometimes found to be effective in keeping the stain from reappearing. Brush or spray the solution of 1 part Brick Klenz to 3 parts water by volume. Do not scrub. Allow it to remain. Other proprietary compounds have been formulated to remove brown stains. Judge their effectiveness only after testing.

Removal of General Externally Caused Stains
These stains are caused by external materials having been spilled on, spattered on, and absorbed by the brick. Each is an individual case and must be treated accordingly.

A large number of external stains can be removed by scrubbing with a kitchen cleanser. Others frequently can be removed by bleaching with a household bleach. A combination such as is found in some kitchen cleansers may prove most effective. Table 14-3 lists sources of some materials suggested for this use.

Agent	Supply Source
1. Aluminum chloride	Pharmacist.
2. Ammonia water	Supermarket. Household ammonia water.
3. Ammonium chloride	Pharmacist. Salt-like substance.
4. Ammonium sulfamate	Nursery and garden stores. Past use was as a base for weed killers. Not now readily available. Substitute any brand weed killer solution.
5. Acetic acid (80%)	Commercial and scientific chemical supply firms.
6. Hydrochloric acid	Hardware stores. Muriatic acid is generally available in 18° and 20° Baume solutions.
7. Hydrogen peroxide (30-35%)	Some commercial and scientific chemical supply firms.
8. Kieselguhr	Commercial, scientific chemical and swimming pool supply firms. Diatomaceous earth.
9. Lime-free glycerine	Drug stores, used as a hand lotion base.
10. Linseed oil	Hardware and paint stores.
11. Paraffin oil	Hardware stores.
12. Powdered pumice	Hardware stores. A sanding or polishing material.
13. Sodium citrate	Pharmacist. Appears like enlarged salt granules.
14. Sodium hydroxide	Supermarket. Available in brand name substances such as Drano.
15. Sodium hydrosulphite	Pharmacist or photographic stores. A white salt or "hypo" of photographic fixing agent.
16. Talc	Drug stores. Inert powder available as "purified talc." Bathroom talcum powder may be substituted.
17. Tricloroethylene	Commercial scientific chemical supply firms and possibly some service stations or supermarkets. A highly refined solvent for dry cleaning purposes.
18. Trisodium phosphate	Paint stores, some hardware stores, supermarkets. Strong base type powdered cleaning material sold under brand names. Also available in brand name substance such as Calgon.
19. Varsol	Service stations. A refined solvent by the brand name Varsol.
20. Whiting	Paint manufacturers, possibly some large paint stores. A powdered chalk. Substitute kitchen flour if purchase is difficult.

Sources of cleaning and masking agents
Table 14-3

Poultice The use of a poultice is included in some of the recommendations that follow. A poultice is a paste, made with a solvent or reagent and an inert material. It works by dissolving the stain and leaching or pulling the solution into the poultice. The powdery substance is simply brushed off. Repeated applications may be necessary. The chief advantages of poultices are that they tend to prevent the stain from spreading during treatment and they pull the stain out of the pores of the brick. Poultices are normally used only for small stain spots.

The inert material may be talc, whiting, fuller's earth, bentonite, or other clay. The solution or solvent used depends on the nature of the stain to be removed. Add enough of the solution or solvent to a small quantity of the inert material to make a smooth paste. Smear the paste onto the stain area with a trowel or spatula and allow it to dry. When dry, scrape off the remaining powder.

If the solvent used in preparing a poultice is an acid, do not use whiting as the inert material. Whiting is a carbonate which reacts with acids to give off carbon dioxide. While this is not dangerous, it makes a foamy mess and destroys the power of the acid.

Paint Stains

To remove fresh paint, apply a commercial paint remover or a solution of trisodium phosphate in water at the rate of 2 pounds (0.91 kilogram) of trisodium phosphate in 1 gallon (3.79 liters) of water. Allow to remain and to soften the paint. Remove with a scraper and wire brush. Wash with clean water. For very old dried paint, organic solvents similar to the above may not be effective, in which case remove the paint by sandblasting or scrubbing with steel wool. There are also new commercial paint removers in the form of a gel solvent. Apply these on a small area for testing.

Iron Stains

Iron stains are quite common and, in some cases, have covered an entire wall. These stains are easily removed by spraying or brushing with a strong solution of 1 pound (0.45 kilogram) per gallon (3.79 liters) of oxalic acid crystals in water. Ammonium bifluoride added to the solution at 1/2 pound per gallon will speed up the reaction. The ammonium bifluoride generates hydrofluoric acid, a very dangerous material, which may etch the brick. Etching will be evident on very smooth brick. In addition, hydrofluoric acid is extremely dangerous, causing severe chemical burns to the skin. Therefore, use this solution only with proper caution.

As an alternate method, mix 7 parts lime-free glycerine with a solution of 1 part sodium citrate in 6 parts lukewarm water, and mix whiting or kieselguhr to make a poultice. Apply a thick paste on the stain with a trowel. Scrape off when dry. Repeat until stain has disappeared and wash thoroughly with clear water. A poultice made with a solution of sodium hydrosulphite and an inert powder (talc) has also been used for the removal of iron-rust stain.

Copper or Bronze Stains

Mix together in dry form 1 part ammonium chloride (sal ammoniac) and 4 parts powdered talc. Add ammonia water and stir until a thick paste is obtained. Place this over the stain and leave until dry. When working on glazed tile, use a wooden paddle to remove the paste. An old stain of this kind may require several applications. Aluminum chloride is sometimes used in the above procedure instead of the sal ammoniac.

Welding Spatter

A problem related to iron staining is welding spatter. When metal is welded too close to a wall or pile of brick, some of the molten metal may splash onto the brick and meld into the surface. The oxalic acid-ammonium bifluoride mixture recommended for iron stains is particularly effective in removing welding stains.

Smoke

Smoke is a difficult stain to remove. Scrub with a scouring powder containing bleach and a stiff brush. Some alkali detergents and commercial emulsifying agents, brush or spray-applied and given sufficient time to work, also perform well. Test them on a small area first. An added advantage is that they can be used in steam cleaners. For small, stubborn stains, a poultice using trichloroethylene will pull the stain from the pores. Exercise caution when using trichloroethylene in confined spaces and provide ventilation for the harmful fumes.

Oil and Tar Stains

Oil and tar stains may be effectively removed by commercial emulsifying agents. For heavy tar stains mix the agent with kerosene to remove the tar, and then mix it with water to remove the kerosene. After application, it can be hosed off. When used with steam-cleaning apparatus, they have been known to remove tar without the use of kerosene.

Where the stain area to be cleaned is small, or where a mess cannot be tolerated, a poultice using benzene, naphtha, or trichloroethylene is most effective in removing oil stains.

Also, dry ice or CO_2 cartridges may be applied to make tar brittle. Light tapping with a small hammer and prying with a putty knife generally is enough to remove thick tar spatters.

Dirt

Dirt is sometimes difficult to remove, particularly from textured brick. Scouring powder and a stiff-bristle brush are effective if the texture is not too rough. Scrubbing with an oxalic acid-ammonium bifluoride solution recommended for iron stains has proven effective on some moderately rough textures. For very rough textures, high-pressure steam cleaning appears to be the most effective method.

Straw and Paper Stains

Straw and paper stains sometimes result from wet packing materials. Not all packing materials stain brick, but those that do can produce a very stubborn stain. Such stains can be removed by applying household bleach. Allow time to dry. Several applications may be required before stains disappear. The solution of oxalic acid-ammonium bifluoride, recommended for iron stain, cleans the stain much more rapidly.

Plant Growth

Occasionally, an exterior masonry surface not exposed to sunlight remains in a consistently damp condition and shows signs of plant growth such as moss. Apply ammonium sulfate or weed killer in accordance with directions furnished with the compound to remove such growths.

Egg Spatter

Brick walls vandalized with raw eggs have been successfully cleaned with a saturated solution of oxalic-acid crystals dissolved in water. Mix in a nonmetallic container and apply with a brush after saturating the surface with water.

White Scum

White scum is a grayish-white film on the face of the brick. It is sometimes mistaken for efflorescence, but technically it is silicic acid scum. This condition results from the failure to saturate the wall with water before application and to thoroughly rinse acid solutions. Generally, it is a film of material that is insoluble in acid solutions except for hydrofluoric acid, which is very dangerous and not generally recommended for this particular use. Test proprietary compounds formulated to remove this condition and judge their effectiveness.

If removal is too difficult, consider masking the film. In time, weathering will remove both the masking solution and the white scum.

Masking solutions may consist of paraffin oil and Varsol, or linseed oil and Varsol. Apply by brush to the affected brick. Linseed oil and Varsol (10 to 25 percent linseed oil) or paraffin oil and Varsol (2 to 50 percent paraffin oil) darkens light-colored brick. Mix and test several batches of solutions with various concentrations. Generally, solutions of 2 to 25 percent paraffin oil are satisfactory. Allow four to five days of warm drying weather, preferably at 70 degrees F (21 degrees C) minimum, before a judgement is made on the effectiveness of the solutions.

Stains of Unknown Origin

Stains of unknown origin can be a real challenge. Appearance may be the first clue. Rust-colored stains may actually be rust. Such stains are quite common and have been known to come from mortar ingredients, welding spatter on the back of the brick, or from something that has been placed on the pile of brick prior to the brick being laid in the wall.

Green stains may be grass, moss, or vanadium efflorescence. Brown stains may also be vanadium efflorescence, or possibly manganese staining. Brown stains can be almost anything.

One test, useful in narrowing down the list of possible causes of a stain, involves a substance ordinarily not placed on a brick-masonry wall. Concentrated sulfuric acid in contact with an organic material turns it black. This is a quick and easy way to identify stains originating from such material. Organic stains can usually be removed with household bleach or oxalic acid.

PAINTING BRICK MASONRY

Although some masonry walls require protective coatings to impart color and to help resist rain penetration, clay masonry requires no painting or surface treatment. Brick are generally selected because, among other characteristics, they have integral and durable color and, when properly constructed, are resistant to rain penetration.

Clay masonry walls may be painted to increase light reflection or for decorative purposes. Most paint authorities agree that, once painted, exterior masonry requires repainting every three to five years.

It is often erroneously assumed that brick masonry walls that are to be painted can be built with less-durable materials and, in some instances, with less than extreme care in workmanship than would normally be used for unpainted brick walls. This is NOT the case. When a brick wall is to be painted, the selection of materials, both brick units and mortar, and the workmanship used in constructing the wall should all be of the highest quality, at least as high in quality as when the walls are to be left exposed. Make sure that joints are properly filled with mortar to avoid the entrance of moisture into the wall, since the moisture may become trapped behind the paint and cause problems. See that there are no efflorescent materials in the wall, whether in the mortar, brick units, or in the backup, since efflorescence beneath the paint film can also cause problems.

Selection of Material

Brick Brick units to be used for walls that are to be painted should conform to the applicable requirements of the ASTM Specification for Building Brick or Facing Brick, C 62 or C 216, respectively. The grades of units (which designate their durability) should not be lower than would be used if the wall were not to be painted. Grade SW is recommended. It may be acceptable to use brick units that are durable but differ in color in a wall to be painted. However, make sure that the units have similar absorption and suction characteristics so that the paint applied will adhere to all of the surfaces and have a uniform acceptable appearance.

Mortar Mortar for brick masonry walls to be painted should conform to the Specifications for Mortar for Unit Masonry, ASTM C 270, Proportion Specifications. The mortar should consist of portland cement and lime. Select the mortar type on the basis of the structural requirements of the wall.

Paint Paint for application to brick masonry walls should be durable, easy to apply, and have good adhesive characteristics. It should be porous if applied to exterior masonry, thereby permitting the wall to breathe and to prevent the trapping of free moisture behind the paint film.

Considerations For Painting Clay Masonry

In selecting a paint system for a brick masonry wall, the primary concern should be the characteristics of the surface and the exposure conditions of the wall. A primer coat may be of particular importance, especially where unusual or severe conditions exist.

Alkalinity of the Masonry The chemical property of masonry which may have a significant effect on paint durability and performance is the alkalinity of the wall. Brick are normally neutral, but are set in mortars which are chemically basic. Paint products based on drying oils may be attacked by free alkali and the oils can become saponified (converted to soap). To prevent this occurrence, use an alkaline-resistant primer.

Efflorescence The deposit of water-soluble salts on the surface of masonry is another factor that can hamper the performance of painted masonry. Remove efflorescence and observe the masonry surface for reoccurrence prior to painting. The section in this chapter on cleaning masonry discusses removal of efflorescence.

Water and Moisture Water or moisture in a masonry system generally hampers the satisfactory performance of the painted surface. Moisture may enter masonry walls in any of several ways: through pores of the material, through incompletely bonded or partially filled mortar joints, copings, sills and projections, through incomplete caulking and improperly installed flashing or where flashing is omitted. In general, brick wall surfaces should be dry for painting.

Surface Preparation

Proper surface preparation is as important as paint selection. Because each coat is the foundation for future coats, success or failure depends largely upon surface preparation. Thoroughly examine all surfaces to determine the required preparation. Previously painted surfaces often require the greatest effort. Before painting, remove all loose matter. Take special care when cleaning surfaces for emulsion paints and primers. They are nonpenetrating and require cleaner surfaces than solvent-based paints. Some paints can or should be applied to damp surfaces. Others must be applied only to dry surfaces. Follow directions accompanying proprietary brands.

New Masonry As a general rule, new clay masonry is seldom painted. It is difficult to justify the extra expenditure for initial and future painting. However, if for any reason the painting of new masonry is desired, there are a few precautions necessary for success.

Do not wash new clay masonry walls with acid cleaning solutions. Acid reactions can result in paint failure. Use alkali-resistant paints. If low-alkali portland cement is not used in the mortar, neutralize the wall to reduce the possibility of alkali-caused failures. Use zinc chloride or zinc sulfate solution, 2 to 3½ pounds per gallon of water.

Existing Masonry Examine older unpainted masonry for evidence of efflorescence, mildew, mold and moss. While these conditions are not common, they all indicate the presence of moisture. Examine all possible entry points for water. Where necessary, repair flashing and caulking, and tuckpoint defective mortar joints.

Remove all efflorescence by scrubbing with clear water and a stiff brush. A wall which has effloresced for a long time may present difficulties. The presence of moisture, the deposition of salts and the probability of alkalies are all factors which may contribute to the deterioration of paints.

If moss has accumulated on damp, shaded masonry, apply an ordinary weed killer. Wet the wall with clear water before applying weed killers to prevent them from being drawn into the wall. Chemical weed killers may contain solubles which can contribute to efflorescence or react unfavorably with the paint, and should be removed after being used by scrubbing the wall with a stiff brush while rinsing with clear water.

Painted Surfaces Previously painted surfaces normally require extensive preparation prior to repainting (refer to Table 14-4 for typical paint failures). Under humid conditions, mildew may have developed. Mildew may feed on a paint film or on particles trapped by the painted surface. Remove it completely before applying paint. Otherwise, growth will continue, damaging new paint. Mildew has been successfully removed by steam cleaning and sandblasting. The following is also effective: 3 ounces trisodium phosphate (Soilax, Spic and Span, etc.) plus 1 ounce detergent (Tide, All, etc.) plus 1 quart 5 percent sodium hyperchloride (Chlorox, Purex, etc.) plus 3 quarts warm water, or enough to make 1 gallon of solution.

Defect	Description
Alligatoring	Wrinkling of the paint surface caused by paint coats of different hardnesses.
Bleeding	The working up of a stain into succeeding coats, imparting a discoloration to the newly applied coat.
Blistering	Bubbles resulting from moisture trapped behind an impermeable paint film.
Chalking	Powdering at or just beneath a paint surface. Slight chalking may be normal due to weathering.
Checking	A defect in organic paints, manifested by slight breaks in the film surface.
Erosion	Wearing away by weathering.
Excessive paint build-up	Result of applying too much paint or coats which are too thick.
Flaking	Detachment of small pieces.
Map checking	Breaks in paint surface extending entirely through the paint film, usually caused by shrinkage.
Mildew	Fungus growth sometimes found feeding on paint of particles adhering to the surfaces in damp places, generally black or gray in color.
Peeling	A partial detachment of paint.
Scaling	An advanced form of flaking.

**Types of paint failures
Table 14-4**

Use this solution to remove mildew and dirt. Scrub with a medium-soft brush until the surface is clean, then rinse thoroughly with fresh water. For small areas, use an ordinary household cleanser. Scrub with a medium-soft brush and rinse thoroughly. Use a mildew proof paint to help prevent molds from recurring.

Remove all peeled, cracked, flaked, or blistered paint by scraping, wire-brushing or sandblasting. In some instances, old paint may be burned off, but this should be done by skilled operators. As with efflorescence, paint blistering is caused by water within the masonry. Search for the source of the water and take the necessary corrective measures to keep the water out of the wall.

If alligatoring exists, remove the entire finish. There is no other means of correction.

If a slight chalking has occurred, brush the surface thoroughly. However, if the chalking is deep, remove it by scrubbing with a stiff fiber brush and a solution of trisodium phosphate and water. Rinse the surface thoroughly afterwards. Use a penetrating primer to improve adhesion of the final coat.

Excessive paint buildup results from too many coats or excessively thick coats. Where it occurs, remove all paint and treat as a new surface.

Completely remove cement-based paints before repainting with other types. An exception to this rule is the use of cement-based paints as primers which will be covered by another paint within a relatively short time. If the wall will be repainted with another cement-based paint, wire brushing and scrubbing will suffice, provided treatments for mildew, efflorescence, etc. are not required.

Masonry Paints
Because all paints have distinct properties and because surfaces vary considerably, even the most experienced paint contractors carefully examine a surface before making recommendations. However, the following will generally indicate the proper use of masonry paints.

Cement-Based Paints For many years, cement-based paints have been satisfactory coatings for masonry surfaces. They achieved popularity because they have relatively good adherence and a tendency to make a wall less permeable to free water. Cement-based paints are permeable, permitting the wall to breathe. Their main components are portland cement, lime and pigments. Additives, binders and sands may be added.

Although cement-based paints are more difficult to apply than other types, good surface protection results when properly applied. While they are not complete waterproofers, cement-based paints help to seal and fill porous areas, excluding large amounts of free water. White and light colors tend to be the most satisfactory. It is difficult to obtain a uniform coating with darker colors. Lighter colors tend to become translucent when wet, and dark colors become darker. Color returns to normal as the wall surface dries. Cement-based paints can provide a good base for other paints within a relatively short time.

The following procedure for applying paint on a properly prepared surface generally applies:

1. Cure new masonry walls for approximately one month before applying cement-based paints.
2. Dampen wall surfaces thoroughly by spraying with water.
3. Cement-based paints are packaged in powdered form. Because their cementitious components begin to hydrate upon contact with water, mix immediately prior to application for optimum results.
4. Apply heavy coats with a stiff brush, allowing at least 24 hours to elapse between coats.
5. During this time, keep the wall damp by periodically spraying it with water.

6. Apply additional coats in the same manner.

7. Keep the final coat damp for several days to cure properly.

Water-Thinned Emulsion Paints Commonly referred to as latex paints, these are relatively easy to apply. Water-thinned emulsions may be brush, roller or spray applied. However, brush application is preferable, especially on coarse-textured masonry. Emulsion paints dry quickly, have practically no odor and present no fire hazard. They may be applied to damp surfaces, permitting painting shortly after a rain or on walls damp with condensation.

As a group, these paints are alkali resistant. Hence, neutralizing washes and curing periods are not necessary before painting. Water emulsion paints possess high water vapor permeability and are known to have performed well on brick substrates that have been properly prepared.

Emulsion paints will not adhere well to moderately chalky surfaces. If possible, repaint before the previous coat chalks excessively. However, specifically formulated latex paints are available containing emulsified oils or emulsified alkyds which facilitate wetting of chalky surfaces. This property enables the paint to bond the chalk both together and to the substrate.

Butadiene-Styrene Paints These relatively low-cost, rubber-based latex paints develop water resistance more slowly than do vinyl or acrylic emulsions. They are most satisfactory in light tints as the chalking rate may be excessive in deep colors.

Vinyl Paints Polyvinyl acetate emulsion paints dry faster, have improved color retention and a more uniform, lower sheen than rubber-based latex paints.

Acrylic Emulsion Paints Acrylic emulsions have excellent color retention permitting recoating in 30 minutes or less and have good alkali resistance. Acrylics have high resistance to water-spotting and may be scrubbed easily.

Alkyd Emulsion Paints Alkyd emulsions are related to solvent-thinned alkyd types, but have all the general characteristics of latex paints. They do have more penetration than most water-thinned emulsions and achieve better adhesion on chalky surfaces. Compared to other emulsion paints, these are rather slow to dry, have more odor, are not as resistant to alkalies, and have poorer color retention. Under normal exposure conditions, alkyd emulsions can serve as a finished coat over a suitable primer.

Multicolored Lacquers A specialized paint group, multicolored lacquers are applied only by spray gun. The finished film appears as a base color with separate dots or particles of contrasting colors. These paints will cover many surface defects and irregularities. However, they must be applied over a base coat of another type such as polyvinyl acetate or acrylic emulsion paints.

Fill Coats These are base coats for exterior masonry. They are similar in composition, application, and use to cement-based paints. However, fill coats contain an emulsion paint in place of some water, giving improved adhesion and a tougher film than unmodified cement paints. Fill coats have greater water retention and give the cement a better chance to cure. This is particularly valuable in arid areas where it is difficult to keep the painted surface moist during the curing period.

Solvent-Thinned Paints Apply solvent-thinned paint only to completely dry, clean surfaces. They produce relatively nonporous films and should be used only on interior masonry walls not susceptible to moisture penetration. The exception to this is special-purpose paint, such as synthetic rubber, chlorinated rubber and epoxy paints.

Oil-Based Paints Oil-based paints have been used for many years. They are relatively nonporous and are recommended for interior use only. Although several coats may be required for uniform color and good appearance, they bind well to porous masonry. As with most solvent-based paints, they have good penetration on relatively chalky surfaces, but are highly susceptible to alkalies. New masonry must be thoroughly neutralized to avoid saponification. Available in a wide color range, oil-based paints are moderately easy to apply. Allow several days of drying time between coats.

Alkyd Paints Alkyd paints are similar to oil-based paints in most general characteristics. They may have slightly less penetration, resulting in somewhat better color uniformity at the cost of adhering power. Alkyd paints are more difficult to brush, dry faster and give a harder film than oil-based paints. These, too, are nonpermeable and are recommended for interior use only.

Synthetic Rubber and Chlorinated Rubber Paints These paints have excellent penetration and good adhesion to previously painted, moderately chalky surfaces as well as to new surfaces. They are reported to be more resistant to efflorescence and are generally good in alkali resistance. They may be applied directly to alkaline masonry surfaces, but are more difficult to brush on than oil paints. Darker-colored synthetic-rubber paints lack color uniformity. Both types have high resistance to corrosive fumes and chemicals. For this reason, they are often specified for industrial applications. Both types require very strong volatile solvents, creating a fire hazard which may prove undesirable.

Epoxy Paints Epoxy paints are of synthetic resins generally composed of two parts: a resin base and a liquid activator. Use them within a relatively short time after mixing. Epoxies can be applied over alkaline surfaces, have very good adhering power, and good corrosion and fume resistance. However, some types chalk excessively if used outdoors. Epoxies are relatively expensive and somewhat difficult to apply.

"High-Build" Paint Coatings These coatings are generally used on interiors to give the effect of glazed brick. Some coatings are based on two-component urethane polyesters and epoxies. Others are of an emulsion-based coat with acrylic lacquer. These paint systems usually include fillers to smooth out surface irregularities.

Other Coatings Heavily applied coatings of the so-called "breathing type" are available with a water or solvent base. They are generally composed of asbestos fiber and sand and applied thickly to hide minor surface imperfections. The presence of moisture on the masonry wall generally will not harm the latex type. Lower application temperatures of 35 degrees F to 50 degrees F less damaging to the solvent type.

For both types, adhesion is mostly mechanical because of low binder and high pigment content. Some coatings require special primers to ensure adhesion. Although these coatings are reported to give good performance on masonry, they tend to show stains where water runoff occurs.

These coatings are capable of allowing passage of water vapor, but they cannot transmit large quantities of water that may enter through construction defects. Failure may occur as a result of water freezing behind the film.

Painting Near Unpainted Masonry

Often windows and trim of masonry buildings are painted with self-cleaning paints to keep the surface fresh and clean. Unfortunately, self-cleaning is generally achieved through chalking. The theory is that rain will wash away chalked paints, constantly exposing a fresh paint surface. The theory works well, but too often no provision is made to keep chalk-contaminated rain water away from masonry surfaces. The result is usually more unsightly than dirty paint on trim or windows. Avoid this staining by choosing nonchalking paints for windows and trim and by providing a means of draining water away from wall surfaces.

Chapter 15
Flashing Clay Masonry

Flashings are usually formed from sheet metals, bituminous membranes, plastics, or a combination of these, the selection being largely determined by cost and suitability. Both installation and material costs vary widely. The differences in total costs reflect both factors and cannot be compared by considering material costs alone. Because replacement costs can exceed the original costs, select a permanent flashing material for the original installation.

Copper
Copper is durable, is available in special preformed shapes, and is an excellent moisture barrier. It is more costly than most other flashing materials. Although exposed copper may tend to stain adjacent masonry, it is not materially affected by the caustic alkalies present in masonry mortars. It can be embedded safely in fresh mortar and will not deteriorate in continuously saturated, hardened concrete unless excessive chlorides are present. When using copper flashing, do not use chloride-based additives in mortar.

Zinc
Galvanized coatings are subject to corrosion in fresh mortar. Although the corrosive products apparently form a very compact film around zinc and a good bond with the mortar, the extent of corrosion cannot be predicted accurately. Some zinc-alloy flashings are available. Like many alloys, these have properties considerably different from those of the pure metal.

Aluminum
The caustic alkalies present in fresh unhardened mortar attack aluminum. Although dry seasoned mortar does not affect aluminum, corrosion can occur if the adjacent mortar becomes wet. Since the purpose of flashing is to direct the flow of water, uncoated aluminum is unsatisfactory as a flashing.

Lead
Lead, like aluminum, is susceptible to corrosion in fresh mortar. Furthermore, when lead is only partially embedded in mortar, in the presence of moisture it develops a differential electric potential, acting as the positive element of an electric cell. The resulting electrolytic action gradually disintegrates the embedded lead.

Plastics
As a group, plastics are becoming one of the most widely used flashing materials. The better plastic flashings are tough, resilient materials which are highly resistant to corrosion. However, because the chemical compositions of plastics are so diverse, it is impossible to place all plastic flashings into one generalized group. Some plastics will not withstand the corrosive effects of masonry mortars. Rely on performance records of materials, the reputation of manufacturers and, where possible, test data to ensure satisfactory performance.

Bituminous Flashing
Fabrics saturated with bitumen are used as damp checks and as low-cost substitutes for metal flashings at the heads of openings at spandrels and at window sills. If they are permanently insoluble in water and are installed with unbroken skins, they can be effective, although not as permanent as a good metal flashing.

Combination Flashings

Often, materials are combined to utilize the better properties of each; for example, sheet metal may be coated with plastic or bituminous materials. This particular combination can provide a lower-cost flashing through reduction in metal gauge or it may permit the use of corrodible metals which would otherwise prove unsatisfactory.

Flashing Placement

Design Considerations Exposure to rain is a basic consideration in design against moisture penetration. Amount of rainfall varies greatly throughout the United States; for example, it is severe along the Gulf Coast, the Atlantic Seaboard, and the Great Lakes; moderate throughout the Midwest, and slight in the far Southwest. Recommendations in this chapter refer to severe or moderate exposures. Where exposure is slight, internal flashing at spandrels, lintels, sills, etc. may be eliminated and external flashing may be reduced to a minimum.

Where flashing extends to the interior, place its end between the furring and the interior finish and turn it up at least 1 inch to collect moisture that may penetrate through the wall.

Wall Base Any moisture which does enter a wall generally travels downward. Place flashing above grade at the wall base to divert this water to the exterior. To stop the upward capillary travel of ground moisture, especially if there is no basement, place damp checks about 6 inches above grade. In areas where first-floor wood joists require protection from termites, use a metal damp check which projects at least 2 inches past the inside face of the wall and which is bent down at an angle. (See Figure 15-1.)

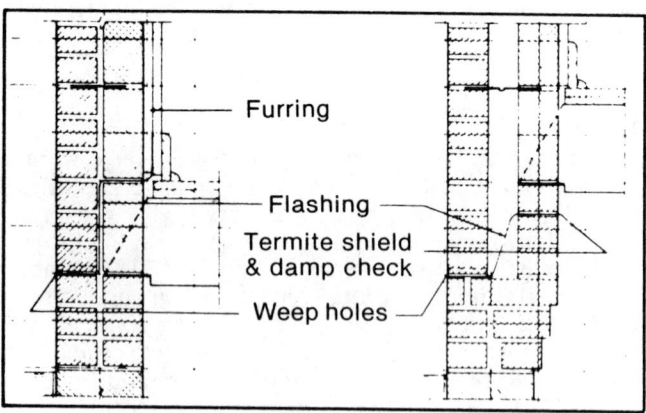

Flashing at bases of masonry walls
Figure 15-1

Window Sills Place through flashing under and behind all sills except impervious monolithic sills. Extend ends of sill flashing beyond the jamb line on both sides and turn them up at least 1 inch into the wall. Slope all sills to drain water away from the building.

Where the undersides of sills do not slope away from the building, provide a drip notch or extend flashing and bend down to form a drip. (See Figure 15-2.) If water which runs down windows and over sills continues down the building face, it is likely to create stains.

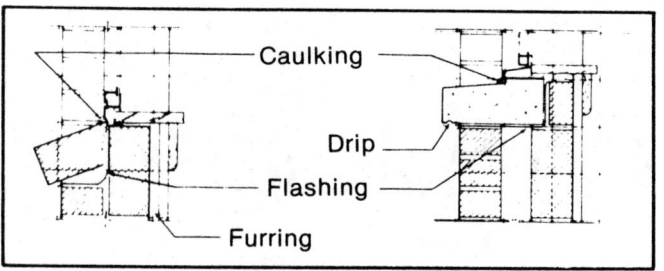

Flashing window sills
Figure 15-2

Opening Heads Install through flashing over all openings except those completely protected by overhanging projections. At steel lintels, place flashing under and behind the facing material and over the top of the lintel and bend its outer edge down to form a drip. (See Figure 15-3.)

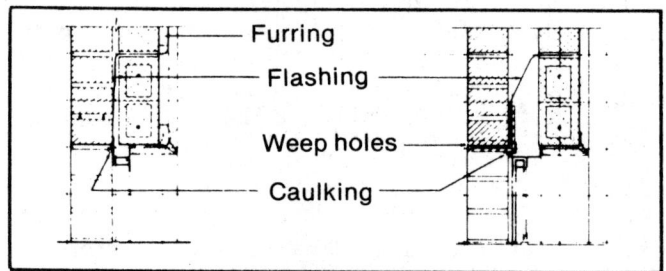

Flashing opening heads
Figure 15-3

Spandrels In skeleton-frame structures, spandrels may be flashed continuously at beams or with reglet. When the entire spandrel is flashed, use two-piece flashings provided they are lapped at least 4 inches. When cavity walls are supported by concrete spandrel beams, place the flashing on the shelf angle and extend it at least 8 inches up the beam and anchor it into a reglet. When galvanized or stainless steel-covered angles are used, flashing may not be necessary except to cover between lengths. (See Figure 15-4.)

Projections and Recesses Projections and recesses tend to hold rainwater and snow. They should have a top slope for drainage and, if possible, a drip to keep the water away from the wall surface below. Place flashing over the top of projections and recesses, with the outer edge bent down to form a drip and the back edge turned up at the inner face of the wall. (See Figure 15-5.)

Roof Flashing Because roof flashing occurs at very

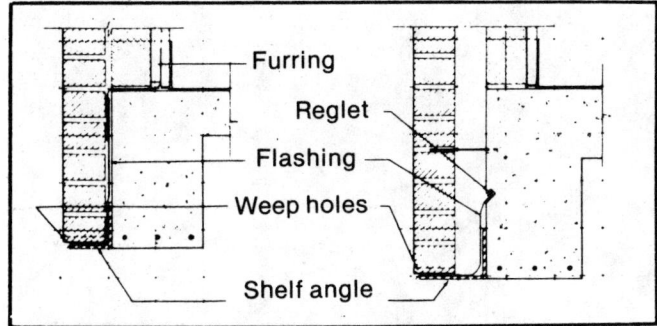

Spandrel flashing
Figure 15-4

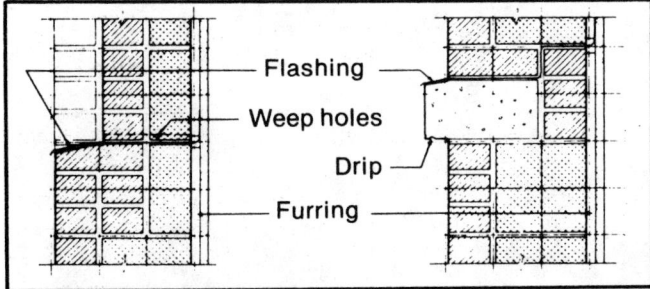

Flashing projections and recesses
Figure 15-5

vulnerable points, design and install it with great care. Base-flashing design depends upon the type of roofing material. Where the base flashing is metal, the counter flashing should also be metal, extending into the wall and overlapping the base metal flashing. At chimney walls, embed counter flashings securely in mortar joints. (See Figures 15-6 and 15-7.)

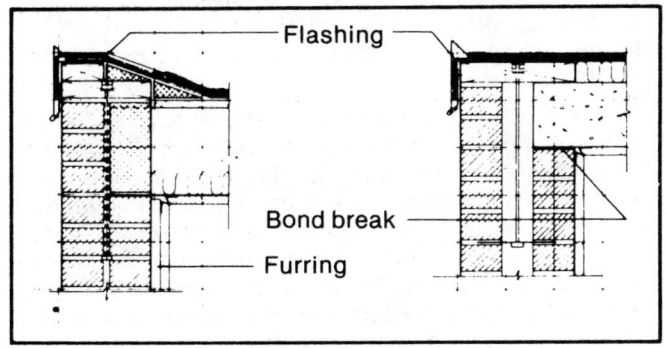

Roof flashing
Figure 15-6

Parapet Walls Face both sides of parapet walls with the same durable masonry materials. Using an inferior material on the side not exposed to view is a frequent cause of future trouble. Do not paint or coat the backs of parapet walls; they must be left free to dry rapidly.

Unless coping is impervious with watertight joints,

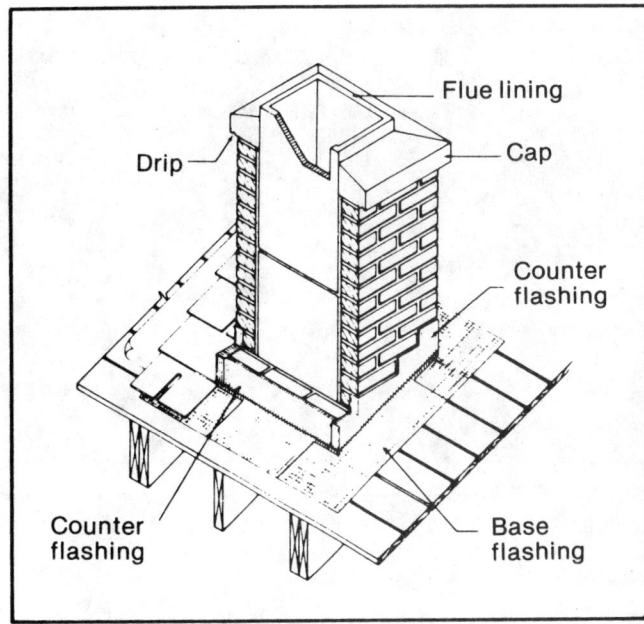

Chimney flashing
Figure 15-7

place through flashing in the mortar bed beneath it. Where the coping provides an adequate drip, the flashing may stop at the wall surface. However, where the copings are flush with the surface of the wall, extend the flashing at least 1/4 inch on both sides and turn it down to provide a drip. Follow the same procedure when topping out piers, pilasters, etc. (See Figure 15-8.)

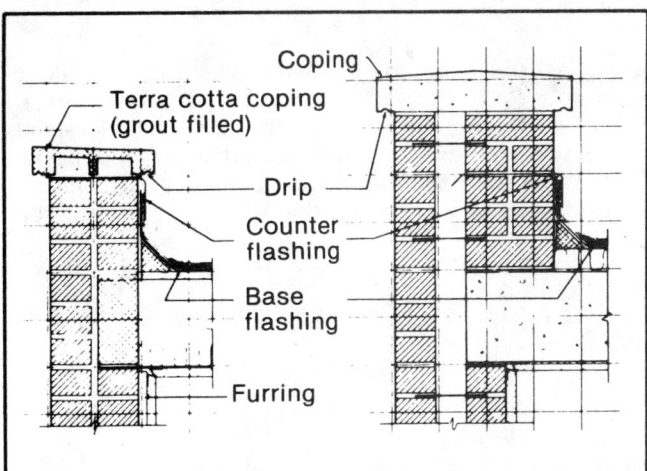

Flashing copings of parapet walls
Figure 15-8

Weep Holes A weep hole is an opening at the level of the flashing that allows the moisture to escape from inside the wall. (See Figure 15-9.) Weep holes are usually made at every third brick. A product that is manufactured for weep holes is simply a plastic tube slightly longer than the width of the brick. It is inserted in the

Masonry & Concrete Construction

**Weep hole
Figure 15-9**

bottom of the weep hole joint to make a more reliable weep hole.

All flashing must be drained to the outside. Tests at the National Bureau of Standards indicate that concealed flashings in tooled mortar joints are not self-draining without weep holes. Rather, they serve as a trap to collect moisture. Every 24 inches, provide weep holes in head joints immediately above all flashings. When building, keep weep holes free of mortar droppings. When wicks of 1/4-inch fiberglass rope or similar materials are used to prevent staining, place weep holes every 16 inches.

Chapter 16
Chimneys and Fireplaces

For hundreds of years the fireplace has been an important feature in homes. Today, a few fireplaces serve as primary sources of heat, but they continue to have an appeal that goes far beyond their utility. And with the rising cost of heating oil, the fireplace offers a way to reduce home-heating expenses. Proper construction of fireplaces and chimneys is essential to ensure their effective operation and safety in the event of a chimney fire.

CHIMNEYS

Flue
The flue is the passage in the chimney through which air, gases, and smoke travel. (See Figure 16-1.) The size (area), shape, height, tightness, and smoothness of the interior determine how effective the chimney is. All of these have an effect on the draft and the release of hot smoke and gases. Overheated or defective flue linings are a major cause of house fires, especially now when so many people are turning to wood for heat.

Most stove manufacturers recommend the size flue to use and offer suggestions for safe hookups.

Flue Linings Chimneys are sometimes built without clay flue linings. This was the way they were built years ago. It has been proven, however, that a chimney with a liner is safer and more efficient.

Lined flues are definitely recommended for use with brick chimneys. When the flue isn't lined, the mortar and the bricks that are exposed to the action of the flue gases disintegrate. This disintegration in addition to that caused by temperature changes can open cracks in the masonry, which reduces draft and increases the fire hazard.

Flue linings must be able to withstand rapid fluctuations in temperature and the action of flue gases. Therefore, they should be made of vitrified fire clay at least 5/8 inch thick. The ASTM specification for Clay Flue Linings C315-78c, defines acceptable flue linings as rectangular, nonmodular, rectangular modular, round or oval.

Rectangular flues are better suited for brick construction. However, a round flue lining is more efficient.

Place each length of lining in position, set in fire clay or strong cement mortar, and strike the inside smooth. In brick construction, lay the lining first and then lay the brick around it. In chimney block construction, lay the blocks one or two ahead of the tile. In brick construction, placing the lining after the brick are laid can cause an air leak—a potential fire hazard. In masonry chimney construction with walls less than 8 inches thick, leave a space between the lining and the chimney walls. Do not fill the space solid with mortar. Use only enough mortar to make good joints and hold the lining in position.

Unless it rests on solid masonry at the bottom of the flue, support the lower section of lining on at least three sides by a brick corbelled to the inside of the chimney.

Always lay flues as plumb as possible. If a projection makes this impossible, then the angle of curve around the projection should never exceed 30 degrees. (See Figure 16-2.) Sharp turns in the flue lining set up eddies which can affect the motion of gas and smoke. When a flue changes directions, make the lining joints as tight as possible by mitering the ends of the tiles. The best way to cut a clay flue lining is with a saw. A diamond blade is best but a small cutoff blade and an electric handsaw will work. Cut the lining before it is built into the chimney; if cut afterwards, it might break and fall into the chimney and be very hard to remove. Another way to cut a lining is to fill the tile full of damp sand. Then tap around the tile with a sharp chisel until the tile breaks.

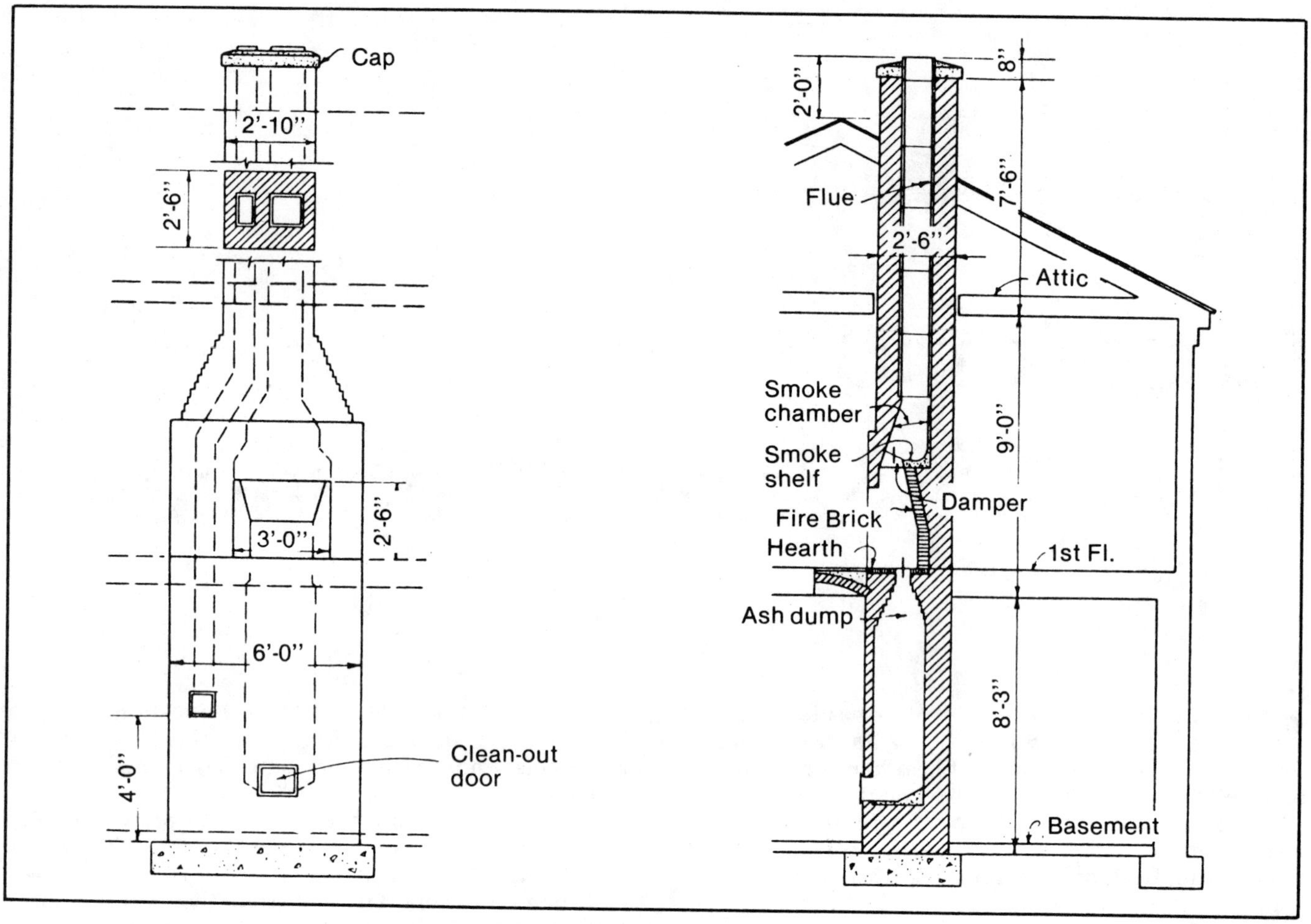

Masonry & Concrete Construction

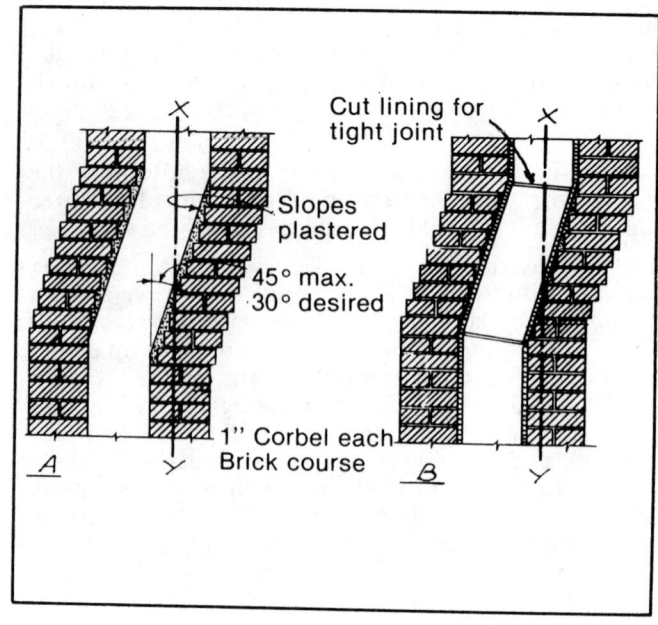

Typical chimney design
Figure 16-1

Offset in a chimney
Figure 16-2

As you lay the tile in either a brick or a block chimney, clean the inside of the chimney as the work progresses to catch the extra material in the inside and to keep the inside as smooth as possible.

A chimney can contain more than one flue. Check the building codes first for this. Building codes generally require a separate flue for each fireplace, furnace, or boiler. If a chimney contains three or more lined flues, separate each group of two flues from the other single flue or group of two flues by a brick division. (See Figure 16-3.) Stagger the lining joints of two flues grouped together without a dividing wall at least 7 inches and completely fill the joints with mortar. If the chimney should contain two or more unlined flues, separate them with a well-bonded wythe at least 8 inches thick.

Height

A chimney should extend about 36 inches above a flat roof and about 24 inches above a peak of a pitched roof. If the chimney is not near the peak, it should be at least 24 inches above any part of the roof that is within 10 feet. A cap or hood should be installed if a chimney can't be built high enough above the ridge of a roof to

Chimneys and Fireplaces

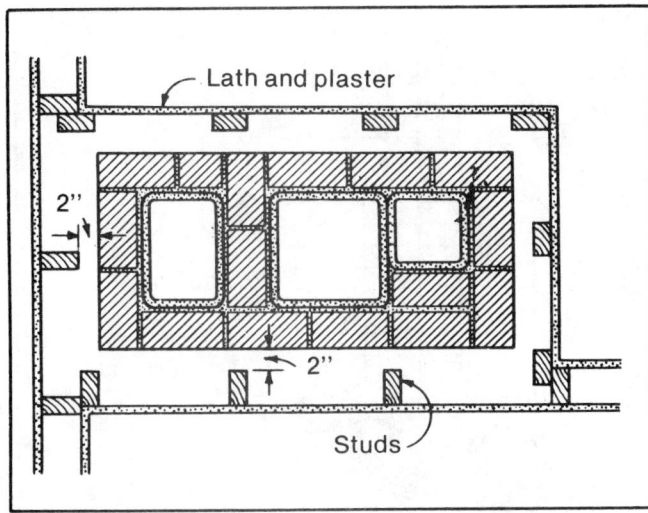

Multiple flue arrangement
Figure 16-3

prevent the wind from deflecting into the chimney. The open end of the hood or cap should be parallel with the ridge of the roof.

Support
A chimney is usually the heaviest part of a building. Always build it on a solid foundation to prevent differential movement in the building. Use concrete footings and design them to distribute the load of the chimney over an area wide enough to avoid exceeding the safe load-bearing capacity of the soil. They should be at least 12 inches thick and should extend about 8 inches past the outer edges of the chimney on all sides.

Construction
Masonry must be relatively smooth and free of projections which might puncture the flashing and destroy its effectiveness. For best results, place through-wall flashing on a thin bed of mortar and place another thin bed of mortar on top to bond the next course.

Thoroughly bond flashing seams in a manner which prevents water penetration. Although most copper sheet metal flashings can be soldered, lockslip joints are required at intervals to permit thermal expansion and contraction. Many plastics can be permanently and effectively joined by heat or appropriate adhesive. The elastic pliability of plastic flashings eliminates the need for expansion and contraction seams.

Effect of Flashing on Wall Strength
The effect of flashing on wall strength depends on mortar placement and bond to both flashing and masonry. Although through-wall flashing does not affect the compressive strength of a wall, it reduces flexural and shearing strengths.

If flashing is placed directly on masonry with no mortar beneath it, the flexural strength is zero at that point. For this condition, the shearing resistance depends on friction. The coefficient of friction between copper

When possible, pour the chimney or fireplace footing at the same time as the foundation
Figure 16-4

flashing and masonry is probably in the order of 0.25 to 0.50.

Limited test data indicate that, where mortar is placed immediately above and below copper flashing, flexural strength is about 30 to 70 percent less than for unflashed walls. There is not sufficient data to permit a similar generalization for shearing strength under similar conditions.

The important consideration is that through-wall flashing does not completely eliminate the resistance of a wall to lateral forces. Fortunately, through-wall flashing generally occurs where flexural resistance is least significant. When the wall ends are restrained, the "arching action" phenomenon assists in resisting bending.

When constructing a new foundation, it is sometimes possible to pour the footing for the fireplace or chimney at the same time. (See Figure 16-4.) It should be about 12 inches thick and reinforced.

If there is no basement in the building, pour the footing for the chimney at a point below the frost line. (The frost line can be obtained from a Cooperative Extension Office or the Department of Agriculture.)

If the house is built of solid masonry (the wall at least 12 inches thick), the chimney can be built integrally with the wall and, instead of carrying it to the ground, it can be offset enough from the wall to provide space for the flue by corbelling. Do not extend the offset more than 6 inches from the face of the wall—each course projecting not more than 1 inch—and do not construct it more than 12 inches high. This type of construction should be limited to single-story construction.

Chimneys in frame buildings should be built from the ground up, or they can rest on the building foundation or basement walls if the walls are at least 12 inches thick and have adequate footings.

Before beginning the construction of a chimney, check local building codes.

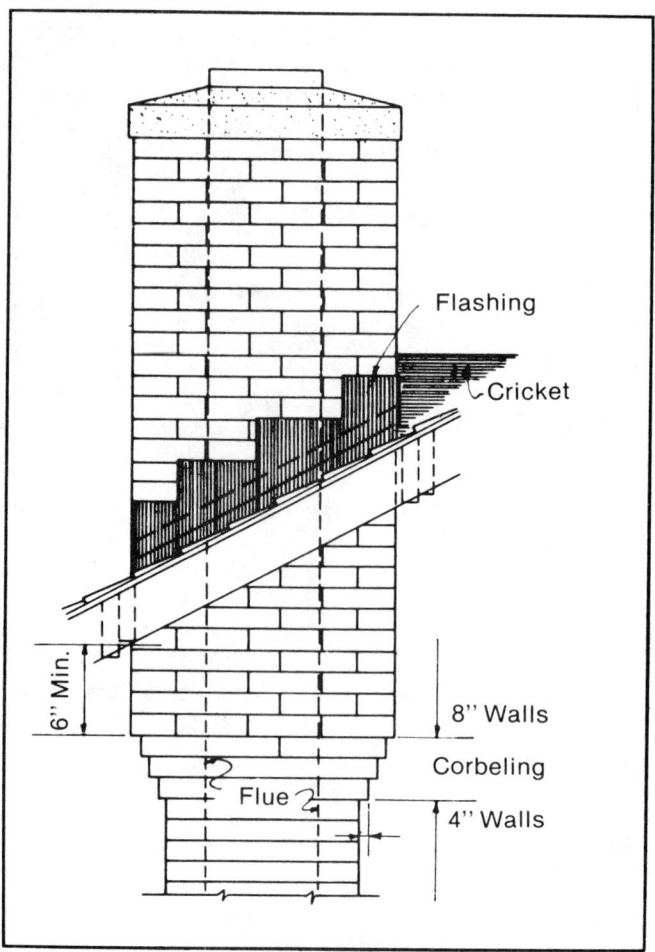

Corbelling of chimney to provide 8-inch walls for the outdoor section
Figure 16-5

Walls

Walls of chimneys with lined flues and not more than 30 feet in height should be at least 4 inches thick if made of brick and at least 12 inches thick if made of stone.

A minimum thickness of 8 inches is recommended for the outside wall of a chimney exposed to the elements. Brick chimneys that extend up through a roof may sway enough in heavy winds to open up mortar joints at the roof line. Openings to the flue at this point are dangerous because sparks from the flue may start fires in the woodwork or roofing. A good practice is to make the chimney wider through corbelling. Make the upper walls 8 inches thick by starting to offset the bricks at least 6 inches below the underside of roof joists or rafters. (See Figure 16-5.)

Soot Pocket and Cleanout

Provide a soot pocket and a cleanout for each flue. (See Figure 16-6.) Deep soot pockets permit the accumulation of an excessive amount of soot which may catch fire. Therefore, make the pocket just deep enough to permit installation of a cleanout door below the smoke-

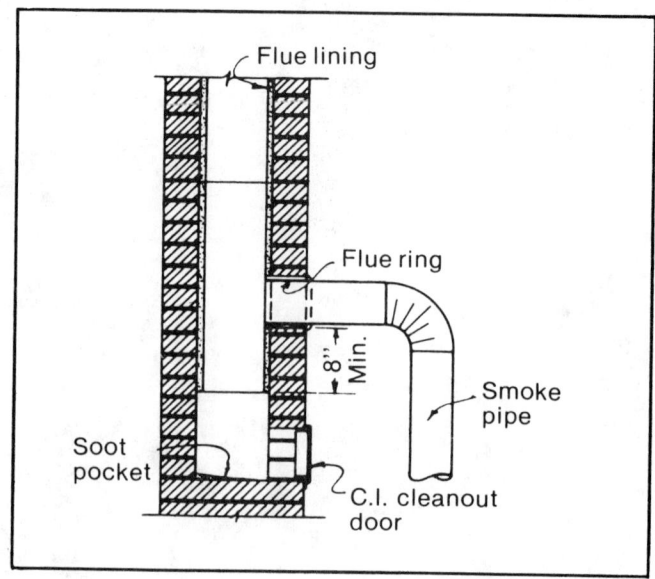

Soot pocket and cleanout
Figure 16-6

pipe connection. Fill the lower part of the chimney—from the bottom of the soot pocket to the base of the chimney—with solid masonry.

The cleanout door should be made of cast iron and should fit snugly and be kept tightly closed to keep air out. A cleanout should serve only one flue. If two or more flues are connected to the same cleanout, air drawn from one to another will affect the draft in all.

Smoke Pipe Connection

Do not connect any range, stove, fireplace, or other equipment to the flue for the central heating unit. In fact, as previously indicated, each unit should be connected to a separate flue, because if there are two or more connections to the same flue, fire may occur from sparks passing into one flue opening and out through another.

A smoke pipe should enter the chimney horizontally and should not extend into the flue. (See Figure 16-6.) Line the hole in the chimney wall with fire clay, or build metal thimbles tightly into the masonry.

The thimble (same material as the flue) should be set level into the flue. Enclose the area surrounding the thimble with brick and fire clay used to cement the thimble to the flue lining. (See Figure 16-7.) To be extra safe, lay fire brick in fire clay around the thimble.

Figure 16-8 shows the inside of the flue lining with the thimble cemented to the lining with fire clay. Note how the thimble does not stick into the chimney. Smooth the fire clay around the thimble for more efficient operation and a better seal.

Many building-supply dealers have flue tile that already have the hole cut for the thimble. (See Figure 16-9.) This is a great timesaver and moneysaver (eliminating breaking tile when cutting). However, it does require more planning. The flue tile has to work out at the right height for the thimble. If you drill a hole through the wall at the center of the thimble and measure down from this point it should make figuring easier.

When building a block chimney in an existing structure, you will have to cut through the soffit and fascia on the edge of a roof. Cutting through the roof is quicker and easier if a chain saw is used. (See Figure 16-10.) However, a chain saw is very dangerous to operate. Never cut with the saw higher than your head. Draw plumb lines on the siding and cut the hole just before you are to lay the blocks.

Connection with Roof

Where the chimney passes through the roof, provide a 2-inch clearance between the wood framing and the chimney for fire protection and to permit expansion owing to temperature changes, settlement, and slight movement during heavy winds.

Flash and counterflash chimneys to make the junction with the roof watertight. (See Figures 16-11 and 16-12.) When the chimney is located on the slope of a roof, build a cricket (Figure 16-13) high enough to shed water around the chimney. Use corrosion-resistant

Thimble set level into the flue, with fire brick and fire clay
Figure 16-7

Inside of flue lining showing entrance of thimble
Figure 16-8

Flue tile with pre-cut thimble hole
Figure 16-9

A chainsaw may be used to cut through the roof to build a chimney
Figure 16-10

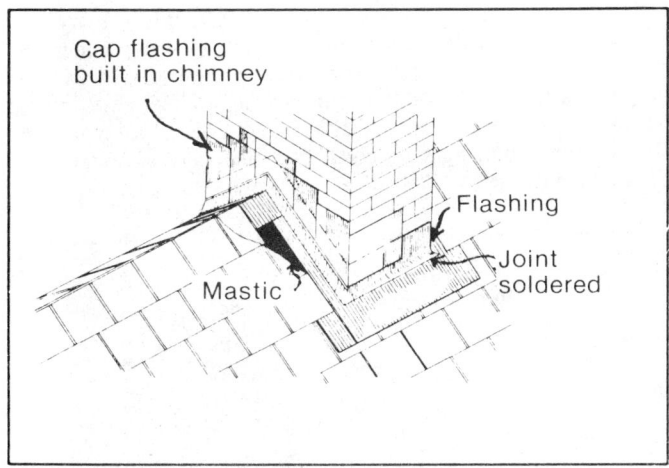

Flashing at a chimney located on a roof ridge
Figure 16-11

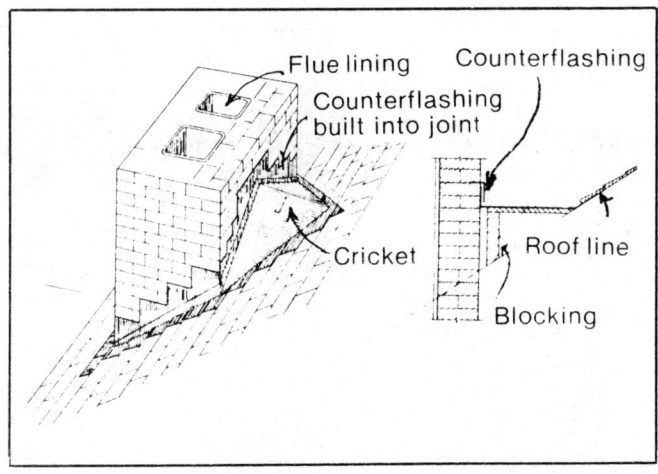

Counterflashing and cricket behind a chimney
Figure 16-12

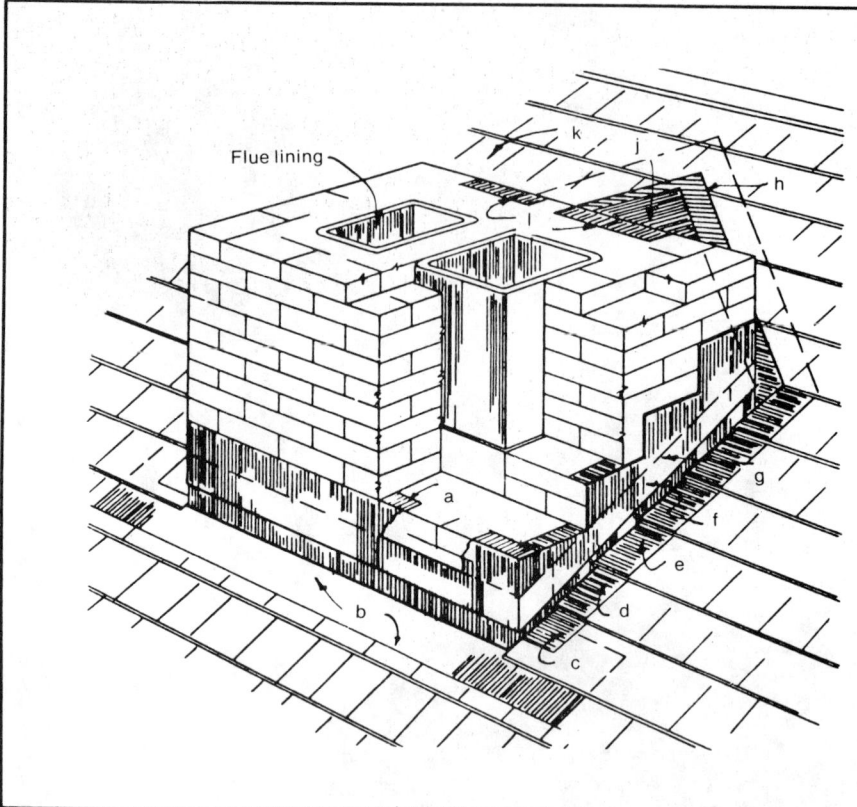

Flashing for a chimney located on the slope of a roof
Figure 16-13

metal, such as copper, zinc, or lead for flashing. Galvanized or tinned sheet steel requires occasional painting.

Top Construction

Figure 16-14(A) shows a good method of finishing the top of the chimney. The flue lining extends at least 4 inches above the cap or top course of brick and is surrounded by at least 2 inches of cement mortar. Finish the mortar with a straight or concave slope to direct air currents upward at the top of the flue and to drain water from the top of the chimney. (See also Figure 16-15.)

Hoods (Figure 16-14(B)) are used to keep rain out of chimneys and to prevent downdrafts caused by nearby buildings, trees, or other tall objects. Common types are the arched brick hood and the flat stone or cast-concrete cap. If the hood covers more than one flue, divide it by wythes so that each flue has a separate section. The area of the hood opening for each flue must be larger than the area of the flue.

Use spark arresters (Figure 16-14 (C)) when burning fuels such as sawdust that emit sparks, or when burning paper or other trash. They may be required when chimneys are on or near combustible roofs, woodland, lumber, or other combustible materials. Do not use them when burning soft coal, because they can become plugged with soot.

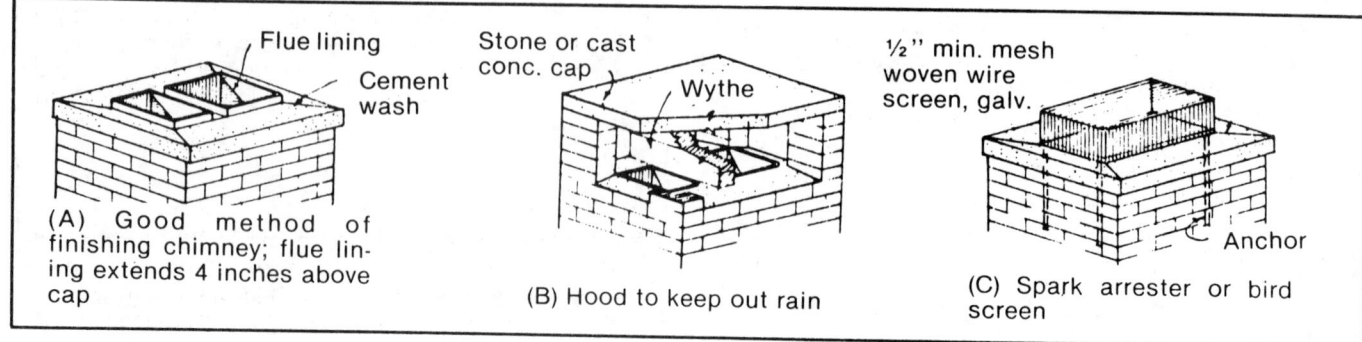

Top construction of chimneys
Figure 16-14

Finished block chimney with a wash around the top to shed water
Figure 16-15

Spark arresters do not entirely eliminate the discharge of sparks, but if properly built and installed, they greatly reduce the hazard. They should be made of rust-resistant material and should have openings not larger than 5/8 inch nor smaller than 5/16 inch. Commercially made screens that generally last for several years are available. Enclose the flue discharge area completely and fasten it securely to the top of the chimney.

FIREPLACES

Design

Varied fireplace designs are possible. Commercial publications frequently feature articles on fireplace design and construction. A fireplace should harmonize in detail and proportion with the room in which it is located, but do not sacrifice safety and utility for appearance.

Location of the fireplace within a room depends on the location of the existing chimney or the best location from the standpoint of safe construction for the proposed chimney. Do not locate a fireplace near doors.

Fireplace openings are usually made from 2 to 6 feet wide. The kind of fuel to be burned can suggest a practical width. For example, where cordwood (4 feet long) is cut in half, an opening 30 inches wide is desirable.

The height of the opening can range from 18 inches for an opening 2 feet wide to 28 inches for a fireplace up to 6 feet wide. The higher the opening, the more chance of a smoky fireplace.

In general, the wider the opening, the greater the depth. A shallow opening throws out relatively more heat than a deep one, but it holds smaller pieces of wood. You have the choice, therefore, between a deeper opening that holds larger, longer-burning logs and a shallower one that takes smaller pieces of wood, but throws out more heat. In small fireplaces, a depth of 12 inches may permit good draft, but a minimum depth of 16 inches is recommended to lessen the danger of sparks falling onto the floor. Place suitable screens in front of all fireplaces to minimize the danger of sparks.

Second-floor fireplaces are generally made smaller than those on the first floor because of the reduced flue height.

Construction

Fireplace construction is basically the same regardless of design. Figure 16-16 shows the construction of a typical fireplace. Table 16-1 gives the recommended dimensions for essential parts or areas of fireplaces of various sizes.

Footings Foundation and footing construction for chimneys with fireplaces is similar to that for chimneys

Masonry & Concrete Construction

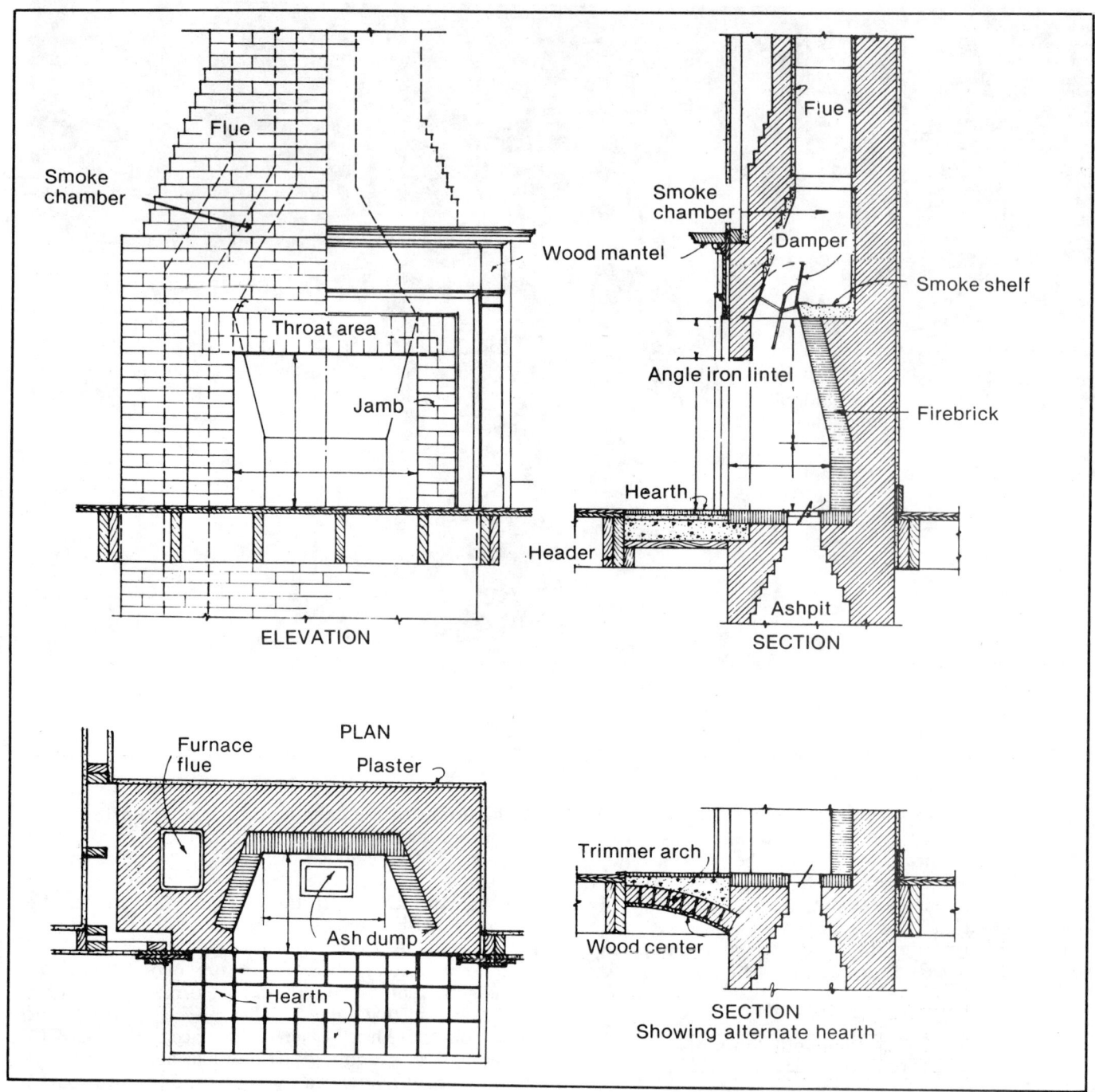

Construction details of a typical fireplace
Figure 16-16

without fireplaces. Always be certain that the footings rest on firm ground since the chimneys are very heavy.

Hearth Make the fireplace hearth of brick, stone, terracotta, or reinforced concrete at least 4 inches thick. It should project at least 20 inches from the chimney breast and should be 24 inches wider than the fireplace opening (12 inches on each side).

The hearth can be flush with the floor so that sweepings can be brushed into the fireplace, or it can be raised. Raising the hearth to various levels and extending the length as desired is presently common practice, especially in contemporary design. If there is a basement, a convenient ash dump can be built under the back of the hearth. (See Figure 16-17.)

In buildings with wooden floors, support the hearth in front of the fireplace by a masonry trimmer arch or

188

Chimneys and Fireplaces

Size of fireplace opening						Size of flue lining required	
Width w inches	Height h inches	Depth d inches	Minimum width of back wall c inches	Height of vertical back wall a inches	Height of inclined back wall b inches	Standard rectangular (outside dimensions) inches	Standard round (inside diameter) inches
24	24	16-18	14	14	16	8½ x 8½	10
28	24	16-18	14	14	16	8½ x 8½	10
30	28-30	16-18	16	14	18	8½ x 13	10
36	28-30	16-18	22	14	18	8½ x 13	12
42	28-32	16-18	28	14	18	13 x 13	12
48	32	18-20	32	14	24	13 x 13	15
54	36	18-20	36	14	28	13 x 18	15
60	36	18-20	44	14	28	13 x 18	15
54	40	20-22	36	17	29	13 x 18	15
60	40	20-22	42	17	30	18 x 18	18
66	40	20-22	44	17	30	18 x 18	18
72	40	22-28	51	17	30	18 x 18	18

Recommended dimensions for fireplaces and size of flue lining required
Table 16-1

other fire-resistant construction. (See Figure 16-16.) Remove wood centering under the arches used during construction when construction is completed. Figure 16-18 shows the recommended method of installing floor around the hearth.

Walls Building codes generally require that the back and sides of fireplaces be constructed of solid masonry or reinforced concrete at least 8 inches thick and that they be lined with firebrick or other approved noncombustible material not less than 2 inches thick or steel lining not less than 1/4 inch thick. Such lining may be omitted when the walls are of solid masonry or reinforced concrete at least 12 inches thick.

Jambs The jambs of the fireplace should be wide enough to provide stability and a pleasing appearance. For a fireplace opening 3 feet wide or less, the jambs can be 12 inches wide if a wood mantle will be used or 16 inches wide if they will be of exposed masonry. For wider fireplace openings, or if the fireplace is in a large room, the jambs should be proportionately wider. Fireplace jambs are frequently faced with ornamental brick or tile.

Do not place woodwork within 6 inches of the fireplace opening. Woodwork above and projecting more than 1½ inches from the fireplace should be placed not less than 12 inches from the top of the fireplace opening.

Lintel Install a lintel across the top of the fireplace opening to support the masonry. For fireplace openings 4 feet wide or less, use 1/2" by 3" flat steel bars or 3½" by 3½" by 1/4" angle irons or specially designed damper frames. Wider openings require heavier lintels. If a masonry arch is used over the opening, the fireplace jambs must be heavy enough to resist the thrust of the arch.

Throat Proper construction of the throat area (Figure 16-16 is essential for a satisfactory fireplace. The sides

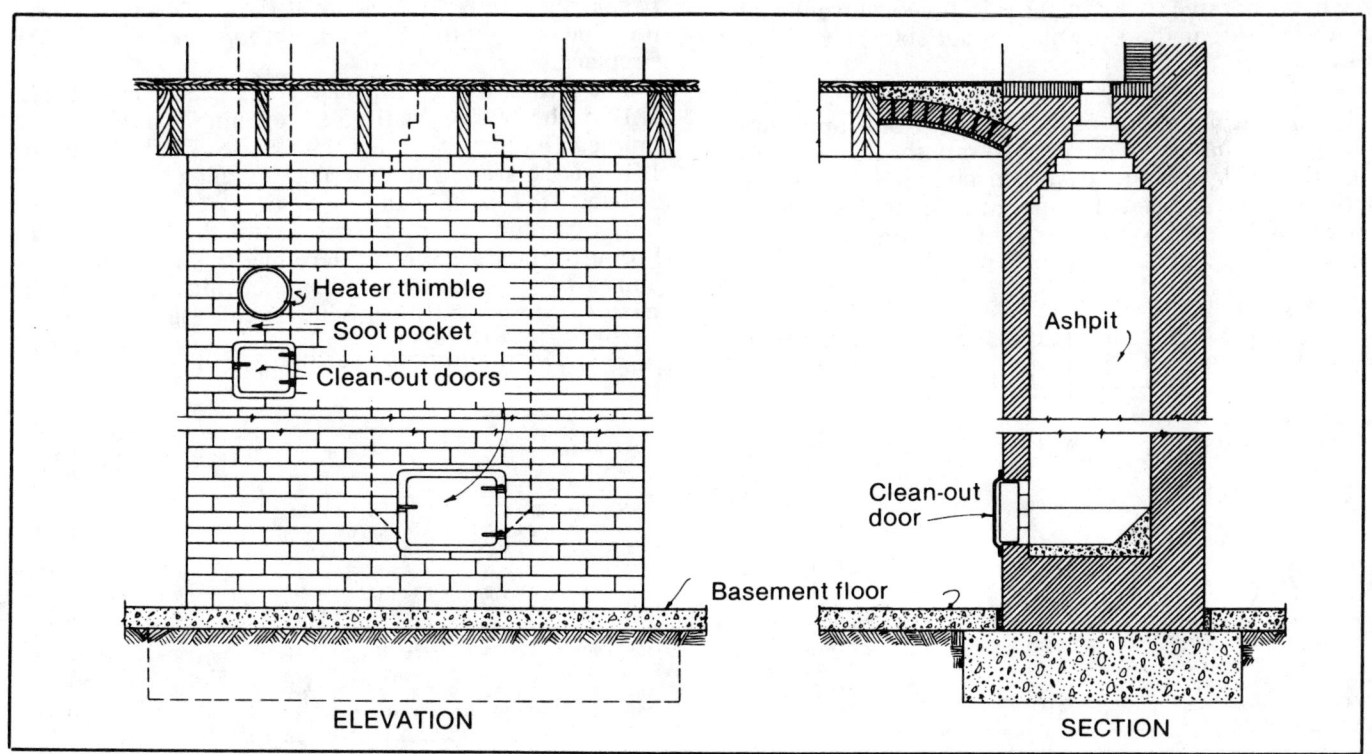

Typical ashpit construction
Figure 16-17

Masonry & Concrete Construction

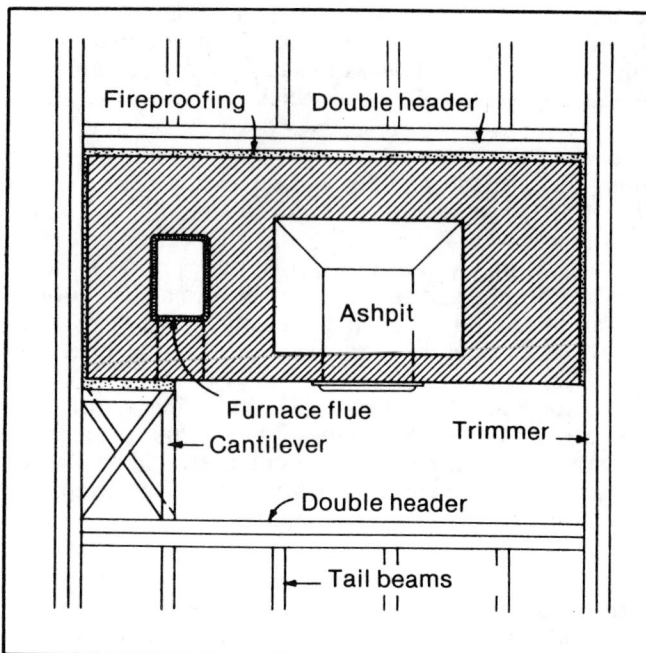

Installation of floor around a chimney and hearth
Figure 16-18

of the fireplace must be vertical up to the throat, which should be 6 to 8 inches or more above the bottom of the lintel. The area of the throat must not be less than that of the flue. The length must be equal to the width of the fireplace opening and the width depends on the width of the damper frame (if a damper is installed). Five inches above the throat the sidewalls should start sloping inward to meet the flue.

Damper A damper consists of a cast-iron frame with a hinged lid that opens or closes to vary the throat opening. Dampers are not always installed, but they are definitely recommended, especially in cold climates. With a well-designed, properly installed damper, you can:

1. Regulate the draft.
2. Close the flue to prevent loss of heat from the room when there is no fire.
3. Adjust the throat opening to the type of fire to reduce the loss of heat. For example, a roaring pine fire may require a full-throat opening, but a slow-burning hardwood log fire may require an opening of 1 or 2 inches. Closing the damper to that opening reduces heat loss up the chimney.
4. Close or partially close the flue to prevent loss of heat from the main heating system. When air heated by the furnace goes up the chimney, you waste fuel.
5. Close the flue in the summer to prevent insects and birds from entering the house.

Dampers of various designs are on the market. Some are designed to support the masonry over fireplace openings, thus replacing ordinary lintels.

Responsible manufacturers of fireplace equipment usually offer assistance in selecting a damper for a given-size fireplace. It is important that the full-damper opening equal the area of the flue.

Smoke Shelf and Chamber A smoke shelf (Figure 16-16) prevents downdraft. It is made by setting the brickwork at the top of the throat back to the line of the flue wall for the full length of the throat. Depth of the shelf may be 6 to 12 inches or more, depending on the length of the fireplace. The smoke chamber is the area from the top of the throat to the bottom of the flue. As indicated under "Throat" above, the sidewalls should slope inward to meet the flue. Plaster the smoke shelf and the smoke chamber with cement mortar at least 1/2 inch thick.

Flue Proper proportion between the size (area) of the fireplace opening, size (area) of the flue, and height of the flue is essential for satisfactory operation of the fireplace.

The area of a lined flue 22 feet high should be at least 1/12 of the area of the fireplace opening. The area of an unlined flue or a flue less than 22 feet high should be 1/10 of the area of the fireplace opening.

Table 16-1 lists dimensions of fireplace openings. The last two columns indicate the size of the flue required. From this table, you can determine the size of lining required for a given-size fireplace opening and also the size of opening to use with an existing flue.

Tables 16-2 through 16-5 show the different sizes of flue liners available and their dimensions.

Outside Dimensions, in. (mm)	Nominal Wall Thickness, in. (mm)	Outside Corner Radius, max. in. (mm)
4½ x 8½ (115 x 215)	⅝ (16)	1 (25)
4½ x 13 (115 x 330)	¾ (19)	1 (25)
8½ x 8½ (215 x 215)	¾ (19)	2 (50)
8½ x 13 (215 x 330)	⅞ (23)	2 (50)
8½ x 17¼ (215 x 450)	1 (25)	2 (50)
13 x 13 (330 x 330)	⅞ (23)	3 (75)
13 x 17¾ (330 x 450)	1 (25)	4 (100)
17¾ x 17¾ (450 x 450)	1¼ (32)	4 (100)
20 x 20 (500 x 500)	1⅜ (35)	5 (125)
20 x 24 (500 x 600)	1½ (38)	5 (125)
24 x 24 (600 x 600)	1⅝ (41)	6 (150)

Rectangular nonmodular clay flue linings - standard dimensions
Table 16-2

Outside Dimensions, in. (mm)	Nominal Dimensions, in. (mm)	Nominal Wall Thickness, in. (mm)	Outside Corner Radius, Max., in. (mm)
3½ x 07½ (90 x 190)	4 x 8 (100 x 200)	⅝ (16)	1 (25)
3½ x 11½ (90 x 290)	4 x 12 (100 x 300)	⅝ (16)	1 (25)
7½ x 7½ (190 x 190)	8 x 8 (200 x 200)	⅝ (16)	1 (50)
7½ x 11½ (190 x 290)	8 x 12 (200 x 300)	¾ (19)	2 (50)
11½ x 11½ (290 x 290)	12 x 12 (300 x 300)	¾ (19)	2 (50)
11½ x 15½ (290 x 390)	12 x 16 (300 x 400)	1 (25)	3 (75)
15½ x 15½ (390 x 390)	16 x 16 (400 x 400)	1⅛ (29)	4 (100)
15½ x 19½ (390 x 490)	16 x 20 (400 x 500)	1¼ (32)	4 (100)
19½ x 19½ (490 x 490)	20 x 20 (500 x 500)	1⅜ (35)	5 (125)
19½ x 23½ (490 x 590)	20 x 24 (500 x 600)	1½ (38)	5 (125)
23½ x 23½ (590 x 590)	24 x 24 (600 x 600)	1⅝ (41)	6 (150)

Rectangular modular clay flue linings - standard dimensions
Table 16-3

Chimneys and Fireplaces

Nominal Inside Diameter, in. (mm)	Permissible Variation in Inside Diameter, ± in. (± mm)	Nominal Wall Thickness, in. (mm)
6(150)	¼(6)	⅝(16)
8(200)	¼(6)	¾(19)
10(250)	⁵⁄₁₆(8)	⅞(23)
12(300)	⅜(10)	1 (25)
15(375)	⅜(10)	1⅛(29)
18(450)	⁷⁄₁₆(11)	1¼(32)
21(525)	⁷⁄₁₆(11)	1⅝(41)
24(600)	½(13)	1⅝(41)
27(675)	⁹⁄₁₆(14)	2 (50)
30(750)	⅝(16)	2⅛(54)
33(825)	¹¹⁄₁₆(17)	2¼(57)
36(900)	1¼(32)	2½(64)

Round clay flue linings - standard dimensions
Table 16-4

Outside Dimensions, in. (mm)	Nominal Wall Thickness in. (mm)	Nominal Outside Corner Radius, in. (mm)
8½, round (215), round	¾(19)	4¼(105)
12¾, round (325), round	1(25)	6⅜(160)
8½ by 12¾(215 by 325)	¾(19)	4¼(105)
8½ by 16¾(215 by 425)	1(25)	4¼(105)
10 by 17¾(250 by 450)	1(25)	5(125)
12¾ by 16¾(325 by 425)	1(25)	6⅜(160)
12¾ by 21(325 by 525)	1⅛(29)	6⅜(160)
16¾ by 16¾(425 by 425)	1(25)	6⅜(160)
16¾ by 21(425 by 525)	1³⁄₁₆(30)	6⅜(160)
21 by 21(525 by 525)	1¼(32)	6⅜(160)

Oval, clay flue linings - standard dimensions
Table 16-5

Chapter 17
Brick Floors and Pavements

Brick can be used to make attractive, long-lasting surfaces for floors and pavements. Care must be taken, however, in selecting materials and making the design practical and appropriate for the intended use.

Rigid Brick Paving
Rigid brick paving consists of units laid in a bed of mortar with subsequent mortar joints between the units. Conversely, flexible brick paving contains no mortar below or between the units.

Rigid-Base Diaphragm
A rigid-base diaphragm is a reinforced concrete slab on grade. Rigid or flexible brick paving may be placed over this type of base.

Semirigid Continuous Base
This type of base usually consists of asphalt or bituminous concrete road pavement. Flexible brick paving is suitable over this type of base.

Flexible Base
A flexible base consists of compacted gravel or a damp, loose, sand-cement mixture tamped in place. Place only flexible brick paving over this type of base.

Suspended Diaphragm
Suspended diaphragm bases are structural roof or floor deck assemblies, the composition of which varies depending upon design. Either flexible or rigid brick paving is suitable for this type of base.

DESIGN CONSIDERATIONS

Traffic
Generally, the first design consideration is to assess the applicable traffic loadings. Heavy vehicular loadings on grade generally require rigid-base diaphragms or semirigid continuous bases. Lighter vehicular traffic, such as for residential driveways, may be supported on flexible bases and flexible paving. Pedestrian traffic can be accommodated over any of the previously mentioned assemblies. Traffic patterns dictating the geometry and size of paved area directly influence the selection of base and cushion material.

Site
A site may involve anything from a small residential patio to a major urban-renewal project encompassing several city blocks. During the planning stages of large projects, give consideration to the location of underground utilities, storm drainage and user convenience.

Successful installations also depend on proper subgrade preparation. Remove all vegetation and organic material from the area to be paved. Remove soft spots containing poor subbase material and refill with suitable material properly compacted.

Drainage
Surface and subsurface drainage are of major importance. Generally, exterior brick paving should be sloped 1/8 to 1/4 inch per foot. Large paved areas for malls and vehicular parking require the larger value. Slope all paving away from buildings, retaining walls and other elements capable of collecting surface water. In areas susceptible to high water tables, use a porous base and cushion of gravel. This type of base serves as a capillary

break, preventing the upward flow of moisture from capillary action. Localities with relatively impervious soils capable of surface water retention may require subsurface drainage systems.

Edging
To prevent horizontal movement of mortarless brick paving units, provide a method of containment around the perimeter of the paved area. This may be a curb of brick soldier coursing set in concrete or mortar. An existing concrete curb, building or retaining wall is sufficient. Construct new edging prior to placement of the paving units.

Joints
There are three basic methods for installing brick paving with mortar joints.

The first method is by the conventional use of mortar and a trowel. Butter brick pavers with mortar and shove them into a leveling base of mortar.

A second method involves placing each unit on a mortar leveling bed leaving 3/8 to 1/2 inch space between the units. Pour a grout mixture into these spaces. Generally, grout proportions of portland cement and sand are the same as for mortar except that hydrated lime may be omitted. When grout is poured into the joints, take special care or protect the units to facilitate cleaning.

A third method involves a dry mixture of portland cement and sand, using the same proportions as for grout. Install brick pavers on a damp cushion of this mixture, and place the same mixture between the units. After cleaning excess material from the paving surface, spray the paving with a fine mist of water until the joints are saturated. Keep the pavement damp for a period of two or three days.

Brick paving without mortar joints may be swept with plain dry sand or a mixture of portland cement and sand. The proper proportions of portland cement and sand are discussed later in this chapter.

Expansion Joints
Give consideration to the potential for thermal and moisture movements, and provide expansion joints to accommodate these movements. A single expansion joint size is not feasible for all types of installations. However, the following location guidelines are offered for large areas. General locations for expansion joints: (1) parallel to curbs and edgings, (2) at 90 degree right turns, and (3) around interruptions.

In addition to expansion joints in the mortared brick surface paving, give consideration to movement joints in the supporting base, which may also be affected by moisture and temperature changes.

Membrane Materials
Membranes are installed for several purposes: (1) to control or reduce the passage of moisture, (2) to discourage weed growth, and (3) to serve as a separating layer to accommodate differential movement. To accomplish these tasks, membranes should be capable of resisting rot and decay. Generally, they are of sheet and liquid materials. Installed properly, fluid types have some advantages over sheet membranes mainly because they are seamless and will conform to irregular surfaces.

Exercise care during construction to avoid membrane damage. Although some membranes resist abrasion better than others, protection should be considered especially for roof-deck construction where resistance to moisture penetration is of primary importance.

Appearance and Esthetics
The visual impact of brick paving units results from the interplay of many factors: shape, size, color, texture, and pattern. Most brick paving consists of solid, uncored units placed flat. However, cored or uncored brick placed on edge are used as paving also.

Bond Patterns An endless variety of patterns can be achieved with brick paving. Therefore, it is important to specify the proper size of the unit, especially for pattern bonds. Two examples of pattern bonds are the "herringbone" and "basket weave" shown in Figure 17-1. Before attempting a paving layout, be familiar with the availability of various types of brick paving units. A recent survey of members of the Brick Institute of America indicates about 38 sizes and shapes of brick paving units are available. The three most common types are shown in Figure 17-2. These include rectangular, square, and hexagonal units. In addition, a few manufacturers make special and moulded shapes.

Color and Texture Brick pavers are available in many colors and textures. Red is the most popular color and is available in several shades. Other colors available include buff, gray and brown.

The texture of a unit affects its performance, installation and maintenance. Therefore, consider a few basic factors. Slip resistance is related to the texture of the unit. The coarser the texture, the better the slip resistance. For interior use, smooth units are more receptive to the application of sealers, coatings and waxes so they are easier to maintain and clean. Generally, smooth low-maintenance units are more desirable in high-traffic areas, such as lobbies and foyers. Coarse-textured units may be ground smooth in a fashion similar to terrazzo floors. Rough units are generally more suitable for exterior use where good slip resistance is desirable.

MATERIAL SELECTION
Brick
In selecting bricks, consider: (1) traffic to which the floor or paving is subjected, (2) exposure to moisture and freezing cycles, and (3) the desired appearance. Also, resistance to chemicals and acids is often a major criterion for industrial paving.

Resistance to Abrasion Resistance to wear and tear is generally associated with dense, hard-burned brick. For floors not subject to heavy wear, such as residential and most nonindustrial floors, a low-absorption, dense brick meeting or exceeding the requirements of ASTM Specification C 62, grade SW should prove satisfactory.

Brick Floors and Pavements

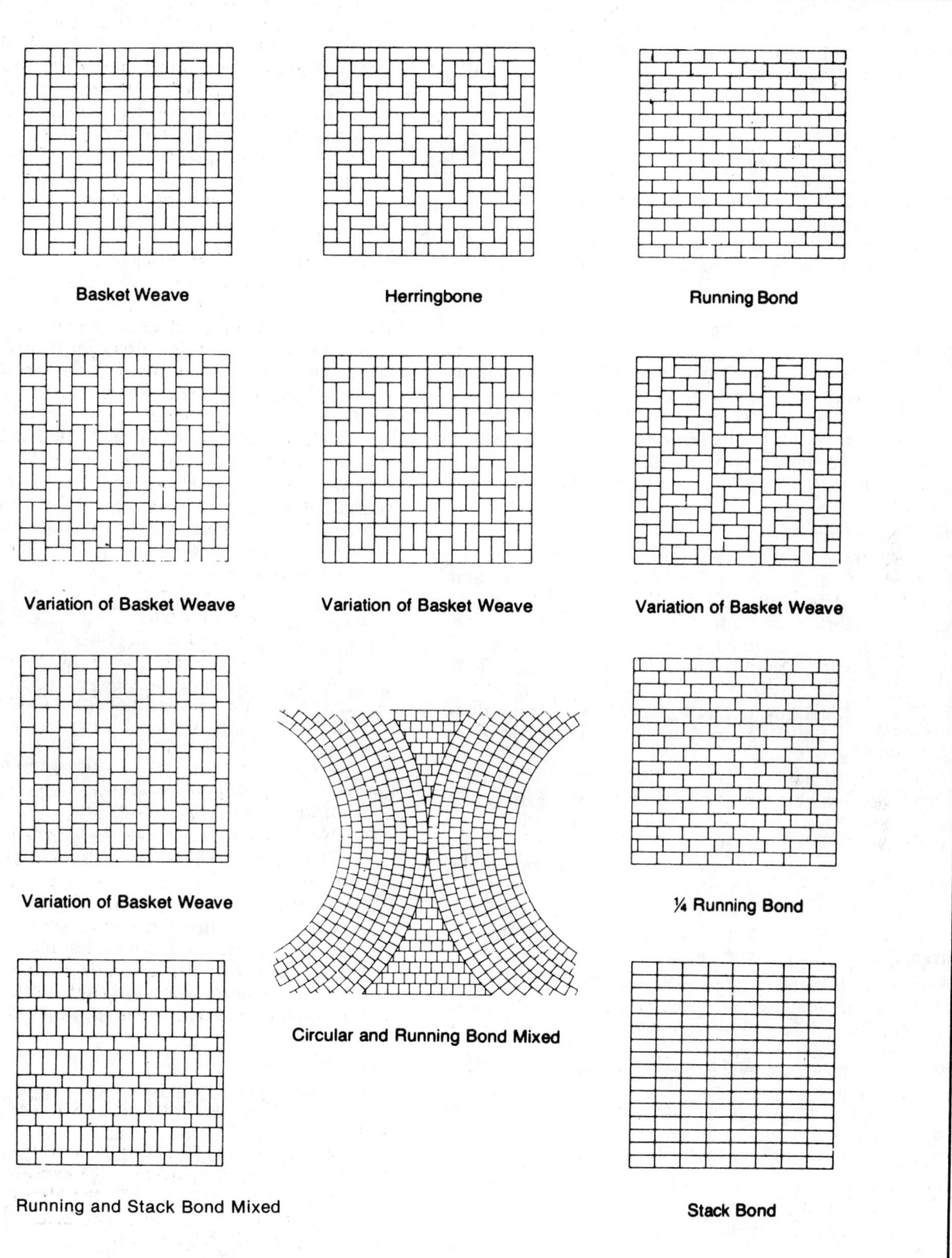

**Brick paving patterns
Figure 17-1**

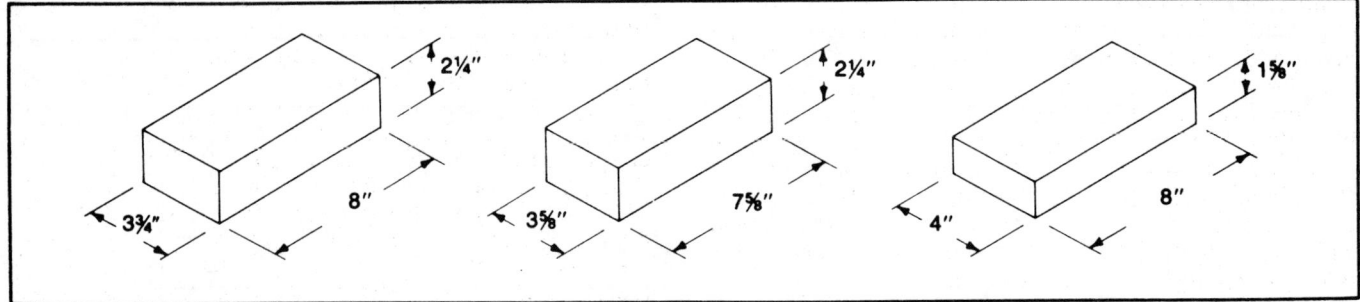

The most common brick paving units
Figure 17-2

In addition, the denser the brick and the lower its absorption, the more resistant it is to discoloring and staining. Consequently, floors of dense, low-absorption brick are more easily cleaned and maintained.

Resistance to Weathering For exterior pavements, resistance to freezing and thawing in the presence of moisture is perhaps more important than resistance to abrasive wear.

Brick Specification

Presently there does not exist any widely accepted standard specification for brick units for floors and pavements. Therefore, care must be exercised in the selection of brick units for this purpose.

In general, dark-colored, hard-burned brick units are associated with durability and good performance. This generalization, however, can be misleading since modern brickmaking technology permits the manufacture of many units that are both light in color and extremely durable in use.

For best results, base the selection of a brick for paving application on the past performance of the units in a similar use. This type of information can usually be obtained from the manufacturer of the brick.

Lacking the specific knowledge of performance, the following physical properties of extruded brick units and molded brick units should be considered minimum.

Extruded Brick Minimum average compressive strength of 8000 psi, maximum average cold-water absorption of 8 percent, and maximum saturation coefficient of 0.78.

Molded Brick Minimum average compressive strength of 4500 psi, maximum average cold-water absorption of 8 percent, and a maximum saturation coefficient of 0.78.

The above physical properties are purposely conservative. There are undoubtedly units that do not conform to these minimums that will perform very satisfactorily. It is therefore most important that performance data be used whenever possible.

If brick paving is placed over existing concrete, the concrete should be sound, with major cracks properly filled.

Cushion Material

Cushion material placed between the base and the paver functions as a leveling layer to help refine the finished grade and compensate for irregularities of the surface and of the units. It may be a 1- or 2-inch layer of sand, pea gravel, stone screenings, or even several layers of roofing felt. Under extremely wet conditions, avoid fine-particle cushions, such as sand or stone screenings, since drainage of moisture may be too slow.

Sand Sand for cushions, bases, joints, and mortar should be specified in accordance with ASTM C 144, Aggregate for Masonry Mortar. Use sand that is free of clay to avoid "scumming" when swept over the face of brick units.

Sand and Cement Mixtures Dry mixtures of sand and cement may be used as bases or cushions. One part portland cement may be mixed with three to six parts damp, loose sand. Bond is not usually achieved between the brick unit and the leveling bed when mixtures high in sand are used.

Roofing Felt Roofing felt, generally of 15- to 30-pound weight, may serve as a cushion between brick pavers and concrete or asphalt paving. This type of cushion can be installed rapidly and provides some compensation for minor irregularities between the surface of the base and the brick pavers. Roofing felt may also impart some resilience to the paving.

Used Brick Do not use salvaged brick for brick paving. The durability of masonry depends on the uniformity of the quality of materials and, generally speaking, used brick are not uniformly durable when exposed to weathering. They may spall, flake, pit, and crack because of freezing in the presence of moisture.

Base Material

Base materials generally have one or more purposes: (1) support, (2) drainage, and (3) ground-swell protection. Cushion materials function as the term implies. They also serve to establish fine grading requirements.

Gravel Bases For maximum drainage efficiency and the prevention of the upward flow of moisture due to capillary action, use clean, washed gravel. The size of the stones will depend on the installation. Pea gravel is self-compacting and can be screeded easily to a finished grade. Where a thick base build-up is desired for drainage purposes, a larger stone size may be more economical.

Bases consisting of unwashed gravel mixed with fine clay and stone dust are popular because they are economical. Referred to as "stone screenings," "bank run," or "bluestone," this material compacts well and is readily available. However, it does present a few problems such as the loss of porosity and drainage characteristics because of hardening from moisture. Brick in direct contact with this material may be susceptible to efflorescence, since moisture in combination with the stone dust particles can deposit water-soluble salts on the face of the brick paver.

Concrete Bases Both existing and new concrete bases may be used for brick paving. Install concrete following recommended concrete practices. If a mortar leveling bed is used, the slab surface should have a raked or floated finish to facilitate good bond. If noncementitious types of leveling beds or cushions are used, the slab need only be screeded, eliminating the need for further finishing.

Asphalt Base New or existing asphalt paving bases may be used to support brick paving. Mortar leveling beds are acceptable with asphalt bases; however, there will be little, if any, bond between the mortar and asphalt. Avoid placing mortar leveling beds on hot asphalt as flash setting of the mortar may occur. Repair major defects in existing asphalt pavement prior to installation of brick paving.

Setting Beds

Mortar setting or leveling beds may be used in conjunction with concrete and asphalt bases. A mortar bed for exterior paving should be a Type M portland cement and lime mortar. For interior applications, use Type S or N. The thickness of the bed may vary from 1/2 inch to 1 inch.

Other mortars using high-bond additives or latex-modified portland-cement mortars are also available. Mix and use these mortars in strict accordance with the manufacturer's recommendations and only after suitable testing or performance documentation. One important advantage of high-bond mortar is its bonding ability to some brick. Since high-bond mortar does not perform well with all brick, exercise caution. Use preproject testing to assure the compatibility of the brick and the mortar combination.

Types of mortar, their uses and specifications are:

1. Portland Cement-Lime Mortars—For exterior mortared paving on grade, use Type M (1:1/4:3) portland cement-lime. It has high durability and is specifically recommended for masonry in contact with the earth. For interior mortared paving, Type S (1:1/2:4¼) or Type N (1:1:6) portland cement-lime mortars are suitable. Use Type S mortar for both reinforced and unreinforced masonry where maximum tensile bond strength is required. It is also suitable for outdoor use, provided the brick slab is not in contact with the ground. Type N is a medium-strength mortar suitable for interior use. Do not use masonry cement mortars.

2. Dry-Mixed or Grout-Type Mortar—Sand and cement may be mixed dry and swept between the brick pavers without mortar joints. The pavement is then fogged down with water to set the mixture. Dry mixtures for exterior use should follow Type M mortar proportions. For interior use, use Type S or N mortar.

Some installers prefer a soupy, grout-type mixture to pour between the mortar joints. The basic difference between the grout and mortar ingredients is the omission of hydrated lime in the grout. To facilitate cleanup, coat the brick units with melted paraffin prior to laying. Thus, grout pours spilling on the face of the coated units will not stain the surface. Make sure the paraffin does not coat the joint side of the brick where grout bond is desirable. After all grouting has been completed, clean off the paraffin coating with a steam jenny. Hot-water hosing or cold-water under pressure are other ways to remove the paraffin coating.

3. Latex-Portland Cement Mortars—Latexes for cement mortar vary among manufacturers. Therefore, carefully follow the manufacturer's directions. Material and installation specifications are contained in ANSI A119.4-1973, "Specification for Latex-Portland Cement Mortar."

4. High-Bond Mortar—SARABOND (R) high-bond mortar additive is a liquid saran polymer that greatly enhances the bonding, compressive and tensile-strength characteristics of the resultant mortar. It permits the designer to take advantage of the higher tensile strength of brick masonry.

According to the Tile Council of America, latex mortars are somewhat more flexible than conventional mortars and have excellent water resistance and cleanup characteristics. This type of mortar is useful for mortared paving supported by a flexural structural system, such as wood-joist floor assembly.

Bituminous

Bituminous setting or leveling beds composed of aggregate and asphaltic cement may be used to support brick paving units. The mix is usually heated at an asphalt plant before being delivered to the job. Typical bases supporting bituminous setting beds are usually concrete slabs or asphalt pavements. Proportions of asphalt and aggregate are generally determined by the specialty contractor and are not explained here.

SUGGESTED BRICK PAVING DESIGN ASSEMBLIES

The supporting base for an on-grade brick paving assembly may be one of three types as discussed earlier; rigid, semirigid, and flexible. The paving assemblies included here are suggested methods, based on experience, for various types of usages. Figures 17-3 to 17-13 show assemblies that make use of all three types of base. Be certain to take the conditions and uses of the work into account when you determine the final design.

Mortared Pavement

This type of installation should be executed by a skilled mason or tile setter.

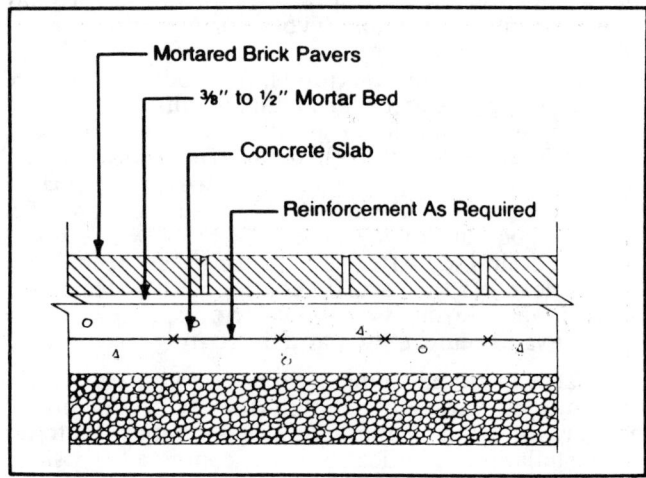

Mortared brick paving for virtually any type design condition for both interior and exterior applications
Figure 17-3

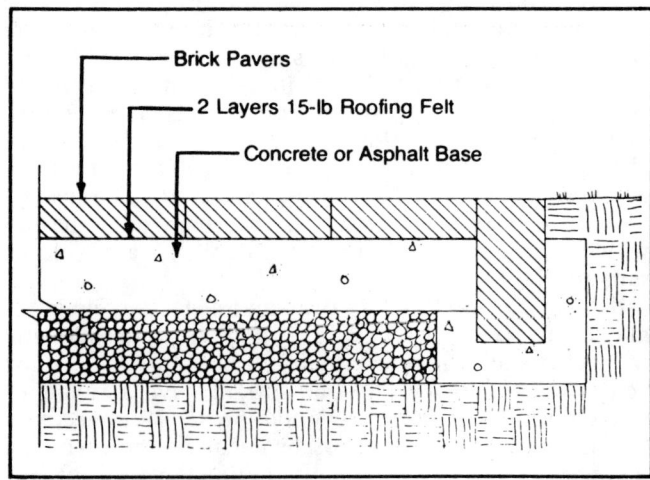

Existing concrete or asphalt base used with flexible brick pavers on an existing rigid or semi-rigid base of a residential driveway
Figure 17-5

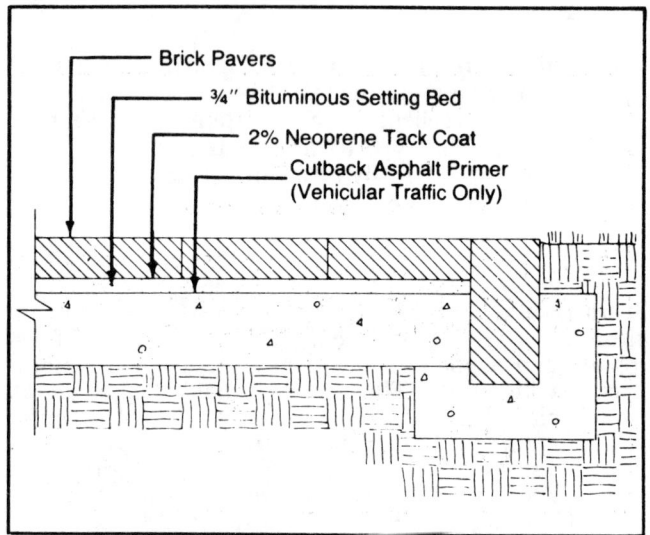

Concrete or asphalt base used with flexible brick paving for pedestrian malls or slow vehicular traffic
Figure 17-4

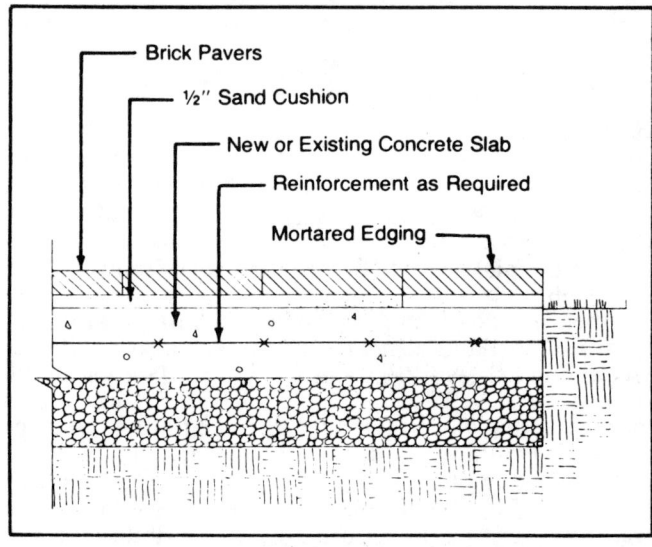

New or existing concrete base used with mortarless pavers on a residential patio
Figure 17-6

Bituminous Setting Bed

Bituminous setting beds are usually placed on rigid or semirigid bases as noted in Figure 17-4. Supporting bases are generally 4 to 6 inches thick, but the thickness of the base is a matter of design and should be sized to accommodate the traffic loading and dead weight of the brick pavers and other materials. Install brick pavers hand-tight on a bituminous setting bed. Apply a tack coat of 2 percent neoprene oxidized asphalt to the setting bed prior to laying brick pavers. The porous bituminous setting bed may consist of approximately 7 percent asphalt and 93 percent graded sand. Delivered hot from the plant, the bitumen material is rolled to a 3/4-inch depth. To accommodate light vehicular traffic, apply a primer coat of rapid-curing cut-back asphalt to the concrete slab or asphalt pavement.

Roofing-Felt Cushion

Where regular flat brick pavers are used, two layers of 15-pound roofing felt over a rigid or semirigid base can suffice as a cushion.

Consider using depressed condrete slabs to avoid an abrupt change in floor finishes (Figure 17-11). An exaggerated vertical projection on one unit above the other can occur as a result of overlapping of felt paper. Avoid this by abutting the edges before the pavers are laid.

Brick Floors and Pavements

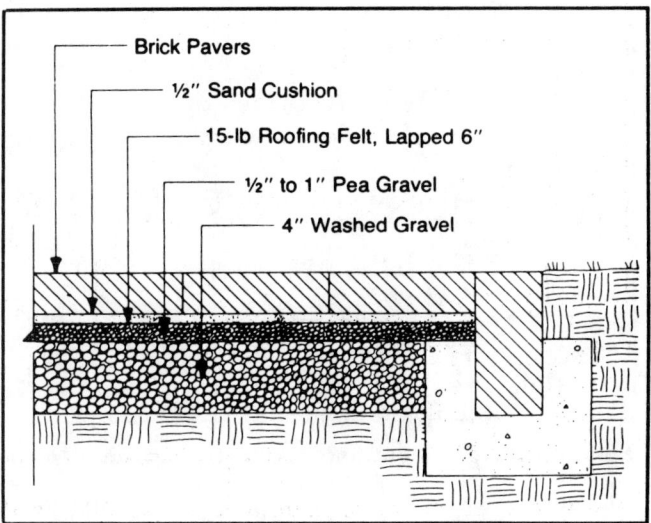

Gravel base for a residential patio in areas that experience severe precipitation, subsurface drainage problems, and high ground water tables
Figure 17-7

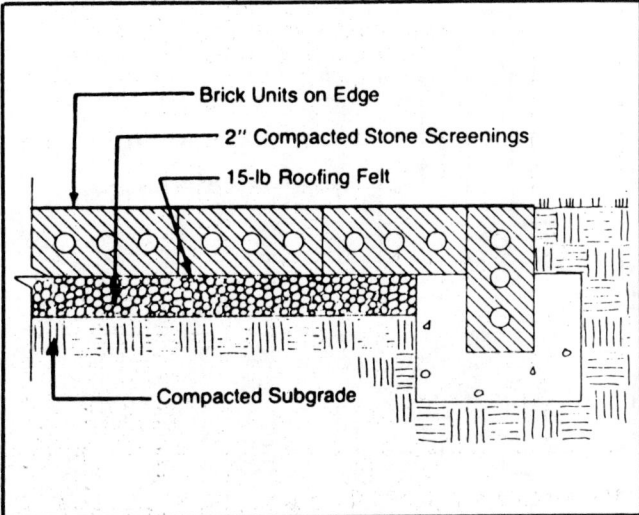

Alternative gravel base suitable for a residential driveway
Figure 17-9

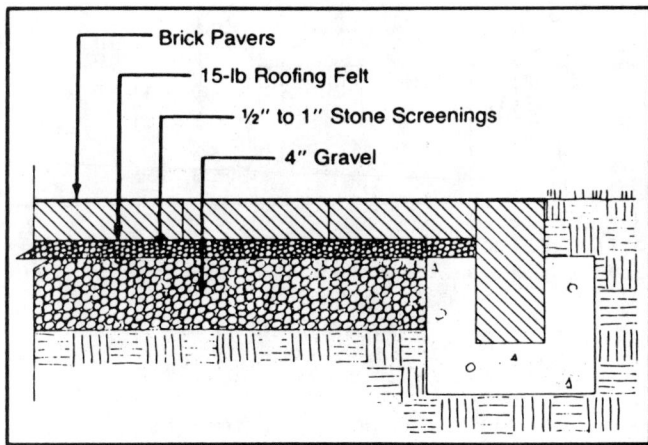

Gravel base for a residential driveway
Figure 17-8

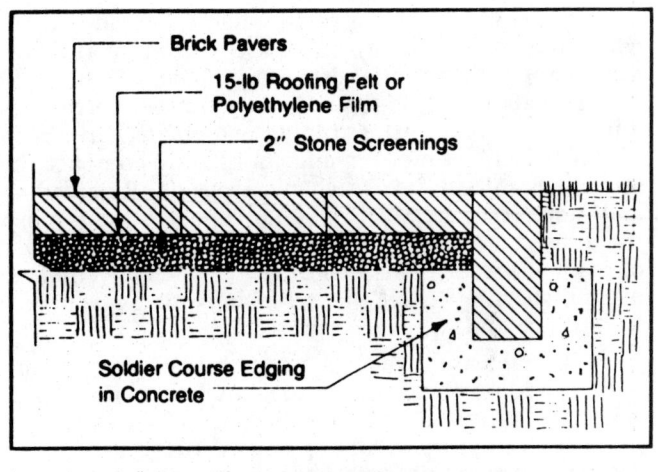

Gravel base good for light pedestrian traffic
Figure 17-10

Flexible Bases

Place only flexible paving over flexible bases. Generally, flexible bases are the most economical type of base to install. They usually consist of layers of gravel, damp loose sand or mixtures of sand and cement.

Porous gravel bases may be of either graded or ungraded stone. Ungraded gravel or stone has the advantage of a better interlocking quality, whereas graded gravel has a tendency to roll under foot. Stone screenings containing finer particles generally compact better than pea gravel. However, pea gravel has better surface-drainage characteristics. As shown in Figure 17-7, a 1/2-inch sand cushion is recommended because it will accomplish two basic purposes. The brick pavers can be laid more efficiently to the desired grade, and the sand cushion lends better stability to the brick paver. In this assembly a membrane must be used or the sand will eventually settle through the pea gravel.

Gravel for the base in Figure 17-8 should contain stones ranging from 1 inch to screenings size. This type is often referred to in many localities as "crusher-run stone."

The assembly shown in Figure 17-9 contains a 2-inch stone base, approximately half of the base thickness shown in Figure 17-8. This reduced base is adequate provided the brick paver thickness is increased. This is easily accomplished by placing units on edge. In this installation, brick units may be cored instead of 100-percent solid. Road construction (Figure 17-8) is generally required to make a smooth, hard, well-bonded support base prior to placement of the thin stone screenings layer. Compaction of the 2-inch stone screenings

199

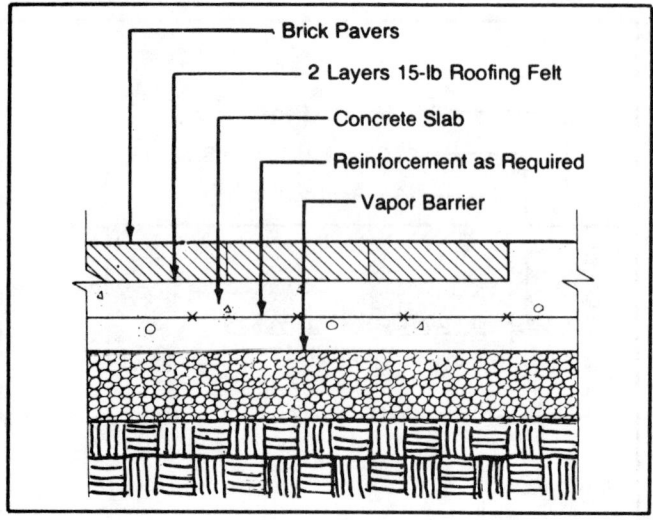

Concrete base for interior flooring. The depressed slab eliminates an abrupt level change from one floor finish to another.
Figure 17-11

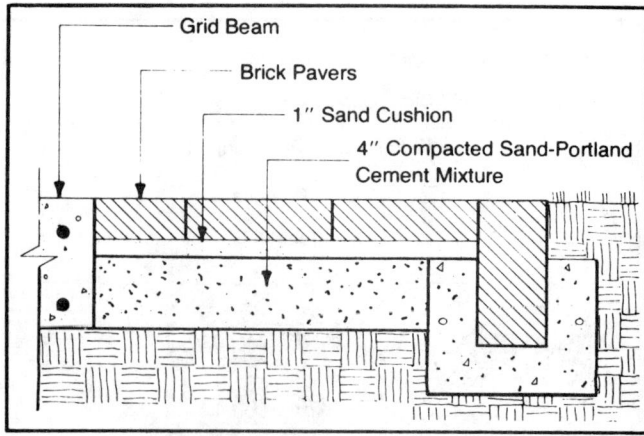

Sand-portland cement base for pedestrian mall traffic
Figure 17-12

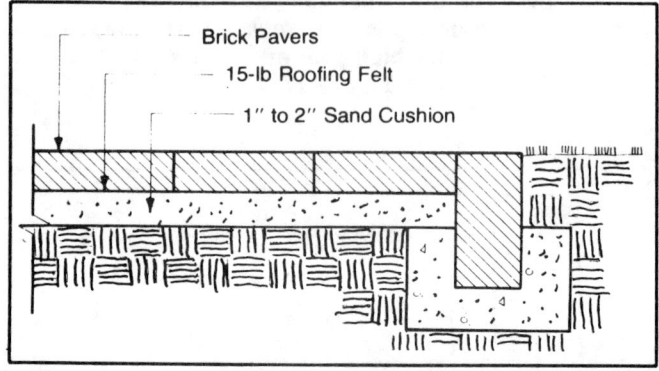

Sand base for residential patios in areas subject to low ground water tables and slight to moderate precipitation.
Figure 17-13

base (Figure 17-9) may be accomplished with less elaborate equipment. It is more suitable for compaction with hand tools. The assemblies in Figures 17-8 and 17-9 accommodate residential driveway traffic.

For walkways (Figure 17-10), spread screenings uniformly in place after grading and edging have been completed. Then moisten and thoroughly compact the screenings by hand or mechanical means. Rescreeding and more compaction may be necessary to accomplish the desired grade as specified. Place a membrane layer of asphalt-impregnated roofing felt or polyethylene plastic sheeting over the screenings to prevent efflorescence and to provide protection.

In special cases where surface drainage slopes are critical, add portland cement to stone screenings to provide stabilization. Add 1 part portland cement to 6 parts stone screenings by volume and mix with only enough water to form a ball. Screed this mixture into place and allow to set up. A stable base is the result and it will permit some drainage through the screenings.

For large areas, such as a pedestrian mall, the installation shown in Figure 17-12 lends itself to soils with generally good subsurface drainage characteristics. Excavate a grid system of earth trenches to accommodate reinforced concrete grade beams. After the concrete has cured, strip the subgrade away and replace it with a sand-cement base (3 to 6 parts sand to 1 part portland cement). Spread a 1-inch sand cushion over the compacted base to facilitate laying the pavers and to separate the pavers from the base.

Figure 17-13 shows the most economical type of installation of residential patio. Use it only over good soils that have good subsurface drainage characteristics.

Chapter 18
Construction Safety

Construction is one of the most hazardous of all occupations. Construction workers work with heavy machinery and often must climb high scaffolds, descend into excavations, and handle acids and other caustic materials.

OCCUPATIONAL SAFETY AND HEALTH ACT

In 1970 Congress passed an act creating the Occupational Safety and Health Act (OSHA) "...to assure so far as possible every working man and woman in the nation safe and healthful working conditions and to preserve our human resources."

In general, coverage of the Act extends to all employers and their employees in the 50 States, the District of Columbia, Puerto Rico, the Canal Zone and all other territories under Federal Government jurisdiction.

As defined by the Act, an employer is "any person engaged in a business affecting commerce who has employees." The Act covers employers and employees in such varied fields as construction, longshoring, and agriculture.

The following are not covered under the act:
1. Self-employed persons.
2. Farms at which only immediate members of the farm employer's family are employed.
3. Workplaces already protected by other federal agencies under federal statutes.

Before the Act became effective, no centralized and systematic method existed for monitoring occupational safety and health problems. Statistics on job injuries and illnesses were collected by some states and by some private organizations; national figures were based on projections not altogether reliable. With OSHA standards came the first basis for consistent nationwide procedures—a vital requirement for gauging problems and solving them.

Employers of eleven or more employees must maintain records of occupational injuries and illnesses as they occur. Employers with ten or fewer employees are exempt from keeping such records unless they are selected by the Bureau of Labor Statistics (BLS) to participate in periodic statistical surveys. The purposes of keeping records are to permit compliance to BLS survey material, to help define high-hazard industries, and to inform employees of the status of their employer's record.

Occupational Injury and Illness

What is an occupational injury or illness? It is any injury such as a cut, fracture, sprain, or amputation that results from a work-related accident or from exposure involving a single incident in the work environment. An occupational illness is any abnormal condition or disorder, other than one resulting from an occupational injury, caused by exposure to environmental factors associated with employment. Included are acute and chronic illnesses which may be caused by inhalation, absorption, ingestion or direct contact with toxic substances or harmful agents.

All occupational illnesses must be recorded regardless of severity.

All occupational injuries must be recorded if they result in:
1. Death (must be recorded regardless of length of time between the injury and death).
2. One or more lost workdays.
3. Restriction of work or motion.
4. Loss of consciousness.
5. Transfer to another job.

6. Medical treatment (other than first aid).

If an on-the-job accident occurs which results in the death of an employee or in the hospitalization of five or more employees, all employers, regardless of size of the company are required to report the accident, in detail, to the nearest OSHA office.

In states with approved plans, employers report such accidents to the state agency responsible for safety and health programs.

Employer Responsibilities and Rights

Employers have certain responsibilities and rights under the Occupational Safety and Health Act of 1970. The checklists that follow provide a review of many of these.

Responsibilities As an employer you must:

1. Meet your general duty responsibility to provide a workplace free from recognized hazards that are causing or are likely to cause death or serious physical harm to employees, and to comply with standards, rules, and regulations issued under the Act.
2. Be familiar with mandatory OSHA standards and make copies available to employees for review on request.
3. Inform all employees about OSHA.
4. Examine workplace conditions to make sure they conform to applicable standards.
5. Minimize or reduce hazards.
6. Make sure employees have and use safe tools and equipment (including personal protective equipment), and that such equipment is properly maintained.
7. Use color codes, posters, labels, or signs to warn employees of potential hazards.
8. Establish or update operating procedures and communicate them so that employees follow safety and health requirements.
9. Provide medical examinations when required by OSHA standards.
10. Report to the nearest OSHA office any fatal accident or one which results in the hospitalization of five or more employees.
11. Keep OSHA-required records of work-related injuries and illnesses, and post a copy of the totals from the last page of OSHA No. 200 during the entire month of February of each year. (This applies to employers of eleven or more employees.)
12. Pay employees for any time spent in taking part in an OSHA inspection.
13. Provide employees, former employees and their representatives access to the log and Summary of Occupational Injuries and Illnesses (OSHA No. 200).
14. Cooperate with OSHA compliance officer by furnishing names of authorized employee representatives who may be asked to accompany the compliance officer during an inspection. (If none, the compliance officer will consult with a reasonable number of employees concerning safety and health in the workplace.)
15. Not discriminate against employees who properly exercise their rights under this Act.
16. Post OSHA citations at or near the worksite involved. Each citation or copy thereof must remain posted until the violation has been abated or for three working days, whichever is longer.
17. Abate cited violations within the prescribed period.

Rights As an employer, you have the right to:

1. Seek advice and off-site consultation as needed by writing, calling or visiting the nearest OSHA office. (OSHA will not inspect merely because an employer requests assistance.)
2. Be active in the involvement of your industry association in job safety and health.
3. Request and receive proper identification of the OSHA compliance officer prior to inspection.
4. Be advised by the compliance office of the reason for an inspection.
5. Have an opening and closing conference with the compliance officer.
6. File a Notice of Contest with the OSHA area director within 15 working days of receipt of a notice of citation and proposed penalty.
7. Apply to OSHA for a temporary variance from a standard if unable to comply because of the unavailability of materials, equipment or personnel needed to make necessary changes within the required time.
8. Apply to OSHA for a permanent variance from a standard if you can furnish proof that your facilities or method of operation provide employee protection at least as effective as that required by the standard.
9. Take an active role in developing safety and health standards through participation in OSHA Standards Advisory Committees, through nationally recognized standard-setting organizations and through evidence and views presented in writing or at hearings.
10. Avail yourself, if you are a small business employer, of long-term loans through the Small Business Administration to help bring your establishment into compliance, either before or after an OSHA inspection.
11. Be assured of the confidentiality of any trade secrets observed by an OSHA compliance officer during an inspection.

OSHA STANDARDS

OSHA standards fall into four major categories, one of which is construction.

The Federal Register is one of the best sources of information on standards, since all OSHA standards are published in the Federal Register when adopted, as are all amendments, corrections, insertions or deletions. The Federal Register is available in many public libraries. Annual subscriptions are available from the Superintendent of Documents, U.S. Government Printing Office, Washington, D.C. 20402.

In masonry construction, you are concerned with the book of standards called 29 CFR PART 1926 SAFETY AND HEALTH REGULATIONS FOR CONSTRUCTION—29 CFR PART 1910 GENERAL INDUSTRY SAFETY AND HEALTH REGULATIONS IDENTIFIED AS APPLICABLE TO CONSTRUCTION.

Here are most of the parts that deal with masonry construction.

1926.20 GENERAL SAFETY AND HEALTH PROVISIONS

(A) Contractor Requirements-Section 107 of the act requires that it shall be a condition of each contract which is entered into under legislation subject to Reorganization Plan No. 14 of 1950 (64 Stat. 1267) as defined in 1926.12, and is for construction, alteration, and/or repair, including painting and decorating, that no contractor or subcontractor for any part of the contract work shall require any laborer or mechanic employed in the performance of the contract to work in surroundings or under working conditions which are unsanitary, hazardous, or dangerous to his health or safety.

(B) Accident-Prevention Responsibility.

(1) It shall be the responsibility of the employer to initiate and maintain such programs as may be necessary to comply with this part.

(2) Such programs shall provide for frequent and regular inspections of the job site, materials, and equipment to be made by competent persons designated by the employers.

(3) The use of any machinery, tool, material, or equipment which is not in compliance with any applicable requirement of this part is prohibited. Such machine, tool, material, or equipment shall be either identified as unsafe by tagging or locking the controls to render them inoperable or shall be physically removed from the place of operation.

(4) The employer shall permit only those employees qualified by training or experience to operate equipment or machinery.

1926.25 HOUSEKEEPING

(A) During the course of construction, alteration, or repairs, form and scrap lumber with protruding nails and all other debris shall be kept clear from work areas, passageways, and stairs, in and around buildings or other structures.

(B) Combustible scrap and debris shall be removed at regular intervals during the course of construction. Safe means shall be provided to facilitate such removal.

(C) Containers shall be provided for the collection and separation of waste, trash, oily and used rags, and other refuse. Containers used for garbage and other flammable, or hazardous wastes, such as caustics, acids, harmful dusts, etc. shall be equipped with covers. Garbage and other waste shall be disposed of at frequent and regular intervals.

1926.26 ILLUMINATION

Construction areas, aisles, stairs, ramps, runways, corridors, offices, shops, and storage areas where work is in progress shall be lighted with either natural or artificial illumination. The minimum illumination requirements for work areas are contained in subpart D of this part.

1926.27 SANITATION

Health and sanitation requirements for drinking water are contained in subpart D of this part.

1926.28 PERSONAL PROTECTIVE EQUIPMENT

(A) The employer is responsible for requiring the wearing of appropriate personal protective equipment in all operations where there is an exposure to hazardous conditions or where this part indicates the need for using such equipment to reduce the hazards to the employees.

(B) Regulations governing the use, selection, and maintenance of personal protective and lifesaving equipment are described under subpart E of this part.

1926.50 MEDICAL SERVICES AND FIRST AID

(A) The employer shall ensure the availability of medical personnel for advice and consultation on matters of occupational health.

(B) Provisions shall be made prior to commencement of the project for prompt medical attention in case of serious injury.

(C) In the absence of an infirmary, clinic, hospital, or physician that is reasonably accessible in terms of time and distance to the worksite, a person who has valid certification in first-aid training from the Bureau of Mines, the American Red Cross, or equivalent training that can be verified by documentary evidence, shall be available at the worksite to render first aid.

(D) First-Aid Supplies.

(1) First-aid supplies approved by the consulting physician shall be easily accessible when required.

(2) The first-aid kit shall consist of materials approved by the consulting physician and placed in a weatherproof container with individual sealed packages for each item. The contents of the first-aid kit shall be checked by the employer before being sent out on each job and at least weekly on each job to ensure that the expended items are replaced.

(E) Proper equipment for prompt transportation of the injured person to a physician or hospital, or a communication system for contacting necessary ambulance service, shall be provided.

(F) The telephone numbers of the physicians, hospitals, or ambulances, shall be conspicuously posted.

1926.51 SANITATION

(A) Potable Water.

(1) An adequate supply of potable water shall be provided in all places of employment.

(2) Portable containers used to dispense drinking water shall be capable of being tightly closed, and equipped with a tap. Water shall not be dipped from containers.

(3) Any container used to distribute drinking water shall be clearly marked as to the nature of its contents and not used for any other purpose.

(4) The common drinking cup is prohibited.

(5) Where single-service cups are supplied, both a sanitary container for the unused cups and a receptacle for disposing of the used cups shall be provided.

(B) Outlets for nonpotable water shall be identified by signs to indicate clearly that the water is unsafe and is not to be used for drinking.

(C) Toilets at construction jobsites.

(1) Toilets shall be provided for employees according to the following table:

Masonry & Concrete Construction

Number of employees	Minimum number of facilities
20 or less	1
20 or more	1 toilet seat & 1 urinal per 40 workers
200 or more	1 toilet seat & 1 urinal per 50 workers

(2) Under temporary field conditions, provisions shall be made to assure not less than one toilet facility is available.

(3) Job sites not provided with a sanitary sewer shall be provided with one of the following toilet facilities unless prohibited by local codes.

(a) Privies (where their use will not contaminate ground or surface water.

(b) Chemical toilets.

(c) Recirculating toilets.

(d) Combustion toilets.

(e) The requirements of this paragraph (C) for sanitation facilities shall not apply to mobile crews having transportation readily available to nearby toilet facilities.

1926.100 HEAD PROTECTION

(A) Employees working in areas where there is a possible danger of head injury from impact, or from falling or flying objects, or from electrical shock and burns, shall be protected by protective helmets.

(B) Helmets for the protection of employees against impact and penetration of falling and flying objects shall meet the specifications contained in American National Standards Institute, Z89.1 1969, Safety Requirements for Industrial Head Protection.

(C) Helmets for the head protection of employees exposed to high-voltage electrical shock and burns shall meet the specification contained in American National Standards Institute.

1926.101 HEARING PROTECTION

(A) Whenever it is not feasible to reduce the noise levels or duration of exposures to those specified . . . ear protective devices shall be provided and used.

(B) Ear protective devices inserted in the ear shall be fitted or determined individually by competent persons.

(C) Plain cotton is not an acceptable protective device.

1926.102 EYE AND FACE PROTECTION

(A) Employees shall be provided with eye and face protection equipment when machines or operations present potential eye or face injury from physical, chemical, or radiation agents.

(1) Eye and face protection equipment required by this part shall meet the requirements specified in American National Standards Institute Z87.1 1968, Practice for Occupational and Educational Eye and Face Protection.

(2) Employees whose vision requires the use of corrective lenses in spectacles, when required by this regulation to wear eye protection, shall be protected by goggles or spectacles of the following types:

(a) Spectacles whose protective lenses provide optical correction.

(b) Goggles that can be worn over corrective spectacles without disturbing the adjustment of the spectacles.

(c) Goggles that incorporate corrective lenses mounted behind the protective lenses.

(3) Face and eye protection equipment shall be kept clean and in good repair. The use of this type equipment with structural or optical defects shall be prohibited.

SUBPART H. MATERIAL HANDLING, STORAGE, USE, AND DISPOSAL
1926.252

(A) All materials stored in tiers shall be stacked, racked, blocked, interlocked, or otherwise secured to prevent sliding, falling or collapsing.

(B) Bagged materials shall be stacked by stepping the layers and cross-keying the bags at least every 10 bags high.

(C) Materials shall not be stored on scaffolds or runways except for supplies needed for immediate operations.

(D) Brick stacks shall not be more than 7 feet in height. When a loose brick stack reaches a height of 4 feet, it shall be tapered back 2 inches in every foot of height above the 4-foot level.

(E) When masonry blocks are stacked higher than 6 feet, the stack shall be tapered back one-half block per tier above the 6-foot level.

SUBPART I. TOOLS—HAND AND POWER
1926.300 GENERAL REQUIREMENTS.

(A) All hand and power tools and similar equipment, whether furnished by the employer or the employee, shall be maintained in a safe condition.

(B) When power-operated tools are designed to accommodate guards, they shall be equipped with such guards when in use.

SUBPART K. ELECTRICAL
1926.401 GROUNDING AND BONDING.

(A) All temporary wiring shall be effectively grounded in accordance with the National Electrical Code. NFPA 70-1971; ANSI CI-1971 (Rev. of CI-1968) Articles 305 and 310.

(B) Construction-site precautions shall be taken to make any necessary open wiring inaccessible to unauthorized personnel.

(C) Temporary lights shall be equipped with heavy-duty electrical cords with connections and insulation maintained in safe condition. Temporary lights shall not be suspended by their electric cords unless cords and lights are designed for this means of suspension.

(D) Splices shall have insulation equal to that of the cable.

SUBPART L. LADDERS AND SCAFFOLDING
1926.450 LADDERS.

(A) Except where either permanent or temporary stairways or suitable ramps or runways are provided, ladders described in this subpart shall be used to give safe access to all elevations.

(B) The use of ladders with broken or missing rungs or steps, broken or split side rails, or other defective construction is prohibited. When ladders with such defects are discovered, they shall be immediately withdrawn from service. Inspection of metal ladders shall include checking for corrosion of interiors of open-end hollow rungs.

(C) Manufactured ladders.

(1) Manufactured portable wooden ladders provided by the employer shall be in accordance with the provisions of the American National Standards Institute. A14.1-1968 Safety Code for Ladders.

(2) Portable metal ladders shall be of strength equivalent to that of wood ladders. Manufactured portable metal ladders provided by the employer shall be in accordance with the provisions of the American National Standards Institute, A 14.2-1956, Safety Code for Portable Metal Ladders.

(3) Fixed ladders shall be in accordance with the provisions of the American National Standards Institute, A 14.3 1956 Safety Code for Fixed Ladders.

(4) Portable ladder feet shall be placed on a substantial base, and the area around the bottom of the ladder shall be kept clear.

(5) Portable ladder shall be used at such a pitch that the horizontal distance from the top support to the foot of the ladder is about one-quarter of the working length of the ladder (the length of the ladder between the foot and the top support). Ladders shall not be used in a horizontal position as platforms, runways, or scaffolds.

(6) Ladders shall not be placed in passageways, doorways, driveways, or any location where they might be displaced by activities being conducted on any other work, unless protected by barricades or guards.

(7) The side rails shall extend not less than 36 inches above the landing. When this is not practical, grab rails which provide a secure grip for an employee moving to or from the point of access shall be installed.

(8) Portable ladders in use shall be tied, blocked, or otherwise secured to prevent their being displaced.

(9) Portable metal ladders shall not be used for electrical work or where they may contact electrical conductors.

(D) Job-Made Ladders.

(1) Job-made ladders shall be constructed for intended use. If a ladder is to provide the only means of access for exit from a working area for 25 or more employees, or if simultaneous two-way traffic is expected, a double-cleat ladder shall be installed.

(2) Double-cleat ladders shall not exceed 24 feet in length.

(3) Single-cleat ladders shall not exceed 30 feet in length between supports (base and top landing). If ladders are to connect different landings or if the length exceeds the maximum length, two or more separate ladders shall be used, offset with a platform between each ladder. Guardrails and toe boards shall be erected on the exposed side of the platforms. (See 1926.451 (E))

(4) The width of single-cleat ladders shall be at least 15 inches, but not more than 20 inches between rails at the top.

(5) Side rails shall be parallel or flared top to bottom by not more than one-quarter of an inch for each 2 feet of length.

(6) Wood side rails of ladders having cleats shall be not less than 1½ inches thick and 3½ inches deep (2 inches by 4 inches nominal).

(7) It is preferable that side rails be continuous. If a splice is necessary to attain the required length, the splice must develop the full strength of a continuous side rail of the same length.

(8) 2-inch by 4-inch lumber shall be used for side rails of single-cleat ladders up to 16 feet long; 3-inch by 6-inch lumber shall be used for single-cleat ladders from 16 feet to 30 feet in length.

(9) 2-inch by 4-inch lumber shall be used for side and middle rails of double-cleat ladders up to 12 feet in length; 2-inch by 6-inch lumber for double-cleat ladders from 12 to 24 feet in length.

(10) Wood cleats shall have the following dimensions:

Length of Cleat (In.)	Thickness (In.)	Width (In.)
Up to and including 20	¾	3
Over 20 and up to and including 30	¾	3¾

(11) Cleats may be made of any wood type similar to the rails.

(12) Cleats shall be inset into the edges of the side rails one-half inch, or filler blocks shall be used on the rails between the cleats. The cleats shall be secured to each rail with three 10d common wire nails or other fasteners of equivalent strength. Cleats shall be uniformly spaced 12 inches top to top.

1926.451 SCAFFOLDING

(A) Scaffolds shall be erected in accordance with requirements of this section.

(B) The footing or anchorage for scaffolds shall be sound, rigid and capable of carrying the maximum load without settling or displacement. Unstable objects such as barrels, boxes, loose brick, or concrete blocks, shall not be used to support scaffolds or planks.

(C) No scaffold shall be erected, moved, dismantled or altered except under the supervision of a competent person.

(D) Guard rails and toe boards shall be installed on all open sides and ends of platforms more than 10 feet above the ground or floor, except on needle-beam scaffolds and floats. Scaffolds 4 feet to 10 feet in height, having a minimum horizontal dimension in either direction of less than 45 inches shall have standard guard rails installed on all open sides and ends of the platform.

(E) Guard rails shall be 2 inches by 4 inches, or the equivalent, approximately 42 inches high, with a midrail, when required. Supports shall be at intervals not to exceed 8 feet. Toe boards shall be a minimum of 4 inches in height.

(F) Where persons are required to pass under the scaffold, scaffolds shall be provided with a screen between the toe board and the guard rail, extending along the entire opening, consisting of no. 18-gauge U.S. Standard wire 1/2-inch mesh or the equivalent.

(G) Scaffolds and their components shall be capable of supporting without failure at least 4 times the maximum intended load.

(H) Any scaffold including accessories such as braces, brackets, trusses, screw legs, ladders, etc. that are damaged or weakened from any cause shall be immediately repaired or replaced.

(I) All load-carrying timber members of scaffold framing shall be a minimum of 1,500-fiber (Stress Grade) construction-grade lumber. All dimensions are nominal sizes as provided in the American Lumber Standards, except that where rough sizes are noted, only rough or undressed lumber of the size specified will satisfy minimum requirements.

(J) All planking shall be scaffold grades, or equivalent, as recommended by grading rules for the species of wood used. The maximum permissible spans for 2-inch by 10-inch lumber or wider planks shall be as shown in the following:

	Full-Thickness Undressed Lumber			Nominal-Thickness Lumber (1)	
Working load (p.s.f.)	25	50	75	25	50
Permissible Span (ft.)	10	8	6	8	6

(1) Nominal-thickness lumber not recommended for heavy-duty use.

(K) The maximum permissible span for 1¼-inch x 9-inch or wider plank of full thickness shall be 4 feet with a medium-duty loading of 50 psf.

(L) All planking of platforms shall be overlapped (minimum of 12 inches) or secured from movement.

(M) An access ladder or equivalent safe access shall be provided.

(N) Scaffold planks shall extend over their end supports not less than 6 inches nor more than 12 inches.

(O) The poles, legs, or uprights of scaffolds shall be plumb and securely and rigidly braced to prevent swaying and displacement.

(P) Overhead protection shall be provided for men on a scaffold exposed to overhead hazards.

(Q) Slippery conditions on scaffolds shall be eliminated as soon as possible after they occur.

(R) No welding, burning, riveting, or open flame work shall be performed on any staging suspended by means of fiber or synthetic rope. Only treated or protected fiber or synthetic shall be used for or near any work involving the use of corrosive substances or chemicals.

(S) Wire, synthetic, or fiber rope used for scaffold suspension shall be capable of supporting at least 6 times the rated load.

(T) The use of shore or lean-to scaffolding is prohibited.

(U) Lumber sizes, when used in this subpart, refer to nominal sizes except where otherwise stated.

SUBPART Q. CONCRETE, CONCRETE FORMS, AND SHORING

1926.700 GENERAL PROVISIONS.

(A) All equipment and materials used in concrete construction and masonry work shall meet the applicable requirements for design, construction, inspection, testing, maintenance and operations described in ANSI A10. 9-1970, Safety Requirements for Concrete Construction and Masonry Work.

(B) Reinforcing steel.
(1) Employees working more than 6 feet above any adjacent working surfaces, placing and tying reinforcing steel in walls, piers, columns, etc., shall be provided with a safety belt, or equivalent device in accordance with Subpart E of this part.
(2) Employees shall not be permitted to work above vertically protruding reinforcing steel unless it has been protected to eliminate the hazard of impalement.
(3) Reinforcing steel for walls, piers, columns, and similar vertical structures shall be guyed and supported to prevent collapse.
(4) Wire mesh rolls shall be secured at each end to prevent dangerous recoiling action.

(C) Bulk storage bins, containers, or silos shall have conical or tapered bottoms with mechanical or pneumatic means of starting the flow of material.

(D) Concrete placement
(1) Concrete mixers. Concrete mixers equipped with 1-yard or larger loading skips shall be equipped with a mechanical device to clear the skip of material.
(2) Guard rails. Mixers of 1-yard capacity or greater shall be equipped with protective guard rails installed on each side of the skip.
(3) Handles on bull floats, used where they may contact energized electrical conductors, shall be constructed of nonconductive material, or insulated with a nonconductive sheath whose electrical and mechanical characteristics provide the equivalent protection of a handle constructed of nonconductive material.
(4) Powered and rotating-type concrete-troweling machines that are manually guided shall be equipped with a control switch that will automatically shut off the power whenever the operator removes his hands from the equipment handles.
(5) Handles of buggies shall not extend beyond the wheels on either side of the buggy. Installation of knuckle guards on buggy handles is recommended.
(6) Pumpcrete or similar systems using discharge pipes shall be provided with pipe supports designed for 100 percent overload. Compressed air hose in such systems shall be provided with positive fail-safe joint connectors to prevent separation of sections when pressurized.
(7) Concrete buckets
(a) Concrete buckets equipped with hydraulic or pneumatically operated gates shall have positive safety latches or similar safety devices installed to

prevent the aggregate and loose material from accumulating on the top and sides of the bucket.

(b) Riding on concrete buckets for any purpose shall be prohibited and vibrator crews shall be kept out from under concrete buckets suspended from cranes or cableways.

(8) When discharging on a slope, the wheels of ready-mix trucks shall be blocked and the brakes set to prevent movement.

(9) Nozzlemen applying a cement, sand, and water mixture through a pneumatic hose shall be required to wear protective head and face equipment as prescribed in Subpart E of this part.

(E) Vertical shoring

(1) General requirements.

(a) When temporary storage of reinforcing rods, material, or equipment on top of formwork becomes necessary, these areas shall be strengthened to meet the intended loads.

(b) The sills for shoring shall be sound, rigid, and capable of carrying the maximum intended load.

(c) All shoring equipment shall be inspected prior to erection to determine that it is as specified in the shoring layout. Any equipment found to be damaged shall not be used for shoring.

(d) Erected shoring equipment shall be inspected immediately prior to, during, and immediately after the placement of concrete. Any shoring equipment that is found to be damaged or weakened shall immediately be reinforced or reshored.

(e) Reshoring shall be provided when necessary to safely support slabs and beams after stripping, or where such members are subject to superimposed loads due to construction work being done.

(2) Tubular welded frame shoring.

(a) Metal tubular frames used for shoring shall not be loaded beyond the safe working load recommended by the manufacturer.

(b) All locking devices on frames and braces shall be in good working order; coupling pins shall align the frame or panel legs; pivoted cross braces shall have their center pivot in place; and all components shall be in a condition similar to that of original manufacture.

(c) When checking the erected shoring frames with the shoring layout, the spacing between towers and cross-brace spacing shall not exceed that shown on the layout, and all locking devices shall be in a closed position.

(d) Devices for attaching the external lateral stability bracing shall be securely fastened to the legs of the shoring frames.

(e) All base plates, shore heads, extension devices, or adjusting screws shall be in firm contact with the footing sill and the form.

Glossary

Absorption: The weight of water a brick unit absorbs, when immersed in either cold or boiling water for a stated length of time, expressed as percentage of the weight of the dry unit. See ASTM Specification C 67.

Abutment: The masonry mass supporting the thrust of an arch or bridge.

Accelerator: Any chemical or other substance added to cement during the mixing process to quicken the setting of the cement.

Addition: A change in the design of a building to increase the overall dimensions; also the original design of a building constructed with connecting parts joined together to make one whole structure.

Admixtures: Materials added to mortar to impart special properties to the mortar.

Aggregate: Various hard materials such as sand, gravel, or crushed stone, added to cement to make concrete.

Air brick: A ceramic or metal unit about the size of a standard brick, open on the ends to permit the entrance of air into the building.

Air-entrained concrete: Portland cement that has had an ingredient added to cause millions of tiny air bubbles to be trapped in the concrete.

Anchor: A piece or assemblage, usually metal, used to attach building parts (plates, joists, trusses, etc.) to masonry or masonry materials.

Anchor bolts: Bolts that are used to fasten sills or plates to masonry.

Angle iron: A piece of iron that forms a right angle and is used to span openings and support masonry at these openings. In brick veneer they are used to secure the veneer to the foundation or to the structure being veneered.

ANSI: American National Standards Institute. This group publishes the American National Standards, which are the approved standards and specifications in all the areas of building construction.

Apex: The architectural term designating the topmost part of a structure.

Apprentice: A person who has entered into an agreement with a trade committee and employers to work for a period of time to learn the trade.

Arch: A curved compressive structural member, spanning openings or recesses; also built flat.

> *back arch:* A concealed arch carrying the backing of a wall where the exterior facing is carried by a lintel.
>
> *jack arch:* One having horizontal or nearly horizontal upper and lower surfaces, also called a flat or straight arch.
>
> *major arch:* Arch with spans greater than 6 feet and equivalent to uniform loads greater than 1000 pounds per foot. Typically known as a Tudor arch, semicircular arch, gothic arch or parabolic arch. Has rise-to-span ratio greater than 0.15.
>
> *minor arch:* Arch with a maximum span of 6 feet and loads not exceeding 1000 pounds per foot. Typically known as a jack arch, segmented arch or multicentered arch. It has a rise-to-span ratio less than or equal to 0.15.
>
> *receiving arch:* One built over a lintel, flat arch, or smaller arch to divert loads, thus relieving the lower member from excessive loading; also known as a discharging or safety arch.
>
> *trimmer arch:* Usually a low-rise arch of brick used for supporting a fireplace hearth.

Asbestos cement: A cement made by combining

portland cement with asbestos. It is fire resistant and waterproof.

Ash dump: A metal frame placed in the floor of a fireplace for the disposing of ashes.

Ashlar masonry: Masonry composed of rectangular units of burned clay, shale, or stone, generally larger in size than brick and properly bonded, having sawed, dressed, or squared beds, and joints laid in mortar.

ASHRAE: The American Society of Heating, Refrigerating and Air-Conditioning Engineers, Inc.

ASTM: The American Society for Testing and Materials. It is a scientific and technical organization formed for "the development of standards and characteristics and performance of materials, products, systems, and services; and the promotion of related knowledge."

Autogenous healing: The longtime (natural) process whereby masonry cracks fill up.

Backfilling: 1. Rough masonry built behind a facing or between two faces. 2. Filling over the extrados of an arch. 3. Brick work in spaces between structural timbers sometimes called nogging. 4. Fill (dirt, stone, or similar material) used to build up the ground between the foundation and the unexcavated part of the surrounding area.

Back-up: The part of a masonry wall behind the exterior facing.

Back hearth: The floor of a fireplace.

Bar chairs: A small device that is used to hold up reinforcing wire or rods in concrete while the concrete is being poured.

Bat: A small piece of brick, usually less than a half.

Batter: Recessing or sloping masonry back in successive courses; the opposite of corbel.

Batterboard: A board set up outside the building line to hold the lines both prior to the excavation and after the excavation to relocate the building lines.

Beam and slab construction: A method of supporting a reinforced concrete floor by a system of reinforced concrete beams or girders.

Bearing plate: A metal plate placed under a beam, girder, or column to spread its weight over a larger area for support.

Bed joint: The horizontal layer of mortar on which a masonry unit is laid.

Bed stone: A large stone sometimes used to support a beam.

Belt course: A narrow horizontal course of masonry, sometimes slightly projected, such as window sills; sometimes called string or sill course.

Bench marks: Permanent marks such as a tree or a fire hydrant used as reference for elevation checks.

Blocking: A method of bonding two adjoining or intersecting walls not built at the same time by means of offsets whose vertical dimensions are not less than 8 inches.

Bluestone: A stone quarried in the southern New York area; commonly used for sills, treads, and lintels.

Bond: 1. Method of tying various parts of a masonry wall by lapping the units one over another or by connecting with metal ties. 2. Patterns formed by exposed parts of units. 3. Adhesion between mortar or grout and masonry units or reinforcement.

Bond beam: Course or courses of masonry wall grouted and usually reinforced in the horizontal direction. Serves as a horizontal tie of the wall, a bearing course for structural members, or as a flexural member itself.

Bond course: The course consisting of units which overlap more than one wythe of masonry.

Bond stone: In stone wall construction, the stones that run through the thickness of the wall to tie the wall together.

Bonder: A bonding unit. (See *header*.)

Breaking joints: Any arrangement of masonry units which prevents continuous vertical joints from occurring in adjacent courses.

Brick: A solid masonry unit of clay or shale formed into a rectangular prism while plastic and burned or fired in a kiln.

   acid-resistant brick: Brick suitable for use in contact with chemicals.

   adobe brick: Large roughly moulded, sun dried clay brick of varying size.

   angle brick: Any brick shaped to an oblique angle to fit a salient corner.

   arch brick: (1.) Wedge shaped brick for special use in an arch. (2.) Extremely hard burned brick from an arch of a stove kiln.

   building brick: Brick for building purposes not especially treated for texture or color. Formerly called common brick. See ASTM Specification C 62.

   clinker brick: A very hard burned brick whose shape is distorted or bloated owing to nearly complete vitrification.

   common brick: See *building brick*.

   dry press brick: Brick formed in molds under high pressures from relatively dry clay (5 to 7 percent moisture content).

   economy brick: Brick whose nominal dimensions are 4" x 4" x 8".

   engineered brick: Brick whose nominal dimensions are 4" x 3.2" x 8".

   facing brick: Brick made especially for facing purposes. Often treated for surface texture and color. See ASTM Specification C 216.

   fire brick: Brick made of refractory ceramic material which will resist high temperatures.

   floor brick: Smooth dense brick, highly resistant to abrasion, used as finished floor surfaces. See ASTM Specification C 410.

   gauged brick: 1. Brick which have been ground to accurate dimensions. 2. A tapered arch brick.

   hollow brick: A masonry unit of clay or shale whose net cross area in any plane parallel to the bearing surface is not less than 60 percent of its gross cross-sectional area measured in the same plane. See ASTM Specification C 652.

   jumbo brick: A generic term indicating a brick larger

in size than the standard.

norman brick: A brick whose nominal dimensions are 4" x 2⅔" x 12".

paving brick: Vitrified brick used where resistance to abrasion is important. See ASTM Specification C 7.

roman brick: Brick whose nominal dimensions are 4" x 2" x 12".

salmon brick: Generic term for underburned brick which are more porous, slightly larger, and lighter in color than hard burned brick.

"SCR brick": Brick whose nominal dimensions are 6" x 2⅔" x 12".

sewer brick: Low absorption abrasion-resistant brick used in drainage structures.

soft mud brick: Brick produced by moulding relatively wet clay (20 to 30 percent moisture).

stiff mud brick: Brick produced by extruding a stiff but plastic clay (12 to 15 percent moisture) through a dye.

Brick and brick: A method of laying brick so that units touch each other with only enough mortar to fill irregularities.

Brick grade: Designation for durability of the unit expressed as SW for severe weathering, MW for moderate weathering, or NW for negligible weathering. See ASTM Specifications C 216 and C 265.

Brick type: Designation for facing brick which controls tolerance, chippage, and distortion. Expressed as FBS, FBX, and FBA for solid brick, and HBS, HBX and HBB for hollow brick. See ASTM Specifications C 216 and C 652.

Brick veneer: A building of masonry in which the brick facing is attached to a surface of a frame with wall ties, and is not bonded to the veneered wall.

Building code: A set of regulations that are adopted by a city or town for the construction of buildings.

Building line: The outside line of the building.

Building permits: A permission form obtained from a state or local government to permit construction of a structure.

Bull float: A wide float with a long handle.

Bull header: A brick having one corner that is rounded; used for window sills and for coping.

Bull nose: A masonry unit that is rounded on one side of one end of the unit for use on corners.

Buttering: Placing mortar on a masonry unit with a trowel.

Buttress: A piece of masonry built against a wall to give the wall more strength.

Caisson: A concrete pile that is constructed for use under water.

Calcium chloride: A granulated salt added to water in mixing concrete and mortar to accelerate the setting time.

Cant brick: A brick that is made with one side beveled.

Capacity insulation: The ability of masonry to store heat as a result of its mass, density and specific heat.

Capping brick: Brick that are made for capping the top of a wall.

C/B ratio: The ratio of the weight of water absorbed by a masonry unit during immersion in cold water to weight absorbed during immersion in boiling water. Also called saturation coefficient. See ASTM Specification C 67.

Centering: Temporary formwork for the support of masonry arches or lintels during construction; also called centers.

Ceramic color glaze: An opaque colored glaze of satin or gloss finish. See ASTM Specification C 126.

Ceramic mosaic: Small ceramic tiles in sheets, usually for use on floors.

Ceramic tile: A flat piece of fired clay in a variety of shapes and compositions.

Chase: A continuous recess built into a wall to receive pipes, ducts, etc.

Clay: A natural, mineral aggregate consisting essentially of hydrous aluminum silicate; it is plastic when sufficiently wetted, rigid when dried and vitrified when fired to a sufficiently high temperature.

Clay mortar mix: Finely ground clay used as a plasticizer for masonry mortars.

Clear ceramic glaze: Same as ceramic color glaze except that it is translucent or slightly tinted, with a gloss finish.

Clip: A portion of a brick cut to length.

Closer: The last masonry unit laid in a course. It may be whole or a portion of a unit.

Closure: Supplementary or short length units used at corners or jambs to maintain bond patterns.

Cold Joint: A joint or a discontinuity formed when a concrete surface hardens before the next pour is placed to it.

Collar joint: The vertical, longitudinal joint between wythes of masonry.

Column: A vertical member whose horizontal dimension measured at right angles to the thickness does not exceed three times its thickness.

Compass brick: A factory-made brick for use in curved work.

Construction Specifications Institute (C.S.I.): An organization that established a format for construction specifications.

Control joint: A groove that is cut or tooled in the surface of concrete to predetermine the place where a crack will occur due to shrinkage of the concrete.

Coping: The material or masonry units forming a cap or finish on top of a wall, pier, pilaster, chimney, etc.

Corbel: A shelf or ledge formed by projecting successive courses of masonry out from the face of the wall.

Counterflashing: The flashing that projects from the masonry wall over the base flashing to protect the upper end of the flashing.

Course: One of the continuous horizontal layers of units bonded with mortar in masonry.

Coursed ashlar: Method of arranging various units in the wall according to height to form courses in stone masonry.

Coursed rubble: Method of laying roughly shaped stones in approximate level beds in stone masonry.

Cricket: A small watershed built behind a chimney to make water run away from the chimney.

CRSI: Concrete Reinforcing Steel Institute.

Culls: Masonry units which do not meet the standards or specifications and have been rejected.

Curing: The process in which mortar and concrete harden.

Damp course: A course or layer of impervious material which prevents capillary action from the ground or lower course.

Dampproofing: Prevention of moisture penetration by capillary action.

Darby: A flat trowel-like tool used to smooth out the surface of concrete soon after the surface has been bull floated.

Datum point: A point that has been established by a city or town to use as a reference point in measuring elevations and distances.

Dog's tooth: Brick laid with their corners projecting from the wall.

Drip: A projecting piece of material, shaped to throw off water to prevent it from running down the face of the wall.

Dry stone wall: A stone wall that is laid without the aid of mortar.

Dutch bond: Brick laid in alternate stretchers and headers.

Edging: The process of finishing the edges of a concrete slab with a tool that has a radius on its edge.

Effective height: The height of a member used for calculating the slenderness ratio.

Effective thickness: The thickness of a member used for calculating the slenderness ratio.

Efflorescence: A powder or stain sometimes found on the surface of masonry resulting from deposition of water soluble salts.

Engineered brick masonry: Masonry in which design is based on a rational structural analysis.

English bond: A masonry bond where courses alternate between headers and stretchers.

English cross bond: Sometimes called Old English bond, it is similar to or the same as Dutch bond.

Expansion joint: See *control joint*.

Exposed aggregate: A concrete finish that has the top of the surface cement washed off to show the stone aggregate.

Face: 1. The exposed surface of a wall or masonry unit. 2. The surface of a unit designed to be exposed in the finished masonry.

Facebrick: Good quality brick used in exposed surface of a brick wall.

Facing: Any material forming a part of the wall used as a finished surface.

False header: A header that does not tie two walls together. As when a half brick is used.

Field: The expanse of wall between openings, corners, etc., principally composed of stretchers.

Filter block: A hollow, vitrified clay masonry unit, sometimes salt glazed, designed for trickling filter floors in sewage disposal plants. See ASTM Specification C 159.

Finish grade: The surface elevation after the grading is completed.

Fire clay: A clay used to make brick which is highly resistant to heat without deforming.

Fire resistant material: A noncombustible material.

Firebrick: A brick that can withstand the effects of heat extremes.

Fireproofing: Any material or combination of materials that increases fire resistance.

Flash set: A process by which concrete sets faster than normal because of too much heat.

Flashing: 1. A thin impervious material placed in mortar joints and through air spaces in masonry to prevent water penetration and/or to provide water drainage. 2. Manufacturing method to produce specific color tones in brick.

Flemish bond: A brick bond in which headers and stretchers alternate on every course.

Form oil: A nonstaining oil usually painted on the forms to act as a nonstick surface between the concrete and the forms.

Frog: A depression on the topside of a brick.

Furring: A method of finishing the interior face of masonry walls to provide a space for insulation, prevent moisture transmittance, or to provide a level surface for finishing.

Gingerbread: Fancy brickwork on old buildings.

Glazed brick: A brick that has a glazing material fused to the face, usually providing a smooth, glassy surface.

Grade: The level of the ground around a building.

Grade beam: A reinforced concrete beam that is used to hold up the walls of a building while supported on piers.

Green cement: Masonry that has not set up yet.

Grounds: Nailed strips placed in masonry walls as a means of attaching trim or furring.

Grout: Mixture of cementitious material, aggregate and water in a pouring consistency.
   high lift grouting: The technique of grouting masonry lifts up to 12 feet.
   low lift grouting: The technique of grouting as a wall is constructed.

Hacking: 1. The procedure of stacking brick on a kiln car for firing. 2. Laying brick so that the bottom edge sets the plane of the wall.

Hard burned: Nearly vitrified clay products which have been fired at high temperatures.

Head joint: The vertical joint between masonry units.

Header: A masonry unit which overlaps two or more adjacent wythes of masonry to tie them together.

> *blind header:* A concealed brick header on the interior of a brick wall.
>
> *clipped header:* A bat placed to look like a header for the purpose of establishing a pattern; also called a false header.
>
> *flare header:* A header of darker color than the rest of the wall.

Heading course: A continuous bonding course of header brick; also called a header course.

Heavyweight concrete: Concrete that is constructed from heavyweight aggregates and weighing about 390 pounds per cubic foot: used in the construction of laboratories as radiation shields.

High chairs: A manufactured product that is used to hold up reinforcing wire in concrete as the concrete is poured.

High early cement: A portland cement sold as Type III; sets up to its full strength faster than other types.

Honeycomb: Method by which concrete is poured and not puddled or vibrated, allowing the edges to have voids or holes after the forms are removed.

Hydrated lime: Material that is left after quicklime is added to water; also called slaked lime.

Hydraulic: The ability of cement to harden when mixed with or under water.

Initial rate of absorption: The weight of water that is absorbed expressed in grams per 30 square inches of contact surface when a brick is partially immersed for one minute. See ASTM Specification C 67.

Isolation joint: A joint that completely separates one piece of concrete from another.

Journeyman: A mason who has learned his trade through an apprenticeship.

Keyway: A recess in one piece of concrete to improve the shear strength of the next pour by tying the pours together mechanically.

Kiln: A furnace oven or heated enclosure used for burning or firing brick.

Kiln run: Brick from one kiln which have not been sorted or graded for size or color variation.

King closer: A brick cut diagonally to have one 2-inch end and one full width end.

Lateral support: Means whereby walls are braced either vertically or horizontally by columns, pilasters, cross walls, beams, floors, roofs, etc.

Lead: The section of wall built up and racked back on successive courses.

Lime, hydrated: Quicklime to which sufficient water has been added to convert the oxides to hydroxides.

Lime putty: Hydrated lime in plastic form ready for addition to mortar.

Lintel: A beam placed over an opening in a wall.

Masonry cement: A mill mixed cementitious material to which sand and water must be added. See ASTM Specification C 91.

Masonry unit: Natural or manufactured building units of burned clay, concrete, stone, glass, gypsum, etc.

> *hollow masonry unit:* One whose net cross-sectional area in any plane parallel to the bearing surface is 74 percent or less of the gross.
>
> *solid masonry unit:* One whose net cross-sectional area in every plane parallel to the bearing surface is 75 percent or more of the gross.

Modular masonry: Masonry in which the materials and dimensions fit the modular grid.

Mortar: A plastic mixture of cementitious materials, fine aggregate and water.

> *fat mortar:* A very sticky mortar containing a high percentage of cementitious components.
>
> *high bond mortar:* Mortar which develops higher bond strengths with masonry units than is normally developed with conventional mortar.
>
> *lean mortar:* Mortar which is deficient in cementitious materials; sandy and difficult to spread.

Nominal dimensions: A dimension, but not more than 1/2 inch greater than a specified masonry dimension by the thickness of the mortar joint.

Nonbearing partition: A wall that does not support the structure; usually a partition wall or a filler wall.

Noncombustible material: Any material which will neither ignite nor actively support combustion in air at a temperature of 1200 degrees F when exposed to fire.

OSHA: Occupational Safety and Health Act of 1970.

Overhand work: Method of laying brick or block while standing inside the finished wall and reaching over the wall.

Pargeting: The process of applying a coat of cement mortar to masonry. Often spelled and/or pronounced parging.

Partition: An interior wall, one story or less in height.

Pick and dip: A method of laying brick whereby the bricklayer simultaneously picks up a brick with one hand and, with the other hand, gathers enough mortar on a trowel to lay the brick.

Pier: An isolated column of masonry.

Pilaster: A wall portion projecting from either or both faces and serving as a vertical column and/or beam.

Plumb rule: A tool of measurement used in a horizontal position as a level and in a vertical position as a plumb rule.

Pointing: Troweling mortar into a joint after masonry units are laid.

Ponding: Curing concrete by flooding the surface with water.

Prefabricated brick masonry: Masonry construction fabricated in a location other than its final in-service location in the structure. Also known as preassembled, panelized and sectionalized brick masonry.

Prestressed concrete: Concrete poured around a steel member that is under tension when the pour is made.

Prism: A small masonry assemblage made with masonry units and mortar used to predict strength of full-scale masonry members.

Pugging: A course of mortar used to deaden sound.

Queen closure: A cut brick having a nominal 2-inch horizontal dimension.

Quoin: A projecting right angle masonry corner.

Racking: A method entailing stepping back successive courses of masonry.

Raggle: A groove in a joint or special unit to receive roofing or flashing, sometimes called a reglet.

RBM: Reinforced brick masonry.

Reinforced masonry: Masonry units, reinforcing steel, grout and/or mortar combined to act together to resist forces.

Return: Any surface turned back from the face of the principal surface.

Reveal: That portion of the jamb or recess which is visible from the face of the wall.

Rowlock: A brick laid on its face edge so that the normal bedding area is visible in the face of the wall.

Salt glaze: A gloss finish obtained by thermochemical reaction between silicates of clay and vapors of salt or chemicals.

Saturation coefficient: See *C/B ratio*.

Scaling: The peeling away of the surface of concrete.

Shale: Clay which has been subjected to high pressure until it has hardened.

Shotcreting: Method of placing concrete on curved surfaces such as swimming pools under pneumatic pressure through a nozzle.

Shoved joints: Vertical joints filled by shoving a brick against the next brick when it is being laid in a bed of mortar.

Slenderness ratio: Ratio of the effective height of a member to its effective thickness.

Slip form: A form that permits constant movement during the placing of concrete; used on dams, towers, etc.

Slump: The stiffness of a concrete mix.

Slushed joints: Vertical joints filled after units are laid by "throwing" mortar in with the edge of the trowel.

Soap: A masonry unit of normal face dimensions, having a nominal 2-inch thickness.

Soft-burned: Clay products that have been fired at low temperature ranges, producing relatively high absorptions and low compressive strengths.

Solar screen: A perforated wall used as a sunshade.

Soldier: A stretcher set on end with face showing on the surface.

Spall: A small fragment removed from the face of a masonry unit by a blow or by action of the elements.

Stack: Any structure or part thereof which contains a flue or flues for the discharge of gases.

Story pole: A marked pole for measuring masonry coursing during construction.

Stretcher: A masonry unit laid with its greatest dimension horizontal and its face parallel to the wall face.

Stringing mortar: The procedure of spreading enough mortar to lay several brick at a time.

Struck joint: Any mortar joint that has been finished with a trowel.

Temper: To moisten and mix clay, plaster or mortar to its proper consistency.

Tie: Any unit of material which connects masonry to masonry or to other materials.

Tooling: Compressing or shaping the face of the mortar joint with a metal tool other than a trowel.

Toothing: Constructing a temporary end of a wall with the end stretcher of every other course projecting.

Traditional masonry: Masonry in which the design is based on empirical rules which control minimum thickness, lateral support requirements, and height without a structural analysis.

Tuck pointing: The filling in with fresh mortar of cutout or defective mortar joints in masonry.

Veneer: A single wythe of masonry, not structurally bonded, for facing purposes.

Vitrification: The condition resulting when kiln temperatures are sufficient to fuse grains and close pores of a clay product, making the mass impervious.

Wall: A vertical member of a structure whose horizontal dimension measured at right angles to the thickness exceeds three times its thickness.

   apron wall: That part of a panel wall between window sill and wall support.

   area wall: 1. The masonry surrounding or partly surrounding an area. 2. The retaining wall around basement windows below grade.

   bearing wall: One which supports a vertical load in addition to its own weight.

   cavity wall: A wall built of masonry units so arranged as to provide a continuous air space within the wall (with or without insulating material), and in which the inner and outer wythes of the wall are tied together with metal ties.

   composite wall: A multiple wythe wall in which at least one of the wythes is dissimilar to the other wythe or wythes with respect to type or grade of masonry unit or mortar.

   curtain wall: An exterior nonload bearing wall not wholly supported at each story.

   dwarf wall: A wall or partition which does not extend to the ceiling.

   enclosure wall: An exterior nonbearing wall in skeleton frame construction.

   exterior wall: Any outside wall or vertical enclosure of a building other than a party wall.

   faced wall: A composite wall in which the masonry facing and backing are so bonded as to exert a common reaction under load.

   fire division wall: A wall which subdivides a building to resist the spread of fire; continuous through all stories from the foundation to the roof.

   fire wall: Any wall which subdivides a building to resist the spread of fire and which extends continuously from the foundation through the roof.

foundation wall: That portion of a load bearing wall below the level of the adjacent grade, or below the first floor beams or joists.

hollow wall: A wall built of masonry units arranged to provide an air space within the wall.

insulated cavity wall: A cavity wall that contains insulation of some kind.

load bearing wall: A wall which supports any vertical load in addition to its own weight.

nonload bearing wall: A wall which supports no vertical load other than its own weight.

panel wall: An exterior, nonload bearing wall wholly supported at each story.

parapet wall: That part of any wall entirely above the roof line.

party wall: A wall used for joint service by adjoining buildings.

perforated wall: One which contains a considerable number of relatively small openings; also called a pierced wall or screen wall.

shear wall: A wall which resists horizontal forces applied in the plane of the wall.

single wythe wall: A wall only one masonry unit in thickness.

solid masonry wall: A wall built of solid masonry units, laid continuously, with mortar joints completely filled with mortar or grout.

spandrel wall: That part of a curtain wall above the top of a window in one story and below the sill of the window in the story above.

veneered wall: A wall having a facing of masonry units or other weather-resistant noncombustible materials securely attached to the backing, but not so bonded as to intentionally exert common action under load.

Wall plate: A horizontal member anchored to a masonry wall to which other structural elements may be attached; also called a head plate.

Wall tie: A bonder or metal piece which connects wythes of masonry to each other or to other materials.

Wall tie, cavity: A rigid, corrosion resistant metal tie which bonds two wythes of a cavity wall.

Wall tie, veneer: A strip of metal used to tie a facing veneer to the backing.

Water retentivity: That property of a mortar which prevents the rapid loss of water to masonry units of high suction.

Water table: A projection of lower masonry on the outside of the wall slightly above the ground to prevent upward penetration of ground water.

Waterproofing: Prevention of moisture flow through masonry due to water pressure.

Weep holes: Openings placed in mortar joints of facing materials at the level of the flashing to permit the escape of water or moisture.

Wythe: 1. Each continuous vertical section of masonry one unit in thickness. 2. The thickness of masonry units separating flues in a chimney.

Index

A

"A" brackets 50-53, 58, 63, 64
Additives 16
Adjustable wall ties 117, 118
Admixtures 16, 41, 42
Aggregate 15, 16
Air-entrained cement 16, 103
Air-entrained concrete 95
American bond 127, 129
American Concrete Institute 60
American Plywood Association 47
American Standard Association 28
Anchoring 35, 36, 138, 139, 148, 153
Architect's rod 11
Architectural concrete 63
Ash pit 189
ASTM 17, 41, 43, 45, 121, 137, 139, 142, 150, 161, 170, 179, 196
Axial loads 146

B

Back-up system 137, 138
Back-up system movements 146
Bank or creek gravel 15
Base shear 77
Basement 7, 116
Batterboard 12, 13, 24
Bearing capacity 9, 10
Bearing partitions 103
Bearing value 8
Benchmark 12
Bending radii 60
Bituminous bed 198
Block 23
Block wall reinforcement 114-116
Blocking 55
Blowouts 91
Bond 26, 27
Bond beams 81, 82, 94, 152
Bond breaks 139, 140
Bonds in brick 121, 122, 130, 132
Boundary lines 11
"Bow" 29
Box system concept 80, 82
Bracing 70
Brick 121, 142, 159, 160
Brick paving patterns 195
Brick sizes 121

Brick veneer 37, 140-149
Brick walls 121
Builder's square 24
Building codes 7, 8, 11, 36, 152, 180
Building frame movements 146
Building inspector 7
Building line 7, 13, 24, 28
Building plan dimensions 11
Building weight 9
Bullfloat 97, 99

C

"C" brackets, 54, 56, 58
Caisson 78
Calcium chloride 16
Calculating footing sizes 8
Cap 180, 181, 186
Caulking 140, 158
Cavity walls 84, 151, 152, 154, 155, 156
Cellars 7, 8
Cement-based paints 171
Chemical compounds 16
Chimney block 179, 183
Chimneys 179, 180, 182, 185, 186
Chlorinated acrylics 17
Chlorinated rubbers 17
Chord plates 61
City engineer's office 7
Clay soils 8
Cleaning agents 168
Cleaning existing masonry 166
Cleaning failures, brick 163
Cleaning guide 164
Cleaning methods 165
Cleaning new masonry 163
Cleanout openings 84, 180, 182, 183
Closure block 33
Coarse aggregate 15
Cold weather work 107
Collar joint 93
Color proportioning 110
Color selection 110
Coloring concrete 109-111
Coloring pigments 110
Colors 110
Columns 33, 60, 82, 94, 153, 156
Common bond 127, 129
Compressive strength 37

Concrete, components 15
Concrete, initial set 18
Concrete, steel reinforcement 18
Concrete base 96
Concrete beams 154
Concrete block 23
Concrete curing 16, 17, 95, 98
Concrete curling 152
Concrete forms 47, 68, 103
Concrete foundations 47
Concrete materials 17
Concrete mixer 19
Concrete placing 18
Concrete pouring 95, 107
Concrete pressures 60
Concrete proportions 17
Concrete reinforcement 37
Concrete segregation 18
Concrete slabs 103
Concrete waxes 18
Continuous wall foundations 35, 36
Control joints 93, 95, 113, 116, 117, 119
Corbelled 179, 182
Corner forms 55, 68, 69, 73
Cornerlocks 55
Crack control joints 91
Cracks in masonry 103
Crawl space 37
Creek gravel 15
Cricket 182, 183, 185
Curing 16, 17, 21, 95, 98
Curling 152
Curtain wall 86, 145
Curved forms 60, 61, 64

D

Dead loads 9, 11, 113
De-aired brick 159
Deflections 78, 113
Differential movements 147, 152, 153, 181
Differential settlements 77, 113
Distance separation 7
Draft 179, 186
Driveways 95
Dumpy level 26, 28
Dur-O-wal reinforcement 114, 115
Dutch bond 130, 132
Dynamic analysis 77

E

"Ears" 28
Eave details 140, 142, 144, 145
Eccentric take-up 51, 53
Edging 95, 97
Efflorescence 42, 43, 111, 167, 170
Elevation 12
Elevation differences 11
Empirical analysis 77
Employer responsibility & rights 202
Engineer's rod 11
English bond 122, 127, 130-132
English cross 130, 132
Expansion joints 95, 103, 139, 140, 153, 156, 157, 194

F

Fill 8
Filled ground 8
Filler panels 55, 68
Filler wall 86
Fine aggregate 15
Fire walls 87
Fireplace construction 187-189
Flash set 46
Flashing 139, 143, 156, 157, 175-178, 181, 183, 185, 186
Flemish bond 122, 127, 128, 130, 132
Flexible base 199
Flexible couplings 79
Flexible joint 79
Flexural deflections 79
Floating slab 38
Flue 179
Flush joints 133, 134
Foot plank 32
Footing 8, 9, 10, 48, 77, 181

Footing plates and clips 50
Form design 60
Form maintenance 66, 67
Forms 47-76
Foundation 13, 23, 28, 30, 35, 77, 139, 143
Foundation bed soil 7
Foundation wall 10, 30, 48, 51, 116
Frost heaving 8, 182
Frost protection 38

G

Gang drilling 49
Gang forms 49, 55
Garden wall bond 130, 132
Grade 12
Grade beams 36, 37
Grade level 8
Graduated rod 11
Gravel base 199
Groover 96, 97
Grouting 84, 93, 98, 101, 152

H

Habitable space 7
Hand float 97, 100
Hand signals 29, 30
Hand tying tool 20
Handsetting forms 69, 70
Hard burned brick 159, 160
Hardened concrete 17
HDO (High Density Overlaid) 47
HDO plyform 47, 67
HDO structural I plyform 49, 61
Headers 84, 127
Hearths 188
Heavyweight concrete block 23
High wall forming 56, 57
High-lift grouting 84, 93
Honeycombing 18
Hood 180, 181, 186
Horizontal angles 13
Horizontal expansion joint 147, 155
Horizontal joint reinforcement 138, 140, 147
Horizontal loads 78
Hydrated lime 41
Hydration 16

I

Inside wall forming 53
Insulation 38
Iron-oxide pigments 109
Isolation joint 95
Ivany block system 118, 119

J

"Jack over jack" 28
Jahn forming system 49
Jamb bars 83
Jambs 140, 141, 144, 145, 189
Jitterbug tamper 103, 107
Joint reinforcement 113

K

Key joint 103-105
Keyway 47

L

Ladder bars 84
Lateral deflections 79
Lateral loads 146, 149, 151
Lateral supports 85, 93
Lava rock 15
Laying block 24
Laying out horizontal angles 13
Laying out the bond 26, 27
Lightweight aggregates 15
Lightweight concrete blocks 23
Lime mortar 41
Lined flues 179, 182, 190, 191
Lintel beams 81, 94
Lintels 139, 141, 189
Live loads 9, 10, 11, 113
Load 9, 11

Load bearing 159
Load-bearing capacities 8, 181
Loads on forms 66, 70
Lot line 7
Low wall forming 53, 54
Low-lift grouting 89, 93

M

Masonry bearing partitions 103
Masonry cement 41, 46
Masonry cleaning 163-168
Masonry wall types 94
Material requirements 45
Measuring diagonals 24, 26
Membrane materials 194
Metal stud wall 146, 147, 149
Metal ties 122, 127
Method of insulation 38
Mica 15
Modular brick 121, 123, 124
Modular construction 28, 81
Modular rule 27
Moisture bridge 152
Moisture content 18
Moisture movements 113, 152
Moisture retention 16
Mortar 41-46
Mortar bed 31
Mortar bond 42
Mortar box 45, 46
Mortar joints 133-136
Mortar mixing 45, 46
Mortar plasticity 42, 46
Mortar types 41, 42
Mortared pavement 197, 198
Movement joints 146, 153

N

Natural cement 41
Natural iron-oxide 109
Non-modular brick 28, 121, 122, 125
Non-structural partition 82, 85
Norman brick 121

O

Occupational injury and illness 201
Occupational Safety and Health Act 201
Offset in chimney 180
Oil-based paints 172
OSHA 201
OSHA standards 202-207
Outside corner forms 55

P

Paint failures 171
Painting brick 169-173
Panel erection 50
Panel stacking 56, 57
Panel walls 145
Parapet walls 147, 150, 151, 155, 156, 157, 177
Partial basement 35
Partitions 33, 81, 84, 103
Patterns in brickwork 121, 122, 132
Paving brick 102, 193-196
Paving brick specifications 196
Paving design assemblies 197
Peaty soil 8
Perforated wall 130, 132
Perimeter wall foundations 38, 39
Pier footing 10
Pier foundations 35
Piers 9, 35, 36, 88, 94
Pilaster forming 76
Pilasters 33, 34, 76, 82, 88, 94
Pile foundation 77, 78
Pivot point 13
Plans 28
Plastic film 16
Plug 28
Plumb bob 13, 24
Plyform 47, 55, 60, 61
Plywood 49, 51, 63
Pony trowel 107
Portland cement 15, 41, 98, 159

Power screed 106
Power trowel 107
Pre-cast concrete 98, 100, 101
Prefabricated forms 47, 48, 67, 68
Premises 7
Properties of mortar 42
Property line 7
Proportioning footings 77
Protection against frost 38
Pumice 15

Q

Queen closure 130

R

Radius wall forms 53, 62, 64
Raft foundations 77
Raked joints 133, 135, 136
"Ranging the wall" 29
Reinforced grouting 84, 89-94
Reinforced masonry 94, 114, 118
Reinforcing 35, 36, 77, 81, 114
Reinforcing rods 18
Removing stains 167-169
Restrained movements 113
Rock 8
Rod 11, 12, 13, 29
Rod reading 12
Rollerbug 103, 106
Roman brick 121
Roof 11
Roofing felt cushion 198, 199
Rough cut joints 133, 134
Running bond 127, 129

S

Salmon brick 160
Salvaged brick 137, 142, 159, 160
Sand 43
Sand base 200
Scabs 53
Scaffold brackets 70, 75
"Scaffold high" 28, 29
Scaffold jacks 56, 57, 58
Scope 26, 29
SCR brick 121
Screed 97, 99
Screed joints 103-105
Screen wall 130, 132
Sealants 140, 158
Sealing 17
Second lifts 56, 59
Second wall forming 54, 73
Segregation 18
Seismic detailing 78
Seismic joint 79
Seismic systems 79
Seismic zones 77, 78
Self-reading rod 11
Separation distance 11
Separation of structures 78
Service walks 95
Setback 11
Shear walls 79, 80, 152
Short-column effects 79
Shrinking 16
Sidewalks 95, 98
Sills 140, 141, 144, 145, 176
Site plans 11
Sizes of brick 121
Skeleton frames 148
Slab floor 37, 95
Slab forms 55, 60, 96, 103
Slab foundations 37, 38, 95
Slab-on-ground 38, 39, 95, 98, 100, 103
Snap ties 50-54
Sodium silicates 17
Soft burned brick 160
Soil 7
Soil analysis 7
Soil bearing capacity 9
Soil characteristics 7
Soil maps 7
Soil survey 7
Soil tests 7

Space frames 79
Spacing reinforcing 116, 144
Spalling 160, 161
Spans for lumber framing 65
Spongy soil 8
Sprinkling 16
Stacked bond 84, 116, 130, 131
Stain removal 167-169
Stairways 79
Staking out a building 12
Standard brick 121
Steel angles 143, 144, 148
Steel beams 154, 155
Steel reinforcement 18, 19
Steel reinforcing rods 20
Step forming 56, 95
Stepdowns 56
Stepped foundation 35
Stepping footings 9
Stiffback 56
Stocking block 33
Storage buildings 9
Story 7
Story drift 79
Story pole 31, 32
Street line 7
Stretcher blocks 28
Stretcher bricks 127
Stringlines 24
Strongbacks 47, 48, 54, 56, 58, 74
Struck joints 133, 134
Structural engineer 8, 10
Structural I plyform 47, 61
Structural load 9
Structural members 94
Structural partition 84
Subgrade 103
Survey 11
Synthetic iron oxide 109, 110

T

Tampers 95, 96
Tamping concrete 95, 103
Target 11
Templates 61
Tensile strength 18
Thermal strains 152
Thimble 183, 184
Three-way forming 56

Tie and hole sizes 50
Tie extenders 56, 58
Ties 71, 117, 138, 142
Tooled joints 133
Transit 11, 13, 26, 27, 28
Trig 31, 32, 33
Troweled joints 133, 134
Type of soil 9

U

Unit masonry 81
Unreinforced footings 10
U.S. Soil Conservation Service 7

V

Veneer 94, 127, 137-139, 142
Veneer movements 146
Vibrate 18
Vibrator 19
Vernier 13, 14
Vertical coursing 125, 126

W

Walers 47, 48, 53, 55, 56, 63, 64, 72
Wall form pressure 60
Wall forms 63
Wall openings 83
Wall piers 83, 88
Wall reinforcement 83, 86
Wall texture 130
Walls and partitions 81, 94, 116
Water 16, 150, 170
Water cement ratio 18
Waterproof paper 16
Weathered joints 133
Weathering 111
Wedge bolts 69
Weep holes 139, 140, 143, 156, 157, 177, 178
Weeping joints 136
Wet blocks 24, 25
Wet burlap curing 16
Windows 148, 158, 176
Wire mesh reinforcement 103, 106
Wire reinforcement 33
Wire ties 20, 84, 115
Workability 16
Workmanship 141
"Wow" 29

Practical References for Builders

Builder's Guide to Accounting Revised
Step-by-step, easy to follow guidelines for setting up and maintaining an efficient record keeping system for your building business. Not a book of theory, this practical, newly-revised guide to all accounting methods shows how to meet state and federal accounting requirements, including new depreciation rules, and explains what the tax reform act of 1986 can mean to your business. Full of charts, diagrams, blank forms, simple directions and examples. **304 pages, 8½ x 11, $17.25**

Building Layout
Shows how to use a transit to locate the building on the lot correctly, plan proper grades with minimum excavation, find utility lines and easements, establish correct elevations, lay out accurate foundations and set correct floor heights. Explains planning sewer connections, leveling a foundation out of level, using a story pole and batterboards, working on steep sites, and minimizing excavation costs. **240 pages, 5½ x 8½, $11.75**

Concrete and Formwork
This practical manual has all the information you need to select and pour the right mix for the job, lay out the structure, choose the right form materials, design and build the forms and finish and cure the concrete. Nearly 100 pages of step-by-step instructions cover the actual construction and erecting of nearly all site fabricated wood forms used in residential construction. **176 pages, 8½ x 11, $10.00**

Reducing Home Building Costs
Explains where significant cost savings are possible and shows how to take advantage of these opportunities. Six chapters show how to reduce foundation, floor, exterior wall, roof, interior and finishing costs. Three chapters show effective ways to avoid problems usually associated with bad weather at the jobsite. Explains how to increase labor productivity. **224 pages, 8½ x 11, $10.25**

Handbook of Construction Contracting Vol. 1 & 2
Volume 1: Everything you need to know to start and run your construction business; the pros and cons of each type of contracting, the records you'll need to keep, and how to read and understand house plans and specs to find any problems before the actual work begins. All aspects of construction are covered in detail, including all-weather wood foundations, practical math for the jobsite, and elementary surveying. **416 pages, 8½ x 11, $21.75**

Volume 2: Everything you need to know to keep your construction business profitable; different methods of estimating, keeping and controlling costs, estimating excavation, concrete, masonry, rough carpentry, roof covering, insulation, doors and windows, exterior finish, specialty finishes, scheduling work flow, managing workers, advertising and sales, spec building and land development and selecting the best legal structure for your business. **320 pages, 8½ x 11, $24.75**

Construction Estimating Reference Data
Collected in this single volume are the building estimator's 300 most useful estimating reference tables. Labor requirements for nearly every type of construction are included: site work, concrete work, masonry, steel, carpentry, thermal & moisture protection, doors and windows, finishes, mechanical and electrical. Each section explains in detail the work being estimated and gives the appropriate crew size and equipment needed. **368 pages, 11 x 8½, $18.00**

National Construction Estimator
Current building costs in dollars and cents for residential, commercial and industrial construction. Prices for every commonly used building material, and the proper labor cost associated with installation of the material. Everything figured out to give you the "in place" cost in seconds. Many time-saving rules of thumb, waste and coverage factors and estimating tables are included. **512 pages, 8½ x 11, $16.00. Revised annually.**

Building Cost Manual
Square foot costs for residential, commercial, industrial, and farm buildings. In a few minutes you work up a reliable budget estimate based on the actual materials and design features, area, shape, wall height, number of floors and support requirements. Most important, you include all the important variables that can make any building unique from a cost standpoint. **240 pages, 8½ x 11, $12.00. Revised annually**

Concrete Construction & Estimating
Explains how to estimate the quantity of labor and materials needed, plan the job, erect fiberglass, steel, or prefabricated forms, install shores and scaffolding, handle the concrete into place, set joints, finish and cure the concrete. Every builder who works with concrete should have the reference data, cost estimates, and examples in this practical reference. **571 pages, 5½ x 8½, $17.75**

Excavation and Grading Handbook
The foreman's and superintendent's guide to highway, subdivision and pipeline jobs: how to read plans and survey stake markings, set grade, excavate, compact, pave and lay pipe on nearly any job. Includes hundreds of practical tips, pictures, diagrams and tables that even experienced "pros" should have. **320 pages, 5½ x 8½, $15.25**

Contractor's Survival Manual
How to survive hard times in construction and take full advantage of the profitable cycles. Shows what to do when the bills can't be paid, finding money and buying time, transferring debt, and all the alternatives to bankruptcy. Explains how to build profits, avoid problems in zoning and permits, taxes, time-keeping, and payroll. Unconventional advice includes how to invest in inflation, get high appraisals, trade and postpone income, and how to stay hip-deep in profitable work. **160 pages, 8½ x 11, $16.75**

Blueprint Reading for the Building Trades
How to read and understand construction documents, blueprints, and schedules. Includes layouts of structural, mechanical and electrical drawings, how to interpret sectional views, how to follow diagrams; plumbing, HVAC and schematics, and common problems experienced in interpreting construction specifications. This book is your course for understanding and following construction documents. **192 pages, 5½ x 8½, $11.25**

Builder's Office Manual
This manual will show every builder with from 3 to 25 employees the best ways to: organize the office space needed, establish an accurate record-keeping system, create procedures and forms that streamline work, control costs, hire and retain a productive staff, minimize overhead, and much more. **208 pages, 8½ x 11, $13.25**

Estimating Tables for Home Building
Produce accurate estimates in minutes for nearly any home or multi-family dwelling. This handy manual has the tables you need to find the quantity of materials and labor for most residential construction. Includes overhead and profit, how to develop unit costs for labor and materials and how to be sure you've considered every cost in the job. **336 pages, 8½ x 11, $21.50**

Berger Building Cost File
Labor and material costs needed to estimate major projects: shopping centers and stores, hospitals, educational facilities, office complexes, industrial and institutional buildings, and housing projects. All cost estimates show both the manhours required and the typical crew needed so you can figure the price and schedule the work quickly and easily. **344 pages, 8½ x 11, $30.00**

Contractor's Year-Round Tax Guide
How to set up and run your construction business to minimize taxes: corporate tax strategy and how to use it to your advantage, and what you should be aware of in contracts with others. Covers tax shelters for builders, write-offs and investments that will reduce your taxes, accounting methods that are best for contractors, and what the I.R.S. allows and what it often questions. **192 pages, 8½ x 11, $16.50**

Wood-Frame House Construction
From the layout of the outer walls, excavation and formwork, to finish carpentry, and painting, every step of construction is covered in detail with clear illustrations and explanations. Everything the builder needs to know about framing, roofing, siding, insulation and vapor barrier, interior finishing, floor coverings, and stairs... complete step by step "how to" information on what goes into building a frame house. **240 pages, 8½ x 11, $11.25. Revised edition**

Roof Framing
Frame any type of roof in common use today, even if you've never framed a roof before. Shows how to use a pocket calculator to figure any common, hip, valley, and jack rafter length in seconds. Over 400 illustrations take you through every measurement and every cut on each type of roof: gable, hip, Dutch, Tudor, gambrel, shed, gazebo and more. **480 pages, 5½ x 8½, $19.50**

Stair Builders Handbook
If you know the floor to floor rise, this handbook will give you everything else: the number and dimension of treads and risers, the total run, the correct well hole opening, the angle of incline, the quantity of materials and settings for your framing square for over 3,500 code approved rise and run combinations—several for every 1/8 inch interval from a 3 foot to a 12 foot floor to floor rise. **416 pages, 8½ x 5½, $12.75**

Computers: The Builder's New Tool
Shows how to avoid costly mistakes and find the right computer system for your needs. Takes you step-by-step through each important decision, from selecting the software to getting your equipment set up and operating. Filled with examples, checklists and illustrations, including case histories describing experiences other contractors have had. If you're thinking about putting a computer in your construction office, you should read this book before buying anything. **192 pages, 8½ x 11, $17.75**

Manual of Professional Remodeling
This is the practical manual of professional remodeling written by an experienced and successful remodeling contractor. Shows how to evaluate a job and avoid 30-minute jobs that take all day, what to fix and what to leave alone, and what to watch for in dealing with subcontractors. Includes chapters on calculating space requirements, repairing structural defects, remodeling kitchens, baths, walls and ceilings, doors and windows, floors, roofs, installing fireplaces and chimneys (including built-ins), skylights, and exterior siding. Includes blank forms, checklists, sample contracts, and proposals you can copy and use. **400 pages, 8½ x 11, $18.75**

Remodeler's Handbook
The complete manual of home improvement contracting: Planning the job, estimating costs, doing the work, running your company and making profits. Pages of sample forms, contracts, documents, clear illustrations and examples. Chapters on evaluating the work, rehabilitation, kitchens, bathrooms, adding living area, re-flooring, re-siding, re-roofing, replacing windows and doors, installing new wall and ceiling cover, re-painting, upgrading insulation, combating moisture damage, estimating, selling your services, and bookkeeping for remodelers. **416 pages, 8½ x 11, $18.50**

Residential Wiring
Shows how to install both rough and finish wiring in both new construction and alterations and additions. Complete instructions are included on troubleshooting and repairs. Every subject is referenced to the 1987 National Electrical Code, and over 24 pages of the most needed NEC tables are included to help you avoid errors so your wiring passes inspection — the first time. **352 pages, 8½ x 5½, $18.25**

Construction Superintending
Explains what the "super" should do during every job phase from taking bids to project completion on both heavy and light construction: excavation, foundations, pilings, steelwork, concrete and masonry, carpentry, plumbing, and electrical. Explains scheduling, preparing estimates, record keeping, dealing with subcontractors, and change orders. Includes the charts, forms, and established guidelines every superintendent needs. **240 pages, 8½ x 11, $22.00**

Process & Industrial Pipe Estimating
A clear, concise guide to estimating costs of fabricating and installing underground and above ground piping. Includes types of pipes and fittings, valves, filters, strainers, and other in-line equipment commonly specified, and their installation methods. Shows how a take-off is consolidated on the estimate form and the bid estimate derived using the complete set of manhour tables provided in this complete manual of pipe estimating. **240 pages, 8½ x 11, $18.25**

Operating The Tractor-Loader-Backhoe
Explains how to get maximum productivity from this highly versatile machine. Describes how experienced operators plan the job before work begins, cut wasted movement to a minimum, work effectively in tight quarters and handle the really difficult jobs safely and efficiently. Covers cutting, filling, compacting, trenching, working around utility lines, craning, tunneling, footings, truck loading, grading, demolition, even stump removal and brush cleaning.

Each task is illustrated with diagrams and photographs. If you own or rent a hoe, you need this practical manual. **192 pages, 8½ x 11, $23.75**

Basic Plumbing with Illustrations
The journeyman's and apprentice's guide to installing plumbing, piping and fixtures in residential and light commercial buildings: how to select the right materials, lay out the job and do professional quality plumbing work. Explains the use of essential tools and materials, how to make repairs, maintain plumbing systems, install fixtures and add to existing systems. **320 pages, 8½ x 11, $17.50**

Cost Records for Construction Estimating
How to organize and use cost information from jobs just completed to make more accurate estimates in the future. Explains how to keep the cost records you need to reflect the time spent on each part of the job. Shows the best way to track costs for sitework, footing, foundations, framing, interior finish, siding and trim, masonry, and subcontract expense. Provides sample forms. **208 pages, 8½ x 11, $15.75**

Estimating Home Building Costs
Estimate every phase of residential construction from site costs to the profit margin you should include in your bid. Shows how to keep track of manhours and make accurate labor cost estimates for footings, foundations, framing and sheathing finishes, electrical, plumbing and more. Explains the work being estimated and provides sample cost estimate worksheets with complete instructions for each job phase. **320 pages, 5½ x 8½, $14.00**

Contractor's Guide to the Building Code
Explains in plain English exactly what the Uniform Building Code requires and shows how to design and construct residential and light commercial buildings that will pass inspection the first time. Suggests how to work with the inspector to minimize construction costs, what common building short cuts are likely to be cited, and where exceptions are granted. **312 pages, 5½ x 8½, $16.25**

Spec Builder's Guide
Explains how to plan and build a home, control your construction costs, and then sell the house at a price that earns a decent return on the time and money you've invested. Includes professional tips to ensure success as a spec builder: how government statistics help you judge the housing market, cutting costs at every opportunity without sacrificing quality, and taking advantage of construction cycles. Every chapter includes checklists, diagrams, charts, figures, and estimating tables. **448 pages, 8½ x 11, $24.00**

Plumbers Handbook Revised
This new edition shows what will and what will not pass inspection in drainage, vent, and waste piping, septic tanks, water supply, fire protection, and gas piping systems. All tables, standards, and specifications are completely up-to-date with recent changes in the plumbing code. Covers common layouts for residential work, how to size piping, selecting and hanging fixtures, practical recommendations and trade tips. This book is the approved reference for the plumbing contractors exam in many states. **240 pages, 8½ x 11, $16.75**

Rough Carpentry
All rough carpentry is covered in detail: sills, girders, columns, joists, sheathing, ceiling, roof and wall framing, roof trusses, dormers, bay windows, furring and grounds, stairs and insulation. Many of the 24 chapters explain practical code approved methods for saving lumber and time without sacrificing quality. Chapters on columns, headers, rafters, joists and girders show how to use simple engineering principles to select the right lumber dimension for whatever species and grade you are using. **288 pages, 8½ x 11, $14.50**

Craftsman BOOK COMPANY

6058 Corte del Cedro
P. O. Box 6500
Carlsbad, CA 92009

In a hurry?
We accept phone orders charged to your MasterCard or Visa. Call (619) 438-7828

We ship promptly and pay the postage when your check covers your order in full. Add 15% for Canadian dollars. California orders add 6%.

Name _____
Company _____
Address _____
City _____ State _____ Zip _____
Send check or money order
Total Enclosed _____ (In California add 6% tax)
If you prefer, use your ☐ Visa or ☐ MasterCard
Card no. _____
Expiration date _____ Initials _____

10 Day Money Back GUARANTEE

☐ 17.50 Basic Plumbing with Illustrations
☐ 30.00 Berger Building Cost File
☐ 11.25 Blueprint Reading for Building Trades
☐ 17.25 Builder's Guide to Accounting Revised
☐ 13.25 Builder's Office Manual
☐ 12.00 Building Cost Manual
☐ 11.75 Building Layout
☐ 17.75 Computers: The Builders New Tool
☐ 10.00 Concrete and Formwork
☐ 17.75 Concrete Construction & Estimating
☐ 18.00 Construction Estimating Reference Data
☐ 22.00 Construction Superintending
☐ 16.25 Contractor's Guide to the Building Code
☐ 16.75 Contractor's Survival Manual
☐ 16.50 Contractor's Year-Round Tax Guide
☐ 15.75 Cost Records for Construction Estimating
☐ 14.00 Estimating Home Building Costs
☐ 21.50 Estimating Tables for Home Building
☐ 15.25 Excavation & Grading Handbook
☐ 21.75 Handbook of Construction Contracting Vol. 1
☐ 24.75 Handbook of Construction Contracting Vol. 2
☐ 18.75 Manual of Professional Remodeling
☐ 16.00 National Construction Estimator
☐ 23.75 Operating the Tractor-Loader-Backhoe
☐ 16.75 Plumber's Handbook Revised
☐ 18.25 Process and Industrial Pipe Estimating
☐ 10.25 Reducing Home Building Costs
☐ 18.50 Remodelers Handbook
☐ 18.25 Residential Wiring
☐ 19.50 Roof Framing
☐ 14.50 Rough Carpentry
☐ 24.00 Spec Builder's Guide
☐ 12.75 Stair Builder's Handbook
☐ 11.25 Wood-Frame House Construction
☐ 13.50 Masonry and Concrete Construction

These books are tax deductible when used to improve or maintain your professional skill.

TH
5311
.N64
1982

TH
5311
.N64
1982

14.16